From High-Temperature Superconductivity to Microminiature Refrigeration

From High-Temperature Superconductivity to Microminiature Refrigeration

Edited by

Blas Cabrera
Stanford University
Stanford, California

H. Gutfreund
The Hebrew University of Jerusalem
Jerusalem, Israel

and

Vladimir Kresin
Lawrence Berkeley Laboratory
University of California
Berkeley, California

Plenum Press • New York and London

Library of Congress Cataloging in Publication Data

William A. Little Symposium (1995: Stanford University)
    From high-temperature superconductivity to microminiature refrigeration / edited by Blas
Cabrera, H. Gutfreund, and Vladimir Kresin.
        p.   cm.
    "Based on the proceedings of the William A. Little Symposium, held September 30, 1995,
in Stanford, California."
    Includes bibliographical references and index.
    ISBN-13: 978-1-4613-8040-5   e-ISBN-13: 978-1-4613-0411-1
    DOI: 10.1007/978-1-4613-0411-1
    1. High-temperature superconductivity—Congresses. 2. Low-temperature engineering—
Congresses. 3. Neural networks—Congresses. 4. Little, William A., 1930–   . I. Cabrera,
Blas. II. Gutfreund, H. III. Kresin, Vladimir Z. IV. Title.
QC611.98.H54W553   1995   96-26470
537.6′ 23—dc20   CIP

Based on the proceedings of the William A. Little Symposium, held September 30, 1995,
in Stanford, California

ISBN-13: 978-1-4613-8040-5

© 1996 Plenum Press, New York

Softcover reprint of the hardcover 1st edition 1996

A Division of Plenum Publishing Corporation
233 Spring Street, New York, N. Y. 10013

When I first recognized the possibility of superconductivity being mediated by an electronic rather than a phonon mechanism and that superconductivity might thus occur near room temperature, I realized that such a discovery would open up what I called "a whole new world of science and technology," in particular, that magnetic levitation might then be used for mass transport — "carrying passengers and cargo above roadways of superconducting sheet, moving like flying carpets without friction and without material wear and tear" [*Scientific American,* **212,** 21 (1965)]. This vision of the "magic carpet" became for my students a shorthand description of our research goal's, and our children's explanation of their Dad's work.

— W. A. Little

PREFACE

This volume, *From High-Temperature Superconductivity to Microminiature Refrigeration*, was compiled as a commemoration to Bill Little's rich scientific career over the past 40 years or more. He has contributed many seminal ideas to such diverse fields of physics as phonon physics at low temperatures, magnetic flux quantization in superconductors, high-temperature superconductivity, neural networks, and microminiature refrigerators.

The first section of the book contains a collection of reprints from Bill Little's most important scientific papers. These papers are preceded by an introduction by Bill himself, which gives many insights into the thinking processes that led him to investigate these diverse topics of research. These publications contain a remarkable range of important contributions from his early treatise on Kapitza boundary resistance for phonons, still his most cited paper, through his introduction of physics techniques to the study of neural networks and to recent important contributions to the understanding of the cuprate high-temperature superconductors.

The second section of the book contains contributions based on the William Little Symposium, which was held at Stanford University on September 30, 1995. These papers cover many subjects, including presentations by Pierre-Gilles de Gennes on polymers, by Vladimir Kresin on recent research on high-temperature superconductors, and by Bill Little himself on the theory and application of microminiature refrigerators. In addition, Walter Meyerhof has contributed a wonderful biography of Bill Little based on the after-dinner talk that he gave at the symposium.

The second section also includes additional contributions that broaden the perspective of Bill Little's impact on physics. These include an historical perspective by Bascom Deaver on the exciting research at Stanford around the time of the discovery of magnetic flux quantization, and contributions by Vitaly Ginzburg on non-phonon models for high-temperature superconductors, by Takehiko Ishiguro on organic superconductivity, and by Lev Gor'kov on the theory of organic conductors. In addition, we have contributions by Roger Freedman on teaching physics and by Gordon Shaw on neural networks.

We conclude by reminding the reader that physics is not the sterile topics of journal publications, which are much like museums of natural history with their dead and stuffed animal exhibits. Rather, it is the lively, engaging, and exciting business of discovery and invention, much like beautiful animals in the wild. For those of us who have had the pleasure of working with Bill Little on an active research project, we have appreciated the excitement, energy, and drive that he has always brought as much as his remarkable insightful physical intuition and creativity. Our aim is that this volume convey some of the excitement of discovery as well as chronicle the achievements themselves. Bill has shown no sign of slowing down, so we look forward to many more contributions and discoveries in the years to come.

Blas Cabrera
Honach Gutfreund
Vladimir Kresin

TABLE OF CONTENTS

Preface

PART ONE: SELECTED PAPERS OF WILLIAM A. LITTLE

PART ONE

Selected Papers of William A. Little

A LAYMAN'S GUIDE TO THE REPRINT COLLECTION

W. A. Little

Physics Department
Stanford University
Stanford, CA 94305

Blas Cabrera and Vladimir Kresin asked me to include in this volume a selection of my papers. Whenever I have seen this done in the past with other authors, I have wanted to know something more of the background to the papers chosen. Why those papers were chosen and what led up to the work reported in them. The papers themselves are largely stripped of history, so to correct for this, I thought it would be interesting to provide something of a commentary, to give a feel for the context in which the work was done and what inspired them. We have lived in exciting times and I would like to convey some of this excitement to the reader, reading this today. In addition, this brief commentary might save the casual reader the necessity of reading all the papers !

The first paper is a section from my Ph.D. thesis presented at Rhodes University in South Africa. John Birks had just taken the chair of physics there and had introduced me to the problem of exciton and photon transfer of energy in organic phosphors. His enthusiasm and excitement were contagious and what I learned from him was to play a role in my later work on superconductivity. My contribution to this paper was the invention of the frequency domain technique for measuring the lifetime of fluorophors. This technique is now used in most commercial fluorometers today.

The second paper on the transport of heat between dissimilar solids, also known as the Kapitza resistance, is, surprisingly, one of my most heavily cited papers. It grew out of months of frustration working on a demagnetisation cryostat as a post doctoral fellow at the University of British Columbia in Vancouver, Canada. I was frustrated because, as others had found before me, there is a real barrier to the flow of heat across an interface at low temperatures and as a result equilibrium at these temperatures is reached very, very slowly. This was a first effort to understand this problem, and though this work was done over thirty five years ago, the problem remains today only partially understood, particularly for the case of an interface between a solid and liquid helium.

I have described in, "Bloch and the New Superconductors" some aspects of the background to the next paper on the, so-called "Little-Parks effect". This experiment done in collaboration with Ron Parks took just 30 days from concept to completion, a fact that colored our views of how experiments should be done for years thereafter, but which we never came close to repeating. It was a great pleasure working with Ron on this project. He had done an elegant piece of work for his thesis on the adiabatic demagnetisation of paramagnetic metals and then worked with me as a post-doc on this effect. He went on to edit the Parks' volumes on Superconductivity, which became one of the most useful collection of work in superconductivity ever published. Tragically he died in the early 80's in the prime of his career.

The background to the next two papers on organic superconductivity I have also given in some detail in "Bloch and the New Superconductors" and in "Organic Superconductivity: the Duke Connection", both of which are included in this collection of papers. I will not repeat this here, however, the description of the pairing interaction in the Scientific American article deserves special comment. It was, I believe, the first attempt to give a layman's interpretation of this interaction and one that seems to have been reasonably successful for it is used today almost universally as a model for the phonon mediated interaction in superconductors. How this came about may be of some interest.

After publishing the paper on the possibility of synthesizing an organic superconductor, I had worked with Harry Aine, an attorney from Varian Associates on a patent application (later abandoned!) on ways to synthesize a high temperature superconductor. This required a description of superconductivity and the pairing interaction, itself. Harry had a good understanding of electronics, a fine legal mind and an excellent memory, but not much knowledge of solid state physics. During the course of the several weeks that we worked together, he drove me to transform my simple, loose, handwaving picture of the superconducting state into a reasonably credible description, free of obvious logical inconsistencies, and one that even freshman can readily grasp. I learned a great deal working with Harry on this and developed at the same time a respect for the legal profession. This work became the basis of the description of superconductivity contained in the Scientific American article.

The Physical Review Letters paper on the measurement of the specific heat of He^3 and He^4 near their critical points was inspired by a conversation with Frank Yang (C. N. Yang), who had visited Stanford's Institute of Theoretical Physics some time earlier. I had shown him the beautiful experimental results of Bagatskii, Voronel, and Gusak referred to in this paper, who had observed a lambda-like singularity of the specific heat at the critical point of argon. Yang had been intrigued by this result and explained to me the mapping he and T.D. Lee had made of the liquid-gas phase transition onto an Ising model and showed how this led one to expect such a singularity along the critical isochore. He encouraged us to extend these measurements to fluids of lower atomic weight, where deviations from the law of corresponding states could be expected from quantum corrections. Mike Moldover, then a student in my group, went ahead with this and obtained the very clean results shown here. This came just at the time that scaling theories of phase transitions were being developed and Mike's results provided an important piece of the puzzle.

The short paper on the decay of persistent currents, while not of earth shaking importance itself, nevertheless was the culmination of almost three years of personal struggle. In the earlier paper on organic superconductivity I had suggested that the well known theorems that prohibited phase transitions in *classical* one-dimensional systems might not apply to a transition to a *superconducting* state. Ferrell and then Rice, showed shortly thereafter that this could not be so, as phase fluctuations in the order parameter would destroy the off-diagonal-long-range-order (ODLRO) in such one-dimensional systems and thus the ordered phase could not exist. I was disturbed by this argument because it implied a knowledge of the relative phase of the order parameter at two points well separated in space, even though no means existed for a measurement of this phase difference. This seemed unphysical. Moreover, the essence of Yang's ODLRO argument had been that the free energy of a ring of a 'superconductor' would depend on the flux through it and whether the ring was open or closed - the Ferrell-Rice argument did not appear to address this issue. This led me to consider a curved 'one-dimensional' filament in the form of a closed ring and to the discovery that phase fluctuations, which are found to preserve the so-called, winding number of the ring, did *not* destroy the free-energy dependence of the ring on the flux. The different circulating current states were separated from one another by free energy barriers which allowed *quasi*-persistent currents to circulate with essentially infinite life. I was able to estimate the height of the barrier but could not come up with an 'attempt' frequency necessary for a complete understanding of the system. Later, Langer and Ambegaokar, and McCumber and Halperin developed an elegant theory that yielded both.

The next paper, a review paper by Hanoch Gutfreund and myself on the, "Prospects of Excitonic Superconductivity" was the result of a ten year collaborative study of the many problems that arose in attempting to devise a superconductor with a pairing interaction mediated by an electronic excitation rather than by phonons. Objections to the mechanism that had been raised earlier by many authors proved upon further study to be invalid, immaterial, or means could be found to minimize their effects. This work is of special interest today, for since the discovery of the high temperature cuprate superconductors many of these issues have been raised again, in spite of the fact that the answers to these questions can generally be found in the literature.

The brief conference paper on "The Missing Link" describes one item that Gutfreund and I had not discusses in depth in the above review, and that was the strength of the excitonic interaction. Here I showed that the magnitude of this interaction, when generated by

4

a virtual transition between a metal and its ligand system, would be a strong function of the difference in valence-state-ionisation-energy between the metal and the ligand. When these are degenerate, the interaction reaches a sharp maximum. This is of interest for the case of the cuprates because such a degeneracy exists between the oxygen and the copper atoms of the copper-oxide planes. Our model suggests that a virtual charge transfer excitation from one to the other would exhibit a strong electronic polarizability with a large transition moment and one that could make a significant contribution to the pairing interaction.

The following paper, a very recent paper in Physical Review Letters by Holcomb, Collman and myself, takes one back to this old problem. We had developed a new technique, Thermal Difference Spectroscopy, for observing over a wide range of photon energies very small changes in the reflectivity of a metal as it enters the superconducting state. These changes can be interpreted as arising from the appearance in the superconducting state of a frequecy dependent gap function, and from them one can identify the energy of the excitations responsible for the pairing interaction. The result of this was that the oxygen-to-copper charge transfer excitation occuring near 1.6 eV was implicated. We suggested that this should be a common feature of all the cuprates if it was responsible for the pairing in these materials. In a subsequent paper in Physical Review we showed that in all three other cuprates studied, this, indeed, was found to be the case. At this time, the Spring of 1996, experiments are on-going to gather further information on this contribution to the pairing.

I have included my earliest paper on neural networks, "The existence of persistent states in the brain" which was published in 1974. In this I was able to establish a mapping between the neural network problem and a generalized Ising spin system. The importance of it lay in the proof that in such a network with arbitrary synaptic connectivity (where a neuron A could affect neuron B but not necessarily the other way round), characteristic patterns of activity could be shown to persist with virtually infinite life even in the presence of noise, imperfect firing, and imperfect connections between the neurons. The proof given in this paper is obscure to say the least, and later a much simpler one was devised, but it is included because it was the first, and it contains the concept of treating the "noise" in the network as an effective temperature, it contains much of the background to the subject, and has, what I believe is an illuminating model in the appendix. Elsewhere, in Concepts in Neuroscience 1,149-164 (1990), in a paper entitled,"The evolution of non-Newtonian views of brain function", I have given a more detailed discussion of this mapping and related it to other development in condensed matter physics.

The next paper published in the American Journal of Physics gives a novel proof of Fermat's theorem of prime numbers. This, somewhat off-the-wall subject arose in attempting to classify the different cyclic states of a neural network, that arise from the lack of symmetry of the synaptic connectivity matrix using the above mapping to a spin 1/2 Ising model. When working on this, I stumbled on this interesting result. Hanoch Gutfreund immediately recognized it as a special case of Fermat's theorem and was able to generalized it to arbitrary spin and obtain the complete theorem, and then went on to derive some further relationships with non-prime numbers.

The short paper on the scaling of miniature cryogenic coolers to microminiature size was the beginning of our work on microminiature refrigerators. Earlier Jim Mercereau, from CalTech had pointed out to me that existing miniature cryocoolers were poorly matched to the cooling needs of the emerging SQUID devices. We had proposed to the Office of Naval Research that this problem of miniaturization should be studied, but the proposal was rejected by the agency. This was done under the auspices of RAI, Inc., a small consulting company we had formed that provided consultation to the government on superconductivity and cryogenics. Some years later, I came up with the idea of the photofabrication of the refrigerators. We were unable to undertake this work at RAI, so turned it over to Stanford University, where eventually we were able to get support from ONR. After the successful development of the first prototype, I licensed back the technology from Stanford and formed MMR Technologies, Inc., where subsequent development of the technology was made. Some of this is discussed in, "Microminiature Refrigeration", Rev. Sci. Instrum. 55, 661 (1984) and in "Recent Advances in Cryogenics" presented at this Symposium.

The final two papers, one honoring William Fairbank in "Near Zero" and the other, Felix Bloch from "Conductivity and Magnetism: the Legacy of Felix Bloch" are themselves commentaries on events, experiences and experiments from the same era spanned by these papers. They are included to provide some perspective on this period.

REPRINTED FROM THE
PROCEEDINGS OF THE PHYSICAL SOCIETY, A,, VOL. LXVI, p. 921, 1953
All Rights Reserved
PRINTED IN GREAT BRITAIN

Photo-Fluorescence Decay Times of Organic Phosphors

BY J. B. BIRKS AND W. A. LITTLE

Physics Department, Rhodes University, Grahamstown, South Africa

MS. received 27th February 1953, and in amended form 6th July 1953

Abstract. A method for measuring photo-fluorescence decay times to approximately 10^{-9} sec is described. Observations are made of the phase and modulation of the light from a 7·5 Mc/s air discharge tube, and of the fluorescence excited by this light, using a photomultiplier modulated at 15 Mc/s. The decay times of anthracene, stilbene, terphenyl and diphenylacetylene crystals of various thicknesses have been measured. Due to the overlap of the emission and absorption spectra, the technical decay time t_f of a thick crystal is greater than the molecular decay time $(t_f)_0$ observed for a microcrystalline specimen. For anthracene $t_f = 14 \pm 2$ mμsec and $(t_f)_0 = 3·5 \pm 1·0$ mμsec. This value of t_f agrees with that derived from $(t_f)_0$ and the comparative technical and molecular emission spectra. For each of the compounds studied the molecular decay time $(t_f)_{00}$, in the absence of internal quenching, is of the order of 3–4 mμsec.

§ 1. INTRODUCTION

T HE decay time of the fluorescence from the organic crystals and solutions, used in scintillation counting, is of the order of 10^{-8} second. Many observations have been reported of the decay time t_I of the fluorescence excited by ionizing radiations for the more important phosphors. Few direct measurements are available however of the decay time t_f of the photo-fluorescence excited by ultra-violet radiation, apart from those of Liebson *et al.* (1950), who found that t_I was greater than t_f, and approximately equal to $2t_f$ for most of the organic crystals investigated.

Previous measurements by Bowen and Lawley (1949) and Little and Birks (1952) have shown that the spectrum of the photo-fluorescence observed in transmission through an organic crystal is critically dependent on crystal thickness, due to the overlap of the absorption and emission spectra. Only a fraction of the molecular emission occurs in a spectral region to which the crystal is transparent, the remainder being absorbed. The absorbed radiation will be re-emitted as fluorescence, this process of emission and absorption recurring until all the initial excitation energy either escapes from the crystal as fluorescence in the transparent region, or is dissipated thermally by internal conversion. Due to this 'photon cascade' process, the technical decay times t_I and t_f observed for thick crystals would be expected to correspond to the sum of the decay times of several molecular emissions (Birks 1953). This distinction between the technical and molecular decay times appears to have been overlooked by previous observers.

An experimental method has therefore been developed for the measurement of photo-fluorescence decay times, and it has been used to study the effect of crystal size on t_f, and to determine the molecular decay time $(t_f)_0$ for various organic compounds of interest.

7

§2. Principle of the Method

When a phosphor of photo-fluorescence decay time t_f is excited by light, whose intensity

$$I = I_c + I_0 \sin^2 \omega t \qquad \ldots\ldots (1)$$

is modulated at a high angular frequency 2ω, the fluorescence emission is also modulated in intensity, but its phase lags behind that of the exciting light, and its degree of modulation differs from that of the exciting light.

The decay of the fluorescence emission from an organic molecule is exponential (Pringsheim 1949). It is shown subsequently that the technical decay process, consisting of several such monomolecular decays, is also exponential. This is confirmed by the shape of the observed modulation curves, and it is in agreement with the exponential scintillation decay pulses, excited in organic crystals by fast electrons and recorded by other observers. A non-exponential scintillation decay may occur with heavy particle excitation, due to bimolecular quenching processes in the primary ionization column (Wright 1953).

The rate of emission of light by a phosphor, excited by a light pulse of intensity I at the time $t = 0$, is at a time t given by

$$\frac{dS}{dt} = \frac{qI}{t_f} \exp\left(-t/t_f\right) \qquad \ldots\ldots (2)$$

where q is the photo-fluorescence efficiency. Hence the intensity of the fluorescence emission, excited by a modulated light source whose intensity is given by (1), is at a time t_0

$$S = \int_{-\infty}^{t_0} \frac{q}{t_f}(I_c + I_0 \sin^2 \omega t) \exp\left\{-(t_0 - t)/t_f\right\}dt$$

$$= qI_c + \tfrac{1}{2}qI_0[1 - \cos\phi \cos(2\omega t_0 - \phi)] \qquad \ldots\ldots (3)$$

where

$$\tan\phi = 2\omega t_f \qquad \ldots\ldots (4)$$

and ϕ is the phase lag between the exciting light and the fluorescence emission.

The degree of modulation of the source is

$$m_s = \frac{I_{max} - I_{min}}{I_{max} + I_{min}} = \frac{I_0}{2I_c + I_0} \qquad \ldots\ldots (5)$$

and of the fluorescence emission is

$$m_f = \frac{S_{max} - S_{min}}{S_{max} + S_{min}} = m_s \cos\phi. \qquad \ldots\ldots (6)$$

Hence the fluorescence decay time t_f can be obtained from observations of either the phase lag ϕ or the relative modulation m_f/m_s of the fluorescence emission.

§3. Experimental Apparatus

3.1. *General Description*

A diagram of the apparatus is shown in fig. 1. The power from a 40 watt 7·5 Mc/s 807 Hartley oscillator is fed via a concentric cable to a resonant circuit across the high-frequency discharge tube, whose light output is thus modulated at 15·0 Mc/s. The oscillator also feeds a frequency doubler stage, and the 15 Mc/s signal is fed via a 'coil line' phase changer, to a constant-output tuned amplifier. The amplifier output, which is of constant amplitude and of variable phase, is applied between the photo-cathode and the first dynode of an RCA 931-A photomultiplier tube, thus modulating the sensitivity of the tube at a frequency of 15 Mc/s. The integrated current i from the photomultiplier is measured with a sensitive galvanometer. The magnitude of i depends on the relative phase of the cathode – first dynode modulating potential and the incident

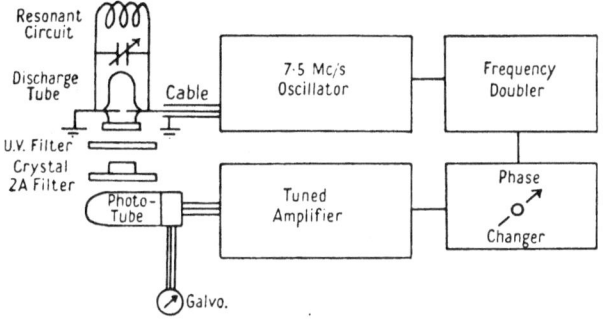

Fig. 1.   Schematic design of apparatus.

modulated light. The current is measured as a function of the phase difference between the potential applied to the discharge tube and that applied to the photomultiplier tube, and hence the degree of modulation and the relative phase of the incident light are obtained. Similar measurements on the fluorescence emission excited in the phosphor by the light from the discharge tube enable ϕ, the phase lag, and m_f/m_s, the relative degree of modulation, to be determined.

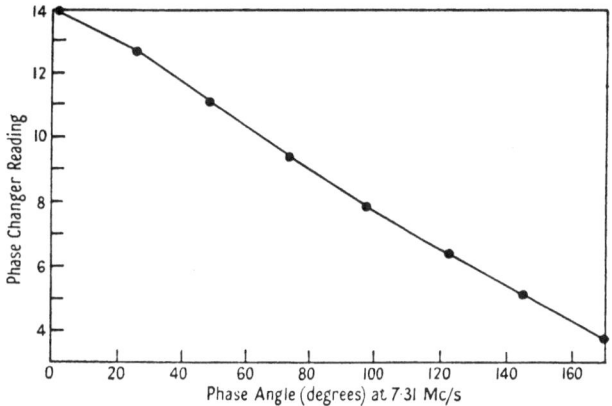

Fig. 2.   Calibration of phase changer.

3.2. *The Design and Calibration of the Phase Changer*

The phase changer (Hund 1936) is an artificial line which consists of ten $2\,\mu\text{H}$ inductances wound on a single bakelite former with each inductance shunted to earth with a $50\,\text{pF}$ condenser, and the line is terminated in a matched resistive load. A movable pick-up coil is mounted inside the coil former on a graduated, adjustable screw. The phase changer was calibrated by comparison with a short length of cable of known electrical length. The 15 Mc/s signal

61–2

from the phase changer was fed to the X deflection plates of a fast oscilloscope, and the 15 Mc/s signal from the frequency doubler was fed via an amplifier through a concentric cable to the Y deflection plates. The phases of the two signals were adjusted to give a linear trace on the oscilloscope screen. The introduction of a short known additional length of cable in the Y-plate channel caused a phase change. The resultant elliptical oscilloscope trace was reduced to a linear trace again by an adjustment of the phase changer, whose reading was equated to the calculated phase change introduced by the cable. The calibration curve of the phase changer is shown in fig. 2. Different calibration runs agreed to within less than 1° of phase, i.e. approximately 2×10^{-10} second. The signal from the phase changer is variable over 300° and is approximately constant in amplitude. It is fed through a limiting tuned amplifier giving an output constant in amplitude to better than 0·1% over the whole range of phase.

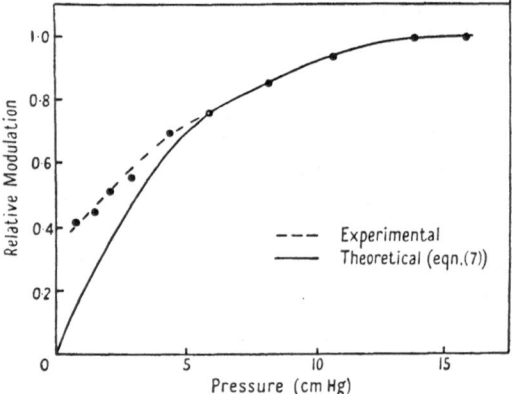

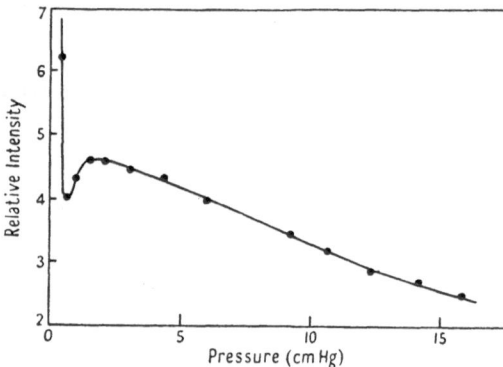

Fig. 3. Degree of modulation plotted against pressure for discharge tube.

Fig. 4. Light intensity plotted against pressure for discharge tube.

3.3. *Design and Studies of Modulated Discharge Tube*

The high-frequency gas discharge tube operates at pressures of up to 15 cm Hg. Air is found to be the most satisfactory and convenient gas. The discharge tube has two needle-point tungsten electrodes with 1 mm separation, sealed into a Pyrex tube which has a flat quartz window. The unperturbed decay times of the excited states of the O_2 and N_2 molecules are relatively long, of the order of 5×10^{-8} sec (Frey 1936), so that at very low pressures the degree of modulation of the light output from the discharge is low though the intensity is high. An increase in the pressure p causes collisional quenching of the excited gas molecules, giving a reduction in intensity, but an increase in the degree of modulation due to the reduced decay time. If it is assumed that the decay time of the excited state of the gas molecule is inversely proportional to the pressure, then the degree of modulation m_s is given theoretically by an equation of the form

$$m_s = (1 + a/p^2)^{-1.2}. \qquad \ldots\ldots(7)$$

The relative degree of modulation of the 7·5 Mc/s air discharge has been measured as a function of p. These measurements are plotted in fig. 3, and they are in good agreement with the theoretical relation (7) for values of p from approximately 3 cm to 15 cm Hg for $a = 35$. The relative intensity of the light output as a function of p is plotted in fig. 4. The large increase in intensity at low pressures is due to a sudden increase in the volume of the discharge.

Although these measurements on the high-frequency gas discharge were primarily conducted to establish the optimum conditions of the modulated light source for the photo-fluorescence studies, the results obtained indicate that the experimental method should be useful in studies of the decay times of the excited states of gas molecules. The degree of modulation of the discharge is simply related to the decay time, and different spectral bands can be isolated and studied by the use of appropriate filters. For the photo-fluorescence studies an air pressure of about 5 cm Hg was used, since this gave a high degree of modulation combined with an adequate intensity.

§4. EXPERIMENTAL RESULTS

4.1. *Decay Times*

A Wood's glass filter, transmitting only ultra-violet radiation, was placed over the window of the discharge tube, and the photomultiplier output current i was measured as a function of the phase difference between the discharge tube potential and the cathode – first dynode potential (fig. 5, curve a). A Wratten 2A

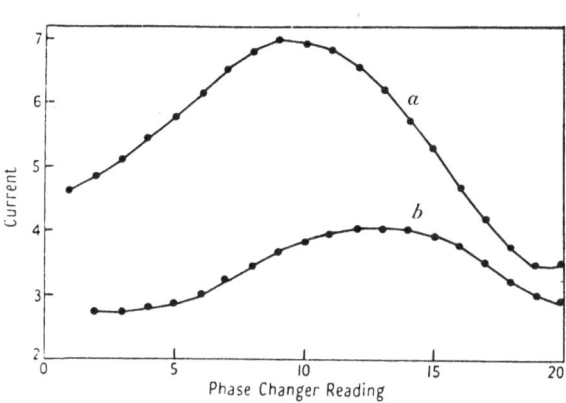

Fig. 5. Photomultiplier current i plotted against phase changer reading; a, light source; b, anthracene crystal.

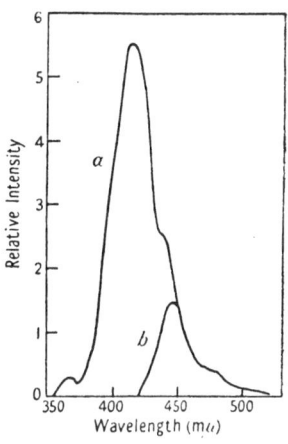

Fig. 6. Fluorescence spectra of anthracene; a, microcrystals (molecular spectrum); b, 1 cm thick crystal (technical spectrum).

filter placed between the Wood's glass filter and the photomultiplier was found to absorb almost all the radiation from the discharge tube. When an anthracene crystal, 1 cm thick, was introduced between the two filters, a strong photo-multiplier signal was obtained due to the anthracene photo-fluorescence, which is transmitted by the Wratten 2A filter. The current i (corrected for the small background current present in the absence of the crystal) was observed as before as a function of phase difference (fig. 5, curve b). The phase lag $\phi = 53° \pm 3°$, corresponding to $t_f = 14\cdot4 \pm 1\cdot5$ mμ sec. A similar value of $\phi = 55° \pm 1\cdot2°$, corresponding to $t_f = 15\cdot5 \pm 0\cdot7$ mμ sec, is obtained from the relative degree of modulation m_f/m_s.

Similar measurements of t_f at room temperature were made on other anthracene single crystals, and also on single crystals of *trans*-stilbene, diphenyl-acetylene and *para*-terphenyl. In addition, observations were made on thin powdered specimens of anthracene and *para*-terphenyl, prepared by grinding,

11

and on microcrystalline layers of each of the four materials formed on a thin glass plate, by evaporation of a dilute solution of the compound in xylene. In the latter case it was only found possible to excite the photo-fluorescence of anthracene and stilbene with sufficient intensity to observe t_f, since the absorption spectra of the other two compounds lie further into the ultra-violet region, where the emission of the source is weak. The mean observed values of t_f for the various specimens studied are listed in the table.

Compound	Description of specimen	t_f (mμ sec)	t_f(L) (mμ sec)
Anthracene	1 cm cube	14 ± 2	17
	Powder	$4\cdot8 \pm 1\cdot5$	—
	Microcrystals	$3\cdot5 \pm 1\cdot0$	—
	$\frac{1}{2}\%$ solution in benzene	—	$2\cdot0 \pm 0\cdot5$
Stilbene	2 mm thick	$3\cdot0 \pm 0\cdot8$	3·1
	Microcrystals	$1\cdot7 \pm 0\cdot6$	—
Diphenylacetylene	2 mm thick	$3\cdot0 \pm 0\cdot8$	2·5
Terphenyl	2 mm thick	$3\cdot8 \pm 1\cdot0$	11
	Fine powder	$3\cdot5 \pm 1\cdot0$	—
	$\frac{1}{2}\%$ solution in benzene	—	$2\cdot5 \pm 0\cdot5$

The values of t_f reported by Liebson (1952) (t_f(L)) for crystals of unspecified size (probably >1 mm thickness) and other specimens are shown in the last column of the table for comparison.

4.2. *Fluorescence Spectra*

The emission spectra of the different anthracene specimens, excited by monochromatic radiation of 254 mμ wavelength, have been measured. The spectra were observed by transmission through the specimens using a modified Cenco-Sheard 'spectrophotelometer' with an EMI 6262 photomultiplier tube as the radiation detector. The intensity readings were corrected for the variation of the sensitivity of the photomultiplier and of the transmission coefficient of the 'spectrophotelometer' with wavelength. The fluorescence spectra from (a) a microcrystalline specimen, and (b) a single crystal, of 1 cm thickness, normalized at a wavelength of 450 mμ, are plotted in fig. 6. The absence of vibrational structure is due to the low resolution of the experimental arrangement.

The spectra of the anthracene powders were intermediate between (a) and (b), extending from a wavelength of about 390 mμ, with an intensity peak at about 425 mμ, and coinciding with the other two spectra at wavelengths greater than 450 mμ. These powder spectra, like the corresponding decay time measurements, are critically dependent on the grain size. For crystal thicknesses greater than 1 mm the spectrum is practically independent of thickness, and similar to that shown in fig. 6, curve b.

The emission spectrum of a 2 mm thick mixed crystal of naphthalene, containing 0·01% of anthracene by weight, and excited by the 254 mμ radiation, was observed to contain two spectral components characteristic of the naphthalene and anthracene molecules respectively. The spectrum of the anthracene emission was found to be similar to that from the pure anthracene microcrystals (fig. 6, curve a). For mixed crystals of greater thickness and higher anthracene concentration, the anthracene emission spectra were intermediate between

(a) and (b), and similar to those from pure anthracene powders. Certain of these spectral measurements have been reported previously (Little and Birks 1952) but have not been published.

§5. Discussion

The dependence of the fluorescence spectra of the anthracene specimens on crystal size is due to the overlap of the absorption and emission spectra. For the microcrystalline specimens the self-absorption is practically eliminated, and the observed spectrum corresponds to the true molecular fluorescence of anthracene in the crystalline state. It is similar to that of anthracene in dilute solid solution in naphthalene, and it also agrees closely with that of anthracene in solid solution in polystyrene, observed by Koski (1951). In each case the complete molecular spectrum is observed because the surrounding medium is transparent to the emission.

In a large crystal, a fraction of the fluorescence emitted by the molecules initially excited by the incident radiation is reabsorbed. This absorbed radiation is subsequently re-emitted as molecular fluorescence, which hence includes both Stokes and anti-Stokes radiation. The process recurs until all the fluorescence emission is in the spectral region to which the crystal is transparent. The spectrum observed (fig. 6, curve b) corresponds to the technical emission spectrum of an anthracene crystal of thickness greater than 1 mm. The probability k_e of the escape of the molecular emission is given by the ratio of the area under the technical spectrum to the area under the molecular spectrum. For the thick anthracene crystals $k_e = 0.2$ from fig. 6. The probability of absorption $k = 1 - k_e = 0.8$.

The effect of the self-absorption on the observed value of t_f may be calculated. Let N be the number of excited molecules at time t, $1/(t_f)_{00}$ be the probability per unit time of molecular emission, and $1/t_i$ be the probability per unit time of internal conversion. Then

$$-\frac{dN}{dt} = N[1/(t_f)_{00} + 1/t_i - k/(t_f)_{00}] = N\lambda, \qquad \ldots\ldots(8)$$

$$N = N_0 e^{-\lambda t}. \qquad \ldots\ldots(9)$$

The intensity of the fluorescence emission escaping from the crystal is

$$J = (1-k)N/(t_f)_{00} = J_0 e^{-\lambda t}. \qquad \ldots\ldots(10)$$

Hence the decay of the emission is exponential, and of technical decay time

$$t_f = \frac{1}{\lambda} = \frac{(t_f)_{00}}{k_e + (t_f)_{00}/t_i}. \qquad \ldots\ldots(11)$$

The molecular decay time ($k=0$, $k_e=1$) is

$$(t_f)_0 = \frac{(t_f)_{00}}{1 + (t_f)_{00}/t_i} = q_0(t_f)_{00} \qquad \ldots\ldots(12)$$

where q_0 is the molecular photo-fluorescence quantum efficiency. In the absence of internal quenching ($q_0 = 1$)

$$t_f = (t_f)_0/k_e. \qquad \ldots\ldots(13)$$

Bowen, Mikiewicz and Smith (1949) have observed $q_0 = 0.9$ for small anthracene crystals at room temperature. This represents a minimum value for q_0, since it is not certain that their specimens were sufficiently thin to eliminate self-absorption completely. Substituting $q_0 = 0.9$, $k_e = 0.2$, $(t_f)_0 = 3.5 \pm 1.0 \, m\mu$ sec in the equations above, we obtain $(t_f)_{00} = 3\cdot9 \pm 1\cdot2$ mμ sec, $t_i = 35 \pm 10$ mμ sec, and $t_f = 12\cdot6 \pm 3\cdot5$ mμ sec. This value of t_f agrees, within the experimental error, with that of $t_f = 14 \pm 2$ mμ sec observed directly.

The photo-fluorescence quantum efficiency q_0 of a solution of anthracene in benzene is $\sim 0\cdot65$ (Bowen 1949). Since self-absorption will be negligible in a relatively dilute solution, we may equate Liebson's value of $t_f = 2\cdot0 \pm 0\cdot5$ mμ sec to $(t_f)_0$, giving $(t_f)_{00} = 3\cdot1 \pm 0\cdot8$ mμ sec in agreement with the value obtained from the crystal measurements.

The smaller difference in the values of t_f observed for thick crystals and microcrystals of *trans*-stilbene is due to the reduced overlap of the absorptino and emission spectra. The reduced self-absorption of stilbene compared with anthracene is confirmed by the observations of Koski (1951), who compared the emission spectrum of a stilbene crystal with that of a dilute solution of stilbene in polystyrene. The data on t_f for stilbene, diphenylacetylene and terphenyl indicate that $(t_f)_{00}$ for each of these materials is of the order of 3–4 mμ sec. This is to be expected since the oscillator strengths of the optical transition from the first electronic excited state to the ground state, responsible for the fluorescence emission, are similar to that of anthracene.

Liebson's value of $t_f = 11$ mμ sec for terphenyl appears anomalous, and disagrees with his value of $t_1 = 6$ mμ sec for the same material. Since $t_1 \sim 2t_f$ we may estimate $t_f \sim 3$ mμ sec from this latter measurement, in agreement with the general results.

§6. Conclusion

The results indicate that the photo-fluorescence molecular decay time $(t_f)_{00}$, in the absence of internal quenching, of the organic compounds studied is of the order of 3–4 mμ sec. This is reduced to $(t_f)_0$ for microcrystals, due to internal conversion, and increased to t_f for large crystals, due to self-absorption. The spectral and decay time measurements on anthracene crystals show clearly that intermolecular energy exchange occurs by photon emission and reabsorption, as proposed by Birks (1953).

Acknowledgment

We wish to thank the South African Council for Scientific and Industrial Research for a research grant in support of this work, and for a bursary awarded to one of us (W. A. L.).

References

Birks, J. B., 1953, *Scintillation Counters* (London : Pergamon Press; New York : McGraw-Hill).

Bowen, E. J., 1949, *The Chemical Aspects of Light* (Oxford : University Press).

Bowen, E. J., and Lawley, P. D., 1949, *Nature, Lond.*, **164,** 572.

Bowen, E. J., Mikiewicz, E., and Smith, F. W., 1949, *Proc. Phys. Soc.* A, **62,** 26.

Frey, A. R., 1936, *Phys. Rev.*, **49,** 305.

Hund, A., 1936, *Phenomena in High Frequency Systems* (New York : McGraw-Hill).

Koski, W. S., 1951, *Phys. Rev.*, **82,** 230.

Liebson, S. H., Bishop, M. E., and Elliot, J. O., 1950, *Phys. Rev.*, **80,** 907.

Liebson, S. H., 1952, *Nucleonics*, **10,** 41.

Little, W. A., and Birks, J. B., 1952, South African Association for the Advancement of Science Congress, Cape Town, South Africa (July 1952).

Pringsheim, P., 1949, *Fluorescence and Phosphorescence* (New York : Interscience).

Wright, G. T., 1953, *Phys. Rev.*, in the press.

THE TRANSPORT OF HEAT BETWEEN DISSIMILAR SOLIDS AT LOW TEMPERATURES[1]

W. A. LITTLE[2]

ABSTRACT

The resistance offered to the flow of heat by the mismatch of the elastic constants at the interface between two materials has been calculated. It is shown that for a perfectly joined interface the heat flow is proportional to the difference of the fourth powers of the temperature on each side of the interface. Deviations from this temperature dependence are to be expected for rough surfaces and for surfaces pressed into contact with one another. The calculated contact resistance between some common solids is given, and graphs are presented from which the heat flow between any two materials may be computed. It is shown too that the spin–phonon, phonon–electron, and phonon–phonon relaxation processes give rise to additional resistive processes in some solids, some of which restrict the heat flow at the surface. The theoretical results compare well with the available experimental data. However, the problem of the contact resistance between helium and metals is still unresolved. A tentative explanation of this is presented which predicts that there should be an appreciable difference between the contact resistance of a metal in the normal and in the superconducting state.

INTRODUCTION

In experiments conducted at very low temperatures it has become well established that a finite difference of temperature will occur at an interface between two dissimilar materials when a heat flux is maintained across this interface. Kapitza (1941) first observed such an effect between helium II and a metal, and since then many investigators have reported similar results with helium II (Fairbank and Wilks 1955) and helium 3 (Fairbank and Lee 1957) and between paramagnetic salts and metallic fins (Goodman 1953). Apart from several attempts to explain the contact resistance in the specific case of a helium–metal interface (Kronig and Thellung 1950; Gorter, Taconis, and Beenakker 1951; and Khalatnikov 1952), no general explanation of the effect has been offered. In view of the fact that this resistance plays a very important role in all experiments using cooling by contact at temperatures below 1° K, a general explanation is long overdue. In this paper such an attempt has been made, firstly, by considering the ideal case of two perfectly joined media and, secondly, by extending this result to several cases of practical importance.

I. PERFECT JUNCTION

Let us consider two semi-infinite isotropic solids joined in the plane zy at $x = 0$. In order to simplify the problem let us suppose that the modulus of rigidity of both materials is zero so that we neglect the transverse phonons in both solids. This greatly simplified the boundary conditions at the interface

[1]Manuscript received November 14, 1958.
Contribution from the University of British Columbia, Vancouver, British Columbia.
[2]Present address: Stanford University, Stanford, California, U.S.A.

and yet allows one to see the trend of the solution in the more general case considered in II.

The number of phonons of energy $h\nu$ incident upon a small area dA of the interface per unit time at an angle of incidence between θ_1 and $\theta_1+d\theta_1$ will be:

$$(1) \qquad \tfrac{1}{2}c_1 N_1(\nu) \cos\theta_1 \sin\theta_1 d\theta_1 dA d\nu,$$

where c_1 is the group velocity of the longitudinal phonons in the first medium and $N_1(\nu)d\nu$ is the number of phonons of energy lying between $h\nu$ and $h(\nu+d\nu)$ per unit volume of the first medium.

Some of these phonons will be reflected at the interface and some will be refracted into the second medium at an angle of refraction between θ_2 and $\theta_2+d\theta_2$. The two angles θ_1 and θ_2 are related by Snell's law of acoustic refraction from which one may obtain the relation:

$$(2) \qquad d\theta_2 = \frac{c_2 \cos\theta_1}{c_1 \cos\theta_2} \cdot d\theta,$$

where c_2 is the group velocity of longitudinal phonons in the second medium. The nett flow of heat across the interface from medium 1 to medium 2 due to these phonons leaving or entering at an angle θ_1 is then:

$$(3) \qquad \frac{dQ}{dt} = \tfrac{1}{2}\int\int\int N_1(\nu)h\nu c_1 \alpha_1(\theta_1) \cos\theta_1 \sin\theta_1 d\theta_1 d\nu dA$$
$$- \tfrac{1}{2}\int\int\int N_2(\nu)h\nu c_2 \alpha_2(\theta_2) \cos\theta_2 \sin\theta_2 d\theta_2 d\nu dA.$$

In this expression $\alpha_1(\theta_1)$ is the transmission coefficient of the interface for phonons incident at an angle θ_1 in medium 1. Because the longitudinal phonons are the quantized longitudinal acoustic waves, this coefficient will be given by the equivalent expression for acoustic waves (Rayleigh 1894):

$$(4) \qquad \alpha_1(\theta_1) = \alpha_2(\theta_2) = \frac{\dfrac{4 \rho_2 c_2}{\rho_1 c_1} \cdot \dfrac{\cos\theta_2}{\cos\theta_1}}{\left(\dfrac{\rho_2 c_2}{\rho_1 c_1} + \dfrac{\cos\theta_2}{\cos\theta_1}\right)^2},$$

where ρ_1 and ρ_2 are the densities of the two media.

The value of $N(\nu)$ is given by the number of degrees of freedom of the lattice lying between ν and $\nu+d\nu$, and the Bose–Einstein energy distribution.

$$(5) \qquad N(\nu)d\nu = \frac{4\pi}{c_1^3} \frac{\nu^2 d\nu}{(e^{h\nu/kT}-1)}.$$

Inserting (5) in (3) and integrating with respect to A one obtains the expression:

$$(6) \qquad \frac{dQ}{dt} = \frac{2\pi k^4 \Gamma A}{h^3 c_1^2} [T_1^4 f(T_1) - T_2^4 f(T_2)] \text{ ergs/second}$$

where

(7)
$$f(T) = \int_0^{h\nu_m/kT} \frac{z^3 dz}{e^z - 1}$$

with

$$z = h\nu/kT$$

and

$$\Gamma = \int_0^{\pi/2} \alpha_1(\theta_1) \sin\theta_1 \cos\theta_1 \, d\theta_1.$$

At low temperatures one may make the usual approximation ($h\nu_m/kT = \infty$) in the integral over z as it occurs in the Debye theory of specific heats, and thus obtain the relation:

(8)
$$\frac{dQ}{dt} = \frac{2\pi k^4 \Gamma A}{h^3 c_1^2} \left(\frac{\pi^4}{15}\right) (T_1^4 - T_2^4)$$

$$= 5.01 \times 10^{16} \frac{\Gamma A}{c_1^2} (T_1^4 - T_2^4) \text{ ergs/second}$$

with c_1 in cm/sec and T in °K.

For small values of the ratio $(T_1 - T_2)/T_1$ we may write this in the convenient form:

(9)
$$\frac{dQ}{dt} = 2.0 \times 10^{17} \frac{\Gamma A}{c_1^2} T_1^3 \Delta T \text{ ergs/second}$$

where

$$\Delta T = T_1 - T_2.$$

The integral Γ is a complicated function of the densities and of the acoustic velocities of the two media. In its explicit form it is difficult to use, so it has been computed for a range of values of the ratio of the densities and the ratio of the acoustic velocities. The integral is given in graphical form in Fig. 1.

This simple acoustic model shows that one should expect a discontinuous change of temperature across the interface and that for small heat fluxes the drop in temperature ΔT should be proportional to the heat flux and to the third power of the mean absolute temperature. From the variation of Γ as shown in Fig. 1, with respect to the relative densities and acoustic velocities in the two media, it can be seen that to obtain the maximum transfer of heat from one medium to another the densities should be similar and the acoustic velocities similar. The strong dependence of Γ on the relative acoustic velocities for $c_2 - c_1$ is due to the total reflection of phonons for θ_1 greater than the critical angle $\sin^{-1}(c_1/c_2)$. It should be noted that the matching of the specific acoustic impedance of each medium (i.e., $\rho_1 c_1 = \rho_2 c_2$) is not the condition for the maximum transfer of heat when $c_2 > c_1$.

The maximum value of Γ is 0.5, which corresponds to a perfect match of the densities and the acoustic velocities. In this case (6) predicts that a

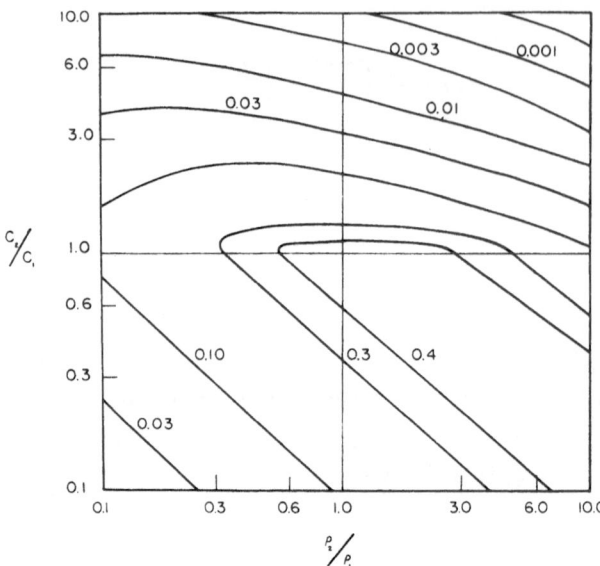

FIG. 1. Contours of Γ for transmission of heat from medium 1 to medium 2 with respect to the ratio of the acoustic velocities c_2/c_1 and the densities ρ_2/ρ_1 of the two media. Poisson's ratio is equal to 0.5 in both media.

temperature difference will occur across the boundary. This apparent paradox is explained by the fact that it was tacitly assumed in (6) that the two media were each at a uniform temperature in the presence of the heat flux. This implies that the region in which the temperature is measured should extend for a distance no greater than the mean free path of the phonons on each side of the barrier. In the particular case of identical media sustaining a heat flux the theory of thermal conductivities requires that there should be a temperature drop over a distance of the mean free path.

That this is so can be shown as follows. If $c_1 = c_2$ and $\rho_1 = \rho_2$ then $\alpha_1(\theta) = 1$ for all values of θ. Let T be the temperature at the barrier and let the mean free path of phonons in this medium be λ then the temperature T_1 of a point where a phonon is last scattered before striking the barrier will be $[T+(\partial T/\partial z)\lambda \cos \theta]$ and likewise T_2 will be $[T-(\partial T/\partial z)\lambda \cos \theta]$. At low temperatures and for small values of T_1-T_2 we obtain

$$\frac{dQ}{dt} = \frac{2\pi k^4 A}{h^3 c_1^2}\left(\frac{\pi^4}{15}\right)4T_1^3 \frac{\partial T}{\partial z}2\lambda \int_0^{\pi/2}\cos^2\theta \sin \theta \, d\theta$$

$$= \frac{1}{3} C_v \lambda c_1 \left(\frac{\partial T}{\partial z}\right) A,$$

where C_v = lattice specific heat unit/volume which gives a thermal conductivity of $\frac{1}{3}C_v\lambda c_1$, which is the usual expression for the conductivity at low temperatures.

The conduction of heat over these small distances at very low temperature is, of course, analogous to that in gases at low pressures. In the case of a gas the heat flow for distance less than the mean free path is limited by the number of molecules, that is, the pressure, while in a solid the heat flow is limited by the number of phonons, which depends upon the temperature. Hence the contact resistance is related to the fact that for distances less than the mean free path the heat flux is not proportional to the temperature gradient but requires a finite temperature difference. The effect of the interface is to reduce this heat flux for a given temperature difference in proportion to the transmission coefficient of the barrier.

It is possible to extend this acoustic model from the ideal case considered above to obtain a qualitative understanding of the effect of roughness and of imperfect contact at the interface upon the temperature dependence and magnitude of the contact resistance. This may be seen by first considering what happens if the effective area of the interface depends upon the frequency of the incident lattice waves. Suppose that the effective area of the interface $A'(\nu)$ for phonons of frequency ν may be written as

$$(10) \qquad A'(\nu) = A_0\nu^n,$$

where n may be 0 or a positive or negative quantity. We will assume that such an expression is possible over the range of frequencies of phonons which contribute appreciably to the heat content of the lattice. Using (10), equation (6) now yields

$$(11) \qquad \frac{dQ}{dt} = B\,(T_1^{n+4} - T_2^{n+4})$$

where B contains terms independent of the temperature.

The heat flux across the interface now will be more or less strongly dependent upon the absolute temperature depending on n being positive or negative.

Let us now consider two cases of practical interest, firstly, an interface which is rough but perfectly joined at all points and, secondly, an interface at which contact between the two media occurs at some points but not at others.

The Rough Interface

Let Fig. 2 represent the interface between the two rough media. Let $\bar{A}$ be defined as the mean amplitude of the roughness. If the mean free path of the

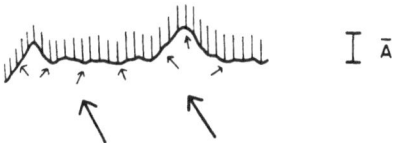

FIG. 2. Diagrammatic representation of a rough interface illustrating the larger effective area of the interface to short lattice waves compared with that for long waves.

phonons is comparable to $\bar{A}$ then the effective area of the interface will differ for phonons of different frequency. The roughness of the interface will occur over a small fraction of a wavelength of a long lattice wave and, consequently, the wave will be transmitted as if the interface was smooth. For these waves the effective area will be the macroscopic area of a plane at the interface. On the other hand the short lattice waves will be transmitted across each individual hill and dale at the interface and the effective area will be the integrated microscopic area of the interface. We should expect, then, the effective area to increase with increasing phonon energy.

$A'(\nu)$ now will be given by (10) with a positive value of n. Also, the transport of heat across the interface will be greater than that given by equation (8) due to the average effective area being greater than the macroscopic area of the interface.

If, however, the mean free path is much greater than $\bar{A}$ the number of phonons of given energy incident upon the interface will be determined by the cross-sectional area of the solid normal to the heat flux and not by the microscopic area of the interface. In this case the effective area of the interface will be the same for all phonons and equation (8) will apply with A given by the macroscopic cross-sectional area of the interface.

Imperfect Contact

Another case of some practical interest is that of the imperfect junction of two media. In many experiments of contact cooling at temperatures below 1° K, the specimen is cooled via a metallic link which is pressed into contact with a powdered paramagnetic salt. Berman (1956) has shown that the amount of heat transported across the interface between two surfaces pressed into contact is proportional to the total force applied. This suggests that the contact occurs at a few points, and that as greater pressure is applied the surfaces deform and more points make contact, the number of point contacts then being proportional to the applied pressure with each carrying approximately the same heat flux. In the case of a paramagnetic salt being pressed in powder form against a metallic fin the contact would occur at a number of points. It is instructive to apply our acoustic model to this case.

We will suppose that the interface can be represented by Figs. 3(a) and 3(b)—a series of point contacts separated by a mean distance, $\bar{B}$. Phonons with wavelengths which are very large compared with $\bar{B}$ will move all the points lying adjacent to one another approximately in phase, and as a result the whole boundary between adjacent points will move in phase (see Fig. 3(a)). The mean displacement at points on the boundary not in contact with the first medium will be only slightly less than that at points of contact. We should expect, therefore, that the strain in the second medium a short distance from the interface will be only slightly less than it would be if the junction were perfect. Hence the energy transported across the interface, which is proportional to the square of the strain, will be only slightly less than that for the ideal case. On the other hand, for phonons with wavelengths which are small compared with $\bar{B}$ the motion of adjacent points of contact will differ by a random phase factor. This factor will be determined for each point by the spatial localization of the phonon and the actual value of the separation of the points of contact. In this case the motion of the barrier *not* in contact with the first medium will tend to zero due to the destructive interference of the surface waves from all the nearby points of contact (see Fig. 3(b)). The mean displacement in the second medium now will be much smaller and the transmitted energy correspondingly reduced.

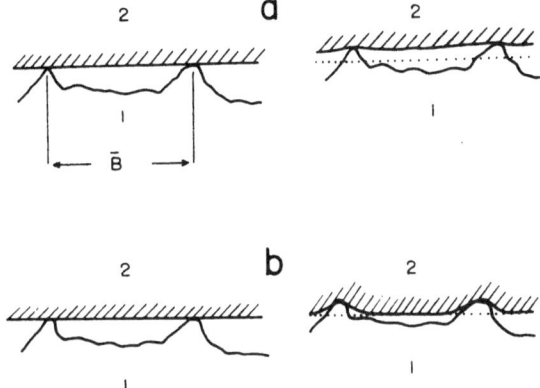

FIG. 3.   Diagrammatic representation of the interface of two solids in imperfect contact.
(a) The surface is shown, firstly, at rest, and, secondly, at the peak of a low frequency lattice wave. The wavelength in this case is much greater than the mean separation of the points of contact $\bar{B}$.
(b) The corresponding picture for a high frequency lattice wave. The dotted line represents the surface of the second medium at rest.

From the above considerations one would expect the effective area of the interface to decrease with increasing phonon frequency, giving (10) with a negative n. Consequently, the transport of heat across the barrier would be less temperature-dependent and the total heat flux less than that given by (8) for the macroscopic area of contact.

II. THE GENERAL CASE

In Section I we have considered the simple case of the transport of heat between two media in which the moduli of rigidity are both zero. This is a somewhat artificial case so it is necessary, for practical reasons, to consider the more general case where both longitudinal and transverse phonons occur. We shall consider the case of a perfect junction. The reasoning for the rough and imperfect junction is equally applicable to this case so there is no need to reconsider these effects.

Equation (8) may be readily generalized to add the contribution from the transverse phonons to the heat flow across the interface. We then obtain:

$$(12) \qquad \frac{dQ}{dt} = 5.01 \times 10^{16} A \left[\frac{\Gamma_1}{c_1^2} + \frac{2\Gamma_t}{c_t^2} \right] (T_1^4 - T_2^4) \text{ ergs/sec cm}^2,$$

where
Γ_1 = transmission coefficient of longitudinal phonons,
Γ_t = transmission coefficient of transverse phonons,
c_1 = velocity of longitudinal phonons,
c_t = velocity of transverse phonons.

In this case the detailed evaluation of Γ_1 and Γ_t is much more difficult. The reason for this is that at the interface each incident wave, whether longitudinal or transverse, in general breaks up into four waves, reflected transverse and longitudinal waves, and refracted transverse and longitudinal waves (Kolsky 1953). Because of the different velocities of the transverse and longitudinal waves, two critical angles of reflection are found and in the interval between these two angles a phase shift occurs in the refracted and reflected waves whose magnitude depends upon the angle of incidence. In this case there is no simple expression corresponding to equation (4). It is easier to calculate the transmission coefficient for each angle of incidence by a numerical solution of the equations determined by the boundary conditions. This has been done and the values of Γ_1 and Γ_t have been computed for a range of relative densities, Poisson's ratio of the media, and the relative velocities of the longitudinal waves.

Method of Calculation

The ALWAC IIIE computer at the U.B.C. Computing Center was programed to solve the eight linear equations given by the boundary conditions for the amplitude and phase of each of the four possible reflected and refracted waves (Kolsky 1953). The fraction of the energy of the incident wave transmitted to the second medium was computed for each angle of incidence θ. The integral was then performed using a Legendre–Gauss five point quadratures routine which approximates the transmission coefficient to a polynomial of ninth degree in θ. The program was checked at several points using the reciprocity relation:

$$(13) \qquad \left[\frac{\Gamma_1}{c_1^2} + \frac{2\Gamma_t}{c_t^2} \right]_{ij} = \left[\frac{\Gamma_1}{c_1^2} + \frac{2\Gamma_t}{c_t^2} \right]_{ji}$$

where the subscripts indicate the values of Γ_1 and Γ_t within the bracket from medium i to medium j.

The contours of Γ are given for a range of relative densities of the two media and the ratios of the velocities of longitudinal waves in the two media. In Figs. 4 and 5 Poisson's ratio σ was chosen to be 0.33 in both media, corresponding to $c_1/c_t = 2$; while in Fig. (6) and (7), $\sigma = 0.33$ for the first medium and 0.44 for the second. Poisson's ratio = 0.44 corresponds to $c_1/c_t = 3$. The values of Γ_t given in Figs. 5 and 7 are the values obtained after integrating over all possible angles of polarization of the transverse waves.

In Table I Γ_t, Γ_1, and dQ/dt are given for some common solids. These have been computed from the room temperature densities and acoustic velocities as given in the *American Institute of Physics Handbook*. In the case of crystalline solids these acoustic velocities are the values averaged over the different crystal directions. It has been assumed that we can approximate these solids to isotropic solids with these average values of the room temperature acoustic velocities and densities. In many cases this is not a very good approximation but the values of Γ still will be correct to within 10 to 20% judging from the variations of Γ in Figs. 4, 5, 6, and 7. This accuracy is sufficient to allow

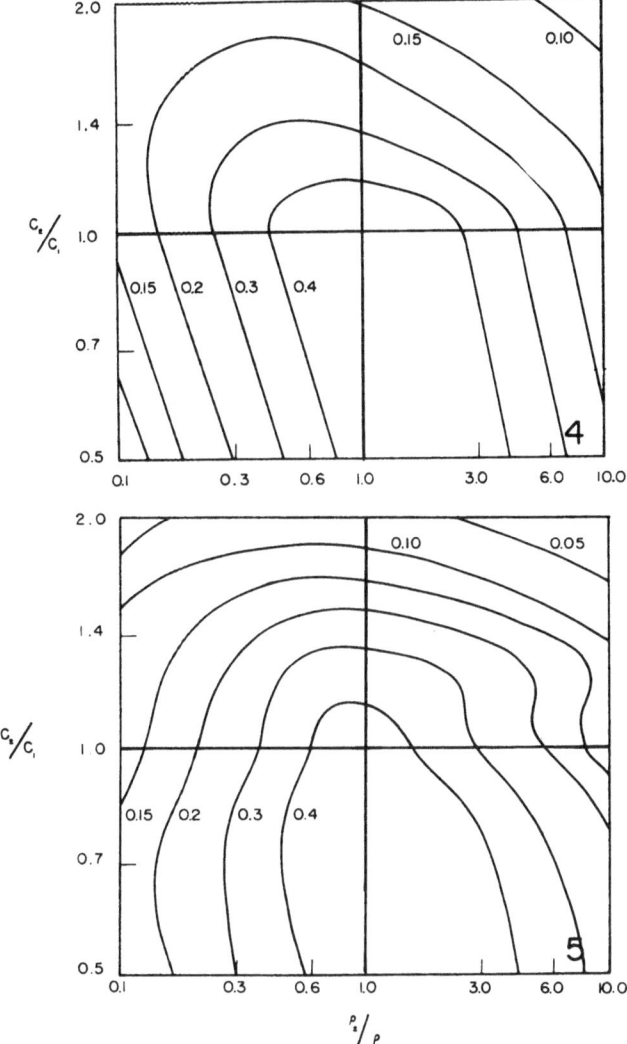

FIG. 4. Contours of Γ_l for the transmission of heat from medium 1 to 2 with respect to the ratio of the velocity of longitudinal waves in the two media and the relative densities. Poisson's ratio is equal to 0.33 in both media.

FIG. 5. Contours of Γ_t for the transmission of heat from medium 1 to 2 with respect to the ratio of the velocities of longitudinal waves in the two media. Poisson's ratio is equal to 0.33 in both media.

these values to be used as a guide in selecting suitable materials for optimum heat transfer.

So far we have only considered the contact resistance due to the imperfect transmission of phonons across the interface. In certain specific cases such as the transport of heat from a metal to a dielectric solid or from a metal to a paramagnetic salt there are additional factors which can contribute to the

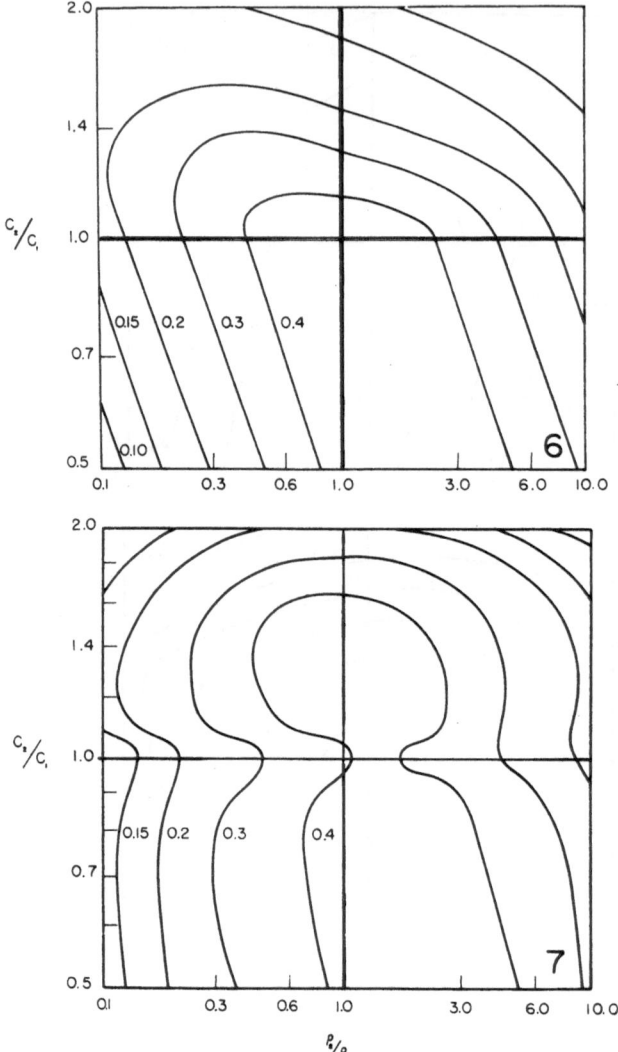

FIG. 6. Contours of Γ_l for Poisson's ratio equal to 0.33 in the first medium and 0.44 in the second.

FIG. 7. Contours of Γ_t for Poisson's ratio equal to 0.33 in the first medium and 0.44 in the second.

apparent contact resistance. This resistance does not necessarily occur at the points of contact of the two media but experimentally one might have difficulty distinguishing these effects from the true contact resistance discussed above. In the case of a metal in contact with a dielectric solid one has to consider the resistive contribution due to the finite electron–phonon relaxation time, whereas in the case of a paramagnetic salt in contact with a metal one has in addition to the phonon–electron relaxation time the spin–lattice and phonon–phonon relaxation times. The contribution of these will be discussed in turn.

III. PHONON-ELECTRON CONTACT RESISTANCE

Let us calculate the rate dQ/dt at which heat may be transported from the system of phonons in a metal to the conduction electrons. Let $L(v)$ be the mean free path of a phonon of energy hv for collisions with the conduction electrons. We will suppose that we are concerned with very low temperatures so that the mean free path for phonon–phonon collisions is very great compared with $L(v)$. In the simple theory of electron–phonon interactions there is no coupling between the transverse phonons and the conduction electrons so we shall calculate the transport of heat from the longitudinal phonons to the conduction electrons alone. This will serve to give us the order of magnitude of this effect. Furthermore, we will assume that on the average a fraction α of the energy of a phonon can be transferred to the electrons at a collision. We obtain the expression:

$$(14) \qquad \frac{dQ}{dt} = \int_v \frac{\alpha h v c_1 N(v) dv}{L(v)}.$$

The mean free path in a metal may be calculated in terms of the electron–phonon interaction and is given by Wilson (1953) as:

$$(15) \qquad \frac{1}{L(v)} = \left(\frac{hv}{k\Theta}\right) \frac{m^{*2} a C_j^2}{h^2 k\Theta M} \left(\frac{4\pi}{3}\right)^{1/3} 6\pi^2$$

where m^* = effective mass of conduction electrons,

$\quad\quad a$ = lattice constant,

$\quad\quad C_j$ = electron–phonon interaction energy $1\ ev < C_j < 10\ ev$,

$\quad\quad \Theta$ = Debye temperature,

$\quad\quad M$ = mass of an atom of the lattice.

Putting (15) in (14) and using the Bose–Einstein distribution for $N(v)$ one obtains the nett transfer of heat from the phonons at temperature T_1 to the electrons at temperature T_2

$$(16) \qquad \frac{dQ}{dt} = 9R\alpha \left[\frac{m^{*2} a^2 C_j^2}{h^2 M\Theta^4}\right] (T_1{}^5 - T_2{}^5)\ 6\pi^2 \left(\frac{4\pi}{3}\right)^{2/3}$$

$$\times \int_0^\infty \frac{x^4}{e^x - 1} dx \text{ ergs/sec g-atom.}$$

This expression has been evaluated for copper on the assumption that $m^* = m$, and $C_j = 3\ ev$, and $\alpha = 0.25$

$$(17) \qquad \frac{dQ}{dt} = 1.0 \times 10^{10}\ (T_1{}^5 - T_2{}^5) \text{ ergs/sec/cc}$$

and the mean free path for the effective transfer of energy to the conduction electrons for phonons of energy $hv = kT$

$$(18) \qquad L = 4 \times 10^{-3} (1/T) \text{ cm.}$$

These calculations serve to illustrate two points. Firstly, it is necessary to have an appreciable volume as well as a large surface area of a metal in contact with a cooling salt in order to effect the maximum transfer of heat at very low temperatures. Secondly, the technique of embedding many fine copper wires in a paramagnetic salt to obtain good thermal contact has only limited value at extremely low temperatures. There is no point in using extremely fine wire in order to increase the surface area alone if the diameter of the wire becomes comparable to the mean free path of the phonons. If the diameter is of this order those phonons which do penetrate the surface will have a good chance of escaping without transferring their energy to the conduction electrons.

IV. CONTACT RESISTANCE WITH PARAMAGNETIC SALTS

Where a paramagnetic salt is used to cool a metal in addition to the contact resistance at the surface and that due to the finite phonon–electron relaxation time, there are other resistive elements due to the spin–lattice and the phonon-phonon relaxation times. It appears to be generally accepted now that in most paramagnetic salts the spin–phonon relaxation times are much shorter than the phonon–phonon relaxation times (Strandberg 1958), with the result that the usual equilibrium distribution of the phonons is distorted in the region of phonon energies close to the spin energy.

This has a very important effect upon the transport of heat between the spin system and a metal. This can be shown quite simply by making the approximation that those phonons lying in the frequency range ν_1 to ν_2 at the electron spin frequency are so strongly coupled to the spin system that their distribution between these frequency limits corresponds to that of the spin temperature, while those phonons outside this range are completely decoupled from the spin system and have a distribution corresponding to the lattice temperature. If this salt is in contact with a metal at a different temperature to the spin temperature those phonons outside the range ν_1 to ν_2 will rapidly come to the temperature of the metal because of the small heat capacity of the lattice at these temperatures, while those phonons inside the range ν_1 to ν_2 will remain at the spin temperature and will be the only means by which energy can be transferred between the spin system and the metal. As a consequence of this the thermal conductivity will be lower than that obtained by assuming an equilibrium phonon distribution and the contact resistance higher. In this case we must use the limits $h\nu_1/kT$ and $h\nu_2/kT$ for the integral occurring in (7). We then obtain for the heat flux at the salt–metal interface:

$$(19) \qquad \frac{dQ}{dt} = 5.01 \times 10^{16} \left(\frac{15}{\pi^4}\right) \left[\frac{\Gamma_1}{c_1^2} + \frac{2\Gamma_t}{c_t^2}\right] A \{T_1^4 f(T_1) - T_2^4 f(T_2)\} \text{ ergs/second}$$

where

$$(20) \qquad f(T) = \int_{h\nu_1/kT}^{h\nu_2/kT} \frac{z^3}{(e^z - 1)} \, dz.$$

There are several cases of interest in which one can approximate to the value of the integral.

(a) *High Temperature Approximation*

If the temperature is high enough for both $h\nu_1/kT$ and $h\nu_2/kT$ to be much less than one and low enough to satisfy the conditions for a non-equilibrium phonon distribution, the exponent may be expanded and the integral becomes $\int z^2\, dz$.

Hence,

$$(21)\qquad \frac{dQ}{dt} = 5.01\times 10^{16}\left(\frac{15}{\pi^4}\right)\left[\frac{\Gamma_1}{c_1^2}+\frac{2\Gamma_t}{c_t^2}\right]\frac{1}{3}\left(\frac{h}{k}\right)^3(\nu_1^3-\nu_2^3)(T_1-T_2).$$

The heat flux is now much less than that given by (8) and dQ/dT is linearly dependent upon the temperature difference across the interface.

(b) *Low Temperature Approximation*

If $\nu_1 = 0$ and $h\nu_2/kT$ is $\gg 1$ we may use the limits 0 and ∞ for z in the integral and obtain the same relation as (8) at very low temperatures. In the region lying between the low and high temperature approximation one should expect the temperature dependence to vary from the fourth power at low temperatures as given by (8) to the linear dependence of (21) at higher temperatures.

(c) *High Field Approximation*

If a large magnetic field is applied to the salt, ν_1 and ν_2 will lie close to the electron Larmor precession frequency $\bar{\nu}$ and we will presume that this width $|\nu_1-\nu_2| = \Delta\nu$ will be much less than $\bar{\nu}$. e^z will be large compared to 1 in the denominator of the integral which may then be approximated to give:

$$(22)\qquad \frac{dQ}{dt} = 5.01\times 10^{16}\left(\frac{15}{\pi^4}\right)\left[\frac{\Gamma_1}{c_1^2}+\frac{2\Gamma_t}{c_t^2}\right]\left(\frac{h\bar{\nu}}{k}\right)^4\left(\frac{\Delta\nu}{\bar{\nu}}\right)\{e^{-h\bar{\nu}/kT_2}-e^{-h\bar{\nu}/kT_1}\}.$$

In this case we should expect a strongly field and temperature-dependent contact resistance.

It should be noted in the foregoing that the thermal conductivity for the paramagnetic salt will also be affected in the same way for each of the above cases. In particular, the thermal conductivity due to those phonons which interact with the spins will be much less than that due to the bulk of the phonons. Because, firstly, there are fewer phonons able to carry the heat from the spins and, secondly, those that are able are severely scattered by the spins themselves and consequently the diffusion of the spin energy to the surface is much slower than one would expect from the gross thermal conductivity.

At very low temperatures one other limitation to the transfer of heat to or from the spin system might be the true spin–lattice relaxation time. One may optimize this heat transfer by the use of ions with strong spin–orbit coupling such as Cr^{++}, Ti^{++}, and Cu^{++} and avoid those ions in an S-state, such as Fe^{+++} and Mn^{+++}.

V. CONTRIBUTION OF SURFACE WAVES

In general one should not expect the surface or Rayleigh waves to play an important role in the transfer of heat from one solid to another. In the first place the total heat flux round the edges of the solid due to these surface waves is much smaller than through the body of the solid. This is because the waves are localized within about a wavelength of the surface and all the heat must be transmitted through this region. The phonons or body waves are not limited in this way and, consequently, they can carry a far greater heat flux. In the second place the surface waves may be excited by the phonons themselves. However, the two additional interactions which are necessary to transfer the heat energy to and from the surface waves must introduce a further resistive contribution in addition to the boundary conditions which must be satisfied by the surface waves themselves. There are, however, a few exceptional cases where the surface waves may play a more important role. These are where: (i) the surface waves may interact directly with the conduction electrons or with the surface energy states in a metal; or (ii) where the acoustic velocities in the two media are so different that almost all the incident phonons are totally reflected. The totally reflected waves give rise to a surface disturbance in the second medium which can interact directly with the phonons or with the conduction electrons. Because of the small coefficient of transmission of the phonons and the large density of surface waves in this case, the role of each could be expected to be reversed and the latter could become then the dominant transfer mechanism.

Further investigation is required to determine when these waves do, in fact, become the dominant transfer mechanism.

DISCUSSION

From the foregoing analysis one can draw the following conclusions regarding the contact resistance.

(a) In general the transfer of heat between two dissimilar materials at low temperatures should not be a more serious problem than the thermal conductivity because the values of Γ seldom are much smaller than 0.2 for common solids. The surface resistance in this case would be equivalent to a little more than the thermal resistance of a thickness of the first medium equal to two phonon mean free paths.

(b) Exceptions to these will occur in the following cases where the contact resistance will be abnormally high:

(i) for materials in imperfect contact where the effective area of the bodies in contact is small;

(ii) between media in which there is a large difference in the densities or acoustic velocities where the Γ are small (see Figs. 4, 5, 6, and 7);

(iii) between the spin system of a paramagnetic salt and a solid in contact with it where only a small fraction of the available phonons are able to carry the energy from the spin system. This should be particularly pronounced in the presence of a strong magnetic field.

These conclusions can be compared with the available experimental results. The contact resistance from liquid helium to copper has been measured by several workers. Ambler (1953) obtained the expression $dQ/dt = 3.4 \times 10^5(T_1^4 - T_2^4)$ ergs/sec cm^2. This is the expected temperature dependence but numerically far greater than the calculated value given in Table I. Clearly

TABLE I

From:	To:	Γ_l	Γ_t	dQ/dT in:	ergs/sec cm²
KCr alum	Copper	0.14	0.19	4.7×10⁶	$(T_1^4 - T_2^4)$
KCr alum	Silver	0.26	0.25	7.3 "	"
KCr alum	Gold	0.17	0.21	5.8 "	"
Pyrex	Copper	0.34	0.36	3.7 "	"
Lucite	Copper	0.11	0.02	2.3 "	"
Garnet	Copper	0.46	0.45	2.3 "	"
Copper	Helium	0.001	0.0008	1.84×10³	"

the transmission of phonons across the interface alone is quite insufficient to account for the experimental results. However, in this case one should expect the surface waves to be of importance because the ratio of the acoustic velocities is greater than 10. In this particular case we have calculated the amplitude of the surface disturbance caused by totally reflected phonons by solving the boundary conditions as described earlier. It was found that the experimental results then can be explained if about 10% of the energy in this disturbance is transferred to the metal. Such a strong interaction suggests that the surface disturbance interacts directly with the conduction electrons. If this view is correct, there should be an appreciable difference in the contact resistance of a metal in the normal and in the superconducting state. Furthermore, if the surface of the metal is covered with an oxide layer or a layer of dirt only the longer wavelength phonons would be able to penetrate to the conduction electrons. This would give rise to a heat flux proportional to the difference of the third powers of the temperatures of the metal and the helium which would explain the results obtained by Fairbank and Wilks (1955).

It is well known that the transfer of heat between bodies merely pressed into contact with one another is extremely small at low temperatures. Moreover, Berman (1956) found the temperature dependence of this heat flux to be proportional to T^2 at helium temperatures. His results are in agreement with the temperature dependence and magnitude predicted for a pair of surfaces in poor contact treated in Section I.

The results of Section IV provide an explanation of the measurements of Goodman (1953) and more recently, of Miedema, Postma, and Steenland (1958) of the contact resistance from copper to a paramagnetic salt. Goodman measured this contact resistance and found dQ/dT proportional to the square of the absolute temperature. The more recent measurements of Miedema et al. on a single crystal of potassium chrome alum cemented to a copper strip show that dQ/dT is proportional to $(T_1^2 - T_2^2)$ at high temperatures and $(T_1^4 - T_2^4)$ at low temperatures. This agrees with the predicted temperature dependence obtained on the assumption of a non-equilibrium phonon spectrum. At low temperatures the numerical value they obtain was $dQ/dt = 1.2 \times 10^5 (T_1^4 - T_2^4)$ergs/sec cm². The calculated value in Table I is about three times greater than this. This is a very satisfactory agreement because the crystal will probably be strained close to the copper surface due to differential contraction and this will impede further the flow of heat.

ACKNOWLEDGMENTS

I wish to thank Dr. J. M. Daniels for helpful advice and criticism during the preparation of this manuscript and Dean G. M. Shrum for the facilities of the Department of Physics during my stay. I am indebted to the staff of the U.B.C. Computing Center for much advice and help in programing the ALWAC IIIE Computer. This work was carried out during the tenure of a National Research Council Postdoctoral Fellowship for which I wish to express my thanks.

REFERENCES

AMBLER, E. 1953. Thesis, University of Oxford.
BERMAN, R. 1956. J. Appl. Phys. 27, 318.
FAIRBANK, H. A. and LEE, D. M. 1957. The Fifth International Conference of Low Temperatures Physics and Chemistry, Wisconsin.
FAIRBANK, H. A. and WILKS, J. 1955. Proc. Roy. Soc. (London), A, 231, 545.
GOODMAN, B. B. 1953. Proc. Phys. Soc. (London), A, 66, 217.
GORTER, C. J., TACONIS, K. W., and BEENAKKER, J. J. M. 1951. Physica, 17, 841.
KAPITZA, P. L. 1941. J. Phys. (U.S.S.R.), 4, 181.
KHALATNIKOV, I. M. 1952. J. Exptl. Theoret. Phys. (U.S.S.R.), 22, 687.
KOLSKY, H. 1953. Stress waves in solids (Oxford University Press, London).
KRONIG, R. and THELLUNG, A. 1950. Physica, 16, 678.
MIEDEMA, A. R., POSTMA, H., and STEENLAND, M. 1958. Kamerlingh Onnes Conference of Low Temperature Physics.
RAYLEIGH, LORD. 1894. Theory of sound (Macmillan & Co., Ltd., London).
STRANDBERG, M. W. P. 1958. Phys. Rev. 110, 65.
WILSON, A. H. 1953. The theory of metals (Cambridge University Press, London).

OBSERVATION OF QUANTUM PERIODICITY IN THE TRANSITION TEMPERATURE
OF A SUPERCONDUCTING CYLINDER*

W. A. Little[†] and R. D. Parks[‡]

Department of Physics, Stanford University, Stanford, California

(Received May 10, 1962; revised manuscript received June 15, 1962)

Deaver and Fairbank[1] and Doll and Näbauer[2] have shown experimentally that the flux which is trapped in a superconducting cylinder is an integral multiple of the unit $hc/2e$. It has been pointed out[3,4] that this result follows because the free energy of the superconducting state is periodic in this unit of the flux if the electrons are paired in the manner described by the Bardeen-Cooper-Schrieffer (BCS) theory.[5] The free energy of the normal state, on the other hand, is virtually independent of the flux. Consequently, the transition temperature T_c, which is the temperature at which the free energy of the normal and superconducting states are equal, must also be a periodic function of the enclosed flux ϕ. The magnitude of the change in T_c was calculated for a thin cylindrical sample using the BCS model in which the possible pairing of particles with net momentum was included. This calculation showed that the binding energy of each pair was reduced by the amount of energy required to provide the center-of-mass motion necessary to maintain the fluxoid,

$$\frac{1}{h} \oint \left(m\vec{v}_s + \frac{2e}{c}\vec{A} \right) \cdot d\vec{s},$$

an integer. Each integer n corresponds to a different superconducting state characterized by a particular pairing arrangement and a different transition temperature. The transition temperature is found to vary as

$$\Delta T_c = \frac{\hbar^2}{16\,m^*R_0^2}\left(\frac{2e}{hc}\phi + n \right)^2.$$

The choice of n which gives the tightest binding and the highest transition temperature switches from 0 to -1, -1 to -2, etc., when ϕ is given by $\frac{1}{2}(hc/2e)$; $\frac{3}{2}(hc/2e)$, etc. We note also that the binding energy of the pair is a minimum at these points and varies periodically with the flux. At the transition temperature the penetration depth becomes infinite and consequently the flux ϕ, enclosed by the cylinder, is determined entirely by the external field. T_c is given by a periodic array of parabolas, each of which is centered on a flux unit (see Fig. 1). One can estimate the expected magnitude of ΔT_c by taking $m^* = m_e$ and a reasonable diameter of say 1 micron for the cylinder. ΔT_c is then approximately 5×10^{-5} K° which is of measurable magnitude in the liquid helium temperature range.

We have observed such an effect with a thin

N522 1-4

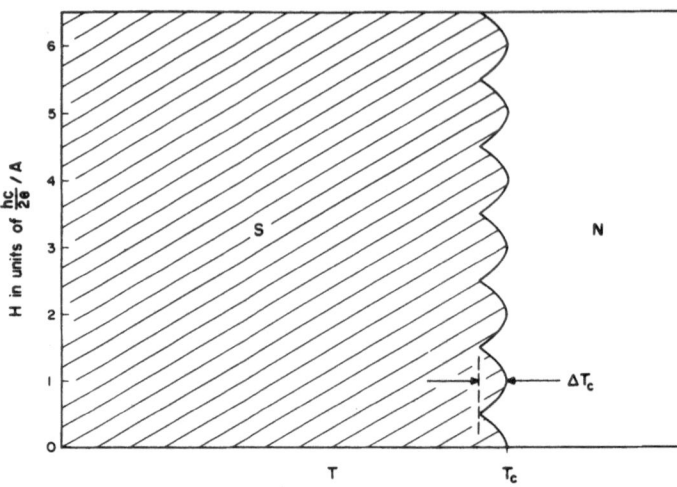

FIG. 1. Phase diagram for a thin cylindrical superconductor in an axial magnetic field. The scalloped edge of the superconducting phase results from the periodicity in the free energy of the bound pairs, in the magnetic flux through the cylinder.

cylinder of tin. A clearly defined series of parabolic variations of the transition temperature was observed as the magnetic field was changed. The parabolas were regularly spaced at intervals of $hc/2e$ in the magnetic flux, in agreement with the results of the experiments on quantized flux by Deaver and Fairbank and Doll and Näbauer.

The cylinder of tin was prepared in the following way. A drop of G.E. 7031 cement was held on the ends of two wires and the wires were then rapidly drawn apart to arm's length. A thin filament of cement was formed which extended from one wire to the other. After some hours of practice we succeeded in drawing a filament of approximately 1 micron in diameter. A portion of this filament was laid carefully onto a glass slide over a slot 3 mm wide by 10 mm long which had been cut in the slide. The filament was cemented in place with a dilute solution of the same cement. The slide was then mounted in a high vacuum evaporator on a rotating "spit" so that the axis of rotation of the "spit" was along that of the filament. In this way it was possible to evaporate metal over the complete perimeter of the filament, the slot in the slide being used to expose the underside of the filament to the evaporating metal. Earlier experience had shown that tin did not adhere well to the filament; indeed, tin films less than 900 Å thick were found to be noncontinuous. For this reason, a thin layer of gold about 25 Å thick was first evaporated onto the filament. This was not electrically continuous but provided a surface to which the tin could adhere. A layer of tin 375 Å thick was then evaporated at a pressure of 3×10^{-6} mm onto the gold substrate. This formed a continuous amorphous film which was electrically conducting.

Electrical contact was made to each end of the cylinder by cementing thin copper wires onto the glass slide and then covering them and the surrounding tin film with a coat of silver paint. The slide was mounted on a bakelite base inside a copper tube which was split lengthwise. The whole assembly was fitted inside a copper solenoid in a glass Dewar system. The system was precooled and then filled with liquid helium. The temperature of the helium bath could be controlled to about 10^{-4} K° with a diaphragm manostat in the pumping line. The resistance of the tin cylinder was measured by passing a constant current of about 10 μA through it and measuring the potential drop across it with a microvoltmeter. The transition from the normal to the superconducting state occurred at about 3.45°K and extended over a temperature range of about 0.05K°. Temperatures in this region were measured with a carbon resistor thermometer placed within one centimeter of the filament and calibrated against the vapor pressure of helium.

If the transition region of a superconductor is spread over a finite temperature interval, there is no unique transition temperature T_c; however, one may be defined by some criterion such as the temperature at which the resistance falls to one-half of its normal value. Alternatively, if the variation in the resistance is measured, one can calculate the change in the transition temperature ΔT_c from the slope of the resistance versus tem-

N522 2-4

perature curve. This is the technique we used. Small changes in the resistance of the tin film were observed as the axial magnetic field was varied and these were interpreted later as changes in the transition temperature.

The results we obtained are shown in the accompanying figures. To obtain these photographs, the magnetic field at the sample was varied sinusoidally at 25 cps and the potential across the filament observed on an oscilloscope while a constant current of 25 μA was flowing through it. Similar results were obtained from dc measurements with the microvoltmeter. Figure 2 is the variation of the resistance with magnetic field and clearly shows a series of parabolas superimposed upon a quadratic background. The upper trace is a measure of the magnetic field, with zero field at the center of the picture. In Fig. 3 an enlarged

view of the parabolas, corresponding to pairing of the particles in the states $n = -1$, 0, and +1, is shown. The minima occur at values of the magnetic field of -14, 0, and +14 gauss. The diameter of the filament had been measured previously with an optical microscope and found to be 1.4 ± 0.1 microns which gives a value of 13 ± 2 gauss for the field which would correspond to one flux unit $hc/2e$. The uncertainty in the diameter results from the difficulty in interpreting the diffraction pattern in the microscope. We could measure no change in this pattern of the filament over its entire length indicating that the filament was of uniform cross section. This suggests that one could usefully determine the diameter to greater precision with an electron microscope and hence determine the flux quantum to a few percent. In Fig. 4 the quadratic background has been reduced to a linear term by electrically differentiating the signal to the oscilloscope. In this way one can see more clearly the parabolas in the region where the quadratic term is large. From this picture one can see the strictly periodic nature of the minima over a region of seven parabolas. From our dc measurements we found that the maximum excursion of the transition temperature ΔT_c was 5×10^{-4} K° (after subtracting the quadratic background). This was somewhat greater than we expected and is probably attributed to the effective mass of the electrons being smaller than that of the true mass, and to some corrections from a more exact theory.

In summary we state the following results and conclusions from this experiment:

1. The quantum periodicity in the free energy of a superconducting pair predicted by Byers and

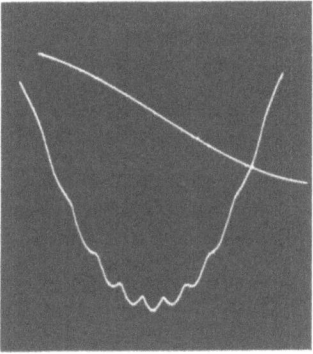

FIG. 2. Lower trace: variation of resistance of tin cylinder at its superconducting transition temperature as a function of magnetic field. Upper trace: magnetic field sweep.

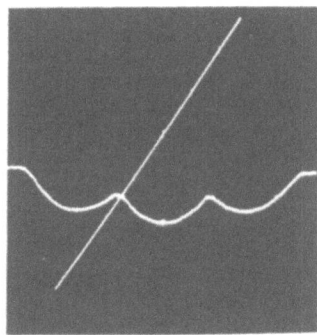

FIG. 3. Enlarged view of parabolic variation of the resistance of tin cylinder for pairs in quantum states -1, 0, and +1. Straight line is magnetic field variation with zero field at the center of the picture.

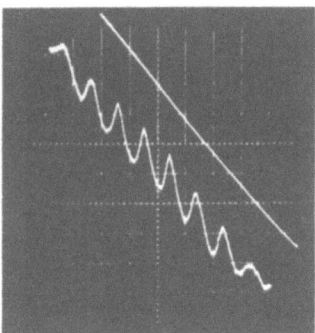

FIG. 4. Lower trace: variation of the resistance of tin cylinder with magnetic field after differentiation which reduces the quadratic background to a linear background. Upper trace: magnetic field sweep.

Yang[4] has been directly observed and this was found to be the same order of magnitude predicted above.

2. Further evidence of electron pairs has been obtained in support of the BCS theory.

3. There was no indication (at least for the tin film measured) of uneven steps in the flux quanta of the type shown by Doll and Näbauer's results on lead.[3]

4. The quadratic background in the variation of the transition temperature gives evidence of the weakening of the binding of the pair in a magnetic field.

5. This type of experiment provides a means of determining to great precision the magnitude of the flux quanta by measuring the minima of the observed parabolas.

We wish to acknowledge the interest and encouragement of B. S. Deaver and W. M. Fairbank in this experiment. We are grateful also to J. R. Schrieffer and W. T. Sommer for stimulating discussions.

*Work supported in part by the Alfred P. Sloan Foundation and the joint program of the Office of Naval Research and the U. S. Atomic Energy Commission.

[†] Alfred P. Sloan Foundation Fellow.

[‡] National Science Foundation Fellow.

[1] B. S. Deaver and W. M. Fairbank, Phys. Rev. Letters 7, 43 (1961).

[2] R. Doll and M. Näbauer, Phys. Rev. Letters 7, 51 (1961).

[3] N. Byers and C. N. Yang, Phys. Rev. Letters 7, 46 (1961).

[4] W. Brenig, Phys. Rev. Letters 7, 337 (1961).

[5] J. Bardeen, L. N. Cooper, and J. R. Schrieffer, Phys. Rev. 108, 1175 (1957).

Reprinted from The Physical Review, Vol. 134, No. 6A, A1416–A1424, 15 June 1964
Printed in U. S. A.

Possibility of Synthesizing an Organic Superconductor*

W. A. Little

Department of Physics, Stanford University, Stanford, California

(Received 13 November 1963; revised manuscript received 27 January 1964)

London's idea that superconductivity might occur in organic macromolecules is examined in the light of the BCS theory of superconductivity. It is shown that the criterion for the occurrance of such a state can be met in certain organic polymers. A particular example is considered in detail. From a realistic estimation of the matrix elements and density of states in this polymer it is concluded that superconductivity should occur even at temperatures well above room temperature. The physical reason for this remarkable high transition temperature is discussed. It is shown further that the superconducting state of these polymers should be distinguished by certain unique chemical properties which could have considerable biological significance.

I. INTRODUCTION

IN the forward to Vol. 1 of his monographs on superfluids, F. London[1] questions whether a superfluid-like state might occur in certain macromolecules which play an important role in biochemical reactions. If this should be the case, an entirely new and important consideration would be added to the problem of understanding living systems. In view of the significance of such an effect, it appears appropriate at this time, when a theory of superconductivity, the Bardeen-Cooper-Schrieffer (BCS) theory[2] has been so remarkably successful in explaining much of the behavior of superconductors, to examine in the light of this whether or not a superconducting state might occur in certain macromolecules. In view of the extreme complexity of biological systems, it would be folly for a physicist to attempt to experiment in such an environment. Instead of attempting this, we shall tackle the problem on our own grounds. The BCS theory, while by no means complete and exact, has succeeded in providing a model with most of the essential features of a superconductor. In particular, it prescribes certain criteria for a system which, if satisfied, should lead to the superconducting state. Our approach is to consider how these criteria might be applied to the design of a particular organic molecule which, if its synthesis is possible, should show some of the essential features of a superconductor and, as we shall show, some remarkable chemical properties as well. One of the interesting features about the particular class of molecules we investigate in detail is that the molecules should be superconducting at room temperature and, indeed, to temperatures well above room temperatures. We can show on simple physical grounds why this is so and perhaps, with hindsight, why this was to be expected.

The idea of superconductivity in organic systems is not a new idea, however, there is a considerable amount

* Supported in part by the National Science Foundation and the U. S. Navy Office of Naval Research.
[1] F. London, *Superfluids* (John Wiley & Sons, Inc., New York, 1950), Vol. 1.
[2] J. Bardeen, L. N. Cooper, and J. R. Schrieffer, Phys. Rev. **108**, 1175 (1957).

of confusion as to the exact meaning of this. The diamagnetic ring currents of aromatic molecules such as benzene, naphthalene, etc., are nondissipative currents similar in many respects to the persistent currents of superconducting rings and, have often been referred to as a form of superconductivity. However, the "superconductivity" of these molecules is not the same as the superconductivity of bulk materials. The reason, I believe, is the following. In macroscopically large superconductors, if superconductivity exists, then a finite fraction of the charge carriers, in general, the BCS pairs are in identically the same center-of-mass momentum state. This state then has a macroscopic occupation. In a magnetic field the canonical momentum of this state remains unchanged, but due to the vector potential term contained in it a current is induced and the energy of the state changes. For a macroscopically large superconductor the kinetic energy of the different center-of-mass momentum states of the pairs lie extremely close to one another, however, because the coherence energy of each state depends upon the square of the number of pairs in that state, the state which is macroscopically occupied is appreciably lower in energy than any of the neighboring states even in a moderate magnetic field. It is only by transitions in which practically all the pairs in the macroscopically occupied state simultaneously move to another state that a lower energy final state can be reached. This is obviously highly forbidden and, consequently, the system of pairs remains in the momentum state into which condensation originally occurred. Thus, it is the coherence energy which prevents the system from freely adjusting itself to take the lowest possible energy. In the aromatic ring compounds practically all the molecules are in their ground states. In a magnetic field the canonical momentum of the electrons in this state remain unchanged and diamagnetic currents flow in the molecule similar to those of a bulk superconductor. The energy of the different momentum states of the electrons in each molecule in this case are well separated though, because the molecules are of microscopic size. Thus, the momenta of the electrons do not change because for fields as large as those available in the laboratory, the state which evolves out of the original ground state still is lower in energy than any other in the presence of the field. If, however, the aromatic system is made arbitrarily large such as in graphite, bulk superconductivity does not result because as the system gets bigger, the different momentum states of the electrons approach each other in energy. Transitions can then occur between states and the induced currents are dissipated. So that in order to get superconductivity in a macromolecule or in a bulk material, something of the nature of a coherence energy is required. In conventional superconductors this is provided by the phonon-induced, electron-electron interaction; in attempting to devise a macromolecule which is to be superconducting

one must provide, therefore, some mechanism similar to this. In our model we do this in the following manner.

II. MODEL SYSTEM

We shall consider a molecule consisting of two parts, a long chain called the "spine" in which electrons fill the various states and may or may not form a conducting system; and secondly, a series of arms or side chains attached to the spine as indicated in Fig. 1. We will show that by appropriate choice of the molecules which constitute the side chains, the virtual oscillation of charge in these side chains can provide an interaction between the electrons moving in the spine. This can be made a sufficiently attractive interaction so that the superconducting state results. We can show further that even if the spine by itself is initially an insulator due to the valence band being full and the conduction band empty, the addition of side chains can increase the electron-electron attraction to the point where it becomes energetically favorable to enter the superconducting state by mixing in states of the conduction band. The spine thus transforms from the insulating or semiconducting state directly to the superconducting metallic state upon the addition of the side chains.

Consider a long chain molecule as shown in the left half of Fig. 1. We will assume this is a conjugated chain of double and single bonds resonating between the two at each link. This corresponds in the band theory of metals to a band which is half filled and ideally is a metallic conductor. (See, however, Sec. III.) At the points P, P', $\cdots$, a regular array of side chain molecules B are attached. The individual side-chain molecules are chosen to have a low-lying excited state such that transitions from the ground state to the excited state correspond classically to an oscillation of charge from end to end of the molecule.

The electrons moving in the spine may be described

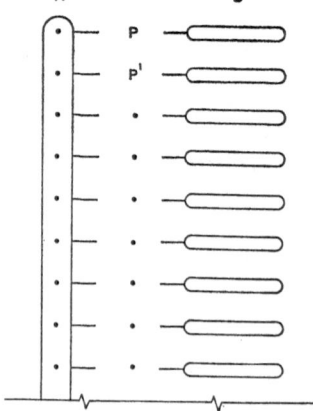

FIG. 1. Proposed model of a superconducting organic molecule. The molecule A is a long unsaturated polyene chain called the "spine." The molecules B are side chains attached to the spine at points P, P', ···.

in the tight-binding approximation[3] by eigenfunctions of the form

$$\phi_{k,m}(r) = \frac{1}{G^{1/2}} \sum_i e^{ikR_i} U_m(r - R_i), \qquad (2.1)$$

where $U_m(r - R_i)$ is the wave function of an electron in a single atom located at R_i. G is the number of links in the chain and ϵ_k is the energy of this electron and we will assume cyclic boundary conditions. To avoid unnecessary complication we shall ignore the electron spin throughout the discussion except where necessary.

The wave function of the electrons in a single side-chain molecule in an eigenstate n we designate by $\psi_n(r_1, r_2, \cdots)$. Due to the interaction of one side chain upon its neighbors, the degeneracy of the levels of the group of side chains will be removed when the side chains are brought together as in Fig. 1. The band of levels of the system of side chains as a whole can be described then by a new wave function similar to (2.1) above.

$$X_{q,n} = \frac{1}{G^{1/2}} \sum_i e^{iqR_i} \psi_n(r_1 - R_j; r_2 - R_j \cdots). \quad (2.2)$$

The Coulomb interaction between the electrons in the spine and those on the side chain will provide an interaction between the electrons in the spine and the side-chain modes. Let this Coulomb interaction be $V_1 = \sum_{i,j} V(r - R_i; r_1 - R_j, r_2 - R_j, \cdots)$ which will give rise to a typical interaction matrix element

$$\int\int \cdots \int \phi_{k'm'}{}^*(r) X_{q'm'}{}^*(r_1, r_2 \cdots)$$

$$\times V_1 X_{qn}(r_1, r_2 \cdots) \phi_{km}(r) d^3r d^3r_1 \cdots. \quad (2.3)$$

In the model we shall consider, screening reduces the range of the Coulomb interaction so that one can safely disregard the contribution to the interaction from any but the neighboring side chains. Secondly, in our model there is negligible overlap between side chains, and also overlap between sites on the spine is relatively small. These features together with the properties of (2.1) and (2.2) and the assumption of regular substitution along the spine allow us to approximate the interaction (2.3) to give the following

$$V_{k'm', q'n', qn, km} = \frac{1}{G} \sum_{n'' = -1, 0, +1} \cos Q n'' a \int \cdots$$

$$\times \int U_{m'}{}^*(r) \psi_{n'}{}^*(r_1 - n'' a, \cdots)$$

$$\times V_1 \psi_n(r_1 - n'' a) U_m(r) \times d^3r d^3r_1 \cdots, \quad (2.4)$$

where $k' = k + Q$, $q' = q - Q$, and a is the spacing between side chains.

[3] See for example N. F. Mott and H. Jones, *The Theory of the Properties of Metals and Alloys* (Clarendon Press, Oxford, 1936).

It is convenient to write this as $(1/G)V(Q)$ for the moment. Then the complete spine side-chain interaction can be described in the representation of second quantization as

$$V_{\text{side chain}} = \frac{1}{G} \sum_{\substack{Q, n, n' \\ k, m, m'}} V(Q) \left(\sum_q c_{q-Q, n'}{}^\dagger c_{q, n} \right) a_{k+Q, m'}{}^\dagger a_{km}, \quad (2.5)$$

where the $a^\dagger$ and a are the creation and destruction operators for the electron in the spine, and $c^\dagger$ and c the corresponding operators for the side-chain modes. The particle-hole operator $\sum_q c_{q-Q, n'}{}^\dagger c_{qn}$ which occurs in (2.5) appears in a similar manner to the phonon creation operator $b_Q{}^\dagger$ in the phonon-electron interaction in a metal. The terms linear in this particle-hole operator which appear through the interaction (2.5) in the total Hamiltonian can be eliminated by the same type of transformation[4] which eliminates the terms linear in the phonon creation operator $b_Q{}^\dagger$ in the conventional theory. This leads to a side-chain induced electron-electron interaction V_2 *between* electrons in the spine of the form

$$V_2 = \frac{1}{2} \sum_{k, k', Q} \left\{ \frac{2V^2(Q)}{G} \frac{\hbar \omega_Q}{(\epsilon_{k+Q} - \epsilon_k)^2 - (\hbar \omega_Q)^2} \right\}$$

$$\times a_{k+Q}{}^\dagger a_{k'-Q}{}^\dagger a_{k'} a_k. \quad (2.6)$$

We have written $\hbar \omega$ for the difference $(E_{n'} - E_n)$ between the energy of the states of the side-chain modes and have considered for simplicity only one excited state n'. This is identical in form to the usual phonon-induced electron-electron coupling. We note that if $|\hbar \omega| \gg |(\epsilon_{k+Q} - \epsilon_k)|$ then the term in brackets reduces to an attractive interaction, $V \approx -2\langle |V(Q)|^2 \rangle_{\text{av}}/G\hbar\omega$, where $\langle |V(Q)|^2 \rangle_{\text{av}}$ is the average of the square of the interaction $V(Q)$.

In addition to this attractive term, the screened Coulomb interaction gives a repulsive term. The sum of these two is the net electron-electron interaction. Let this sum be $\langle V(Q) \rangle_{\text{av}}$; then the total Hamiltonian for the electrons in the spine is

$$\mathcal{H} = \sum_k (\epsilon_k - \mu) a_k{}^\dagger a_k$$

$$+ \frac{1}{2} \sum_{k, k', Q} \langle V(Q) \rangle_{\text{av}} a_{k+Q}{}^\dagger a_{k'-Q}{}^\dagger a_{k'} a_k, \quad (2.7)$$

where $\langle V(Q) \rangle_{\text{av}}$ is given above and μ is the Fermi energy.

The BCS theory then shows that a superconducting state is possible if the following equation can be satisfied for a nonzero gap, Δ:

$$-1 = \sum_k \frac{V \tanh \frac{1}{2} \beta E_k}{2E_k}, \quad (2.8)$$

[4] J. Bardeen, *Encyclopedia of Physics*, edited by S. Flügge (Springer-Verlag, Berlin, 1956), Vol. 15, p. 352.

where $E_k = [(\epsilon_k - \mu)^2 + \Delta^2]^{1/2}$ and the sum is limited by $|\epsilon_k - \mu| < \hbar\omega$. V is the average over the region of Q where $\langle V(Q) \rangle_{av}$ is attractive. The critical temperature is given by

$$kT_c = 1.14\hbar\omega \exp[-1/N(0)V], \qquad (2.9)$$

where $N(0)$ is the density of states of one spin at the Fermi surface.

It is appropriate to dispose of a difficulty here which will arise shortly in considering a particular example of a molecule. In long-chain conjugated molecules the double and single bonds do not resonate freely between each position in the chain, but tend to localize so that a stationary periodic charge distribution and periodic bond length is established. This introduces a periodic potential of twice the atomic spacing in the chain which in turn produces a gap in the density of states halfway up the band. As there is one electron per atom in the band, this band is filled up to the new gap. The conjugated system is then no longer a metallic conductor but a semiconductor or insulator, i.e., the density of states at the Fermi surface $N(0)$ would be zero.[5] However, it is incorrect to interpret Eq. (2.9) as indicating that $T_c = 0$ in this case, for Eq. (2.9) is only an approximation. Instead, one must go back to (2.8) and examine this to see whether in this case a solution is possible. In the next section we consider this in detail.

III. SUPERCONDUCTIVITY IN A SEMICONDUCTOR

Let us consider a somewhat idealized case of a semiconductor with a band structure as shown in Fig. 2. Before the introduction of the periodic potential which generates the gap, the energy of the states would have been given by that shown by the dashed line in the figure. Let us take the band gap as δ and for convenience we will assume the density of states near the band edge is the same in the two bands. If the lower band is completely filled and the upper empty, then the Fermi level will be halfway between the two bands. We will assume that $\hbar\omega$ is greater than δ.

Let $N(0)$ be the density of states at the Fermi surface for the system prior to the introduction of the periodic potential and let ϵ be the energy of these states measured with respect to the Fermi surface. The semiconductor band gap can then be conveniently introduced by changing the energy of each state ϵ to $\epsilon' = [\epsilon^2 + (\delta/2)^2]^{1/2}$. We can now use Eq. (2.8) to see if the superconducting state can occur at any temperature. At $T = 0°K$ a

---

[5] Where one does not have alternating double and single bonds, but a double bond separated from the next by two single bonds, the band theory does not appear to work and one finds that the chain is an insulator instead of a metal. This empirical fact does not appear to have been explained, but seems to be a consequence of the same considerations of the metallic and insulating states discussed by N. F. Mott [Phil. Mag. 6, 287 (1961)], and W. Kohn [Phys. Rev. 133, A171 (1964)]. As this question is not quite settled we have limited ourselves to a conjugated chain which is known to behave as one would expect from the band theory.

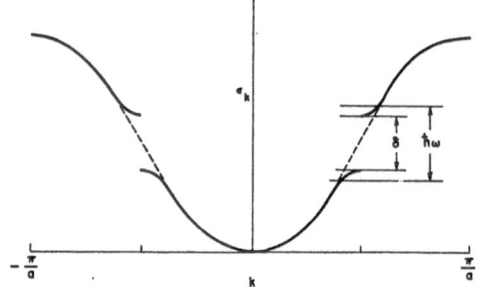

Fig. 2. Band structure of a conjugated chain semiconductor caused by the localization of the double bond. The energy $\hbar\omega$ is the energy of the transitions of the side chains.

superconducting state is possible if V is negative and

$$1 < \sum_\epsilon \frac{|V|}{2[\epsilon^2 + (\delta/2)^2]^{1/2}}, \qquad (3.1)$$

where the sum is limited by $|\epsilon'| < \hbar\omega$.

Transforming to an integral, we require

$$\frac{1}{N(0)|V|} < \int_0^x \frac{d\epsilon}{[\epsilon^2 + (\delta/2)^2]^{1/2}}, \qquad (3.2)$$

where $x^2 = (\hbar\omega)^2 - (\delta/2)^2$ and $N(0)$ is defined above prior to the introduction of the gap.

$$\frac{1}{N(0)|V|} < \ln\left[\frac{[(\hbar\omega)^2 - (\delta/2)^2]^{1/2} + \hbar\omega}{\delta/2}\right]. \qquad (3.3)$$

If $(\delta/2)^2 \ll (\hbar\omega)^2$ then the criterion for obtaining the superconducting state is that

$$\delta < 4\hbar\omega \exp\left[\frac{-1}{N(0)|V|}\right]. \qquad (3.4)$$

If this criterion can be satisfied, then the transition temperature can be obtained from the expression

$$\frac{1}{N(0)|V|} = \int_0^x \frac{\tanh\frac{1}{2}\beta_c\epsilon'}{\epsilon'} d\epsilon, \qquad (3.5)$$

which is the same expression for determining the temperature at which the energy gap of a superconductor becomes equal to $\delta/2$. The form of this expression is given graphically in the BCS paper. We only need note that T_c rapidly approaches the transition temperature for $\delta = 0$ as $|V|N(0)$ is increased beyond that necessary to satisfy the criterion. So that if the criterion can be satisfied, then the transition temperature generally will be of the same order as the transition temperature of the metallic superconductor. We see then that a gap in the band structure does not necessarily exclude the superconducting state. Incidentally, this conclusion does not violate Yang's[6] statement that superconductivity or off-diagonal-long-range order (ODLRO) cannot occur in an insulator because his definition of an insulator is one in which there are no available empty states. Our point is that these states are, in fact, available in our model.

It is perhaps appropriate to note that a gap in the density of states such as that considered above would make the average side-chain induced interaction more strongly attractive because for interband transitions $(\epsilon_{k+Q} - \epsilon_k)$ would be at least δ and as can be seen from (2.6), the attractive term would thus become larger provided that $\delta < \hbar\omega$. If one takes the variation of $\langle V(Q) \rangle_{av}$ with Q into account, the details of the above inequality are, of course, changed, but the general feature that the existence of superconductivity is determined by some such criterion remains.

IV. PARTICULAR EXAMPLE

As a particular example of a molecule of the type considered in Sec. II, we will consider in detail the molecule illustrated in Fig. 3. The spine is a conjugated chain of alternating double and single bonds. To this is attached a series of side-chain molecules as shown. Because of the great thickness of the benzene rings in the side chains, compared to the carbon-carbon spacing in the spine, it is not possible to attach a side chain to every carbon atom on the spine. This can be seen more clearly in Fig. 4 where the molecule is drawn to scale using the known values for the van der Waals radii of the constituent atoms. The changed periodicity of the new structure necessitates a slight modification of the wave function used in (2.1) and (2.2) above, but the modification is an obvious one which we will handle later. The side-chain molecule is part of a well-known dye molecule used for sensitizing photographic plates in the red, a diethyl-cyanine iodide, and it has been

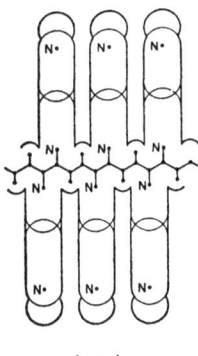

FIG. 4. Approximate scale drawing of the proposed superconducting organic polymer. The plane of the benzene rings in the side chains are oriented at right angles to the spine. The two nitrogen sites on each side chain are indicated, but the iodine site has been omitted for the sake of clarity.

4 Å

chosen because its absorption spectrum is known, and its ground and excited states are well understood.[7]

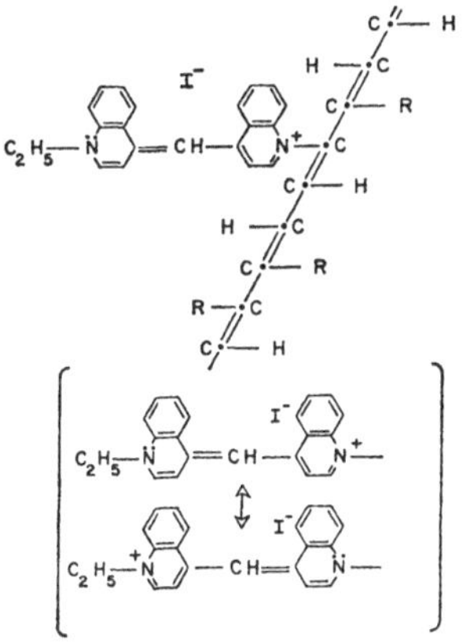

Before estimating the matrix elements of the interaction of the side chains with the electrons in the spine, it is useful to consider first what effect screening will have upon the Coulomb interaction.

Screening

The Coulomb potential at a distance r from a charge which is placed in a conducting medium is screened by the rearrangement of the charges of the medium. The potential is then given approximately by $(e/r)e^{-\lambda r}$. The screening length $1/\lambda$ in a free-electron gas can be estimated by using the Thomas-Fermi method as shown by Mott and Jones.[4] From this one finds that $\lambda^2 = (4me^2/\hbar^2)(3N_0/\pi)^{1/3}$ where N_0 is the number of electrons per unit volume. For the carbon atoms on the spine, one valence electron per atom is relatively free, and from the known size of the atom one can estimate N_0 and thus λ. We find that $1/\lambda$ is approximately 0.5 Å. This value varies extremely slowly with the number of valence electrons so that our choice of one free valence electron is not critical. We note then that the Coulomb interaction is screened out in an extremely short distance.

It should be noted too that screening can occur only where the charges of the medium are free to move, as for example, within an atom or along the conjugated series of atoms of the spine. Where the charges are not free to move indefinitely, then only a limited displacement of the charges occurs which merely modifies the Coulomb interaction by the dielectric constant of the medium, but does not screen it. These two cases must be born in mind in considering the Coulomb interaction and the side-chain interaction. Because the electrons are free to move in the spine one must use the screened Coulomb interaction for computing the Coulomb repulsion between the electrons in the spine. However,

FIG. 3. Chemical structure of the proposed superconducting organic polymer. At each point R on the spine a similar side chain to the one shown is attached. These side chains are resonating hybrids of the two extreme structures shown in the inset. The positive charge resonates between the two nitrogen sites as illustrated.

[6] C. N. Yang, Rev. Mod. Phys. **34**, 694 (1962).

[7] K. Mees, *The Theory of the Photographic Process* (The Macmillan Company, New York, 1942), p. 987.

the interaction between the electrons in the spine and the charges on the side chain is only partially screened because the side chain is insulated from the spine and a free movement of charge between the two cannot occur, i.e., the conjugation of the side chain does not extend to the spine. Here one must use the Coulomb interaction modified by the dielectric constant of the medium together with some screening due to the induced movement of charge in the spine.

Coulomb Repulsion

For steric reasons, it is not possible to attach a side chain to each carbon atom of the spine. Consequently, the Hamiltonian of the spine is not invariant under a displacement from one atom to the next as assumed in our earlier discussion. In our example, the unit cell is repeated only after four carbon atoms and thus, we are dealing here with a lattice with a basis of four atoms. The wave function of the spine is then

$$\phi_{km} = \frac{1}{(G)^{1/2}} \sum_i e^{ikR_i} \sum_{j=1}^{4} \alpha_{kj} U_m(r - r_{j,i}). \quad (4.1)$$

R_i is now the position of the lattice point measured along the zig-zag line joining the carbon atoms and $r_{j,i}$ the position of the jth atom in the unit cell measured with respect to R_i. α_{kj} is a phase factor which one would expect would be very nearly $(1/4^{1/2})e^{ik(r_{j,i})}$, which is the value it would have if the Hamiltonian was perfectly invariant under a $C-C$ displacement. The number of unit cells, G is now a quarter of the number of carbon atoms in the spine.

Let us now calculate the Coulomb repulsion between electrons in states given by (4.1):

$$V(Q)_{\text{Coulomb}} = \iint \phi_{k-Q}^*(r_1)\phi_{k'+Q}^*(r_2) V(r_{12})$$
$$\times \phi_{k'}(r_2)\phi_k(r_1)d^3r_1 d^3r_2. \quad (4.2)$$

Using (4.1) and the fact that the Coulomb interaction is screened from all except immediate neighboring ions, we obtain in the long-wave limit, i.e., $Q \approx 0$

$$V(0)_{\text{Coulomb}} = \frac{1}{4G}\left[\int |U(r_1-r_j)|^2 V(r_{12})|U(r_2-r_j)|^2 d^3r_{12} \right.$$
$$\left. + \sum_{k=j\pm1} \int |U(r_1-r_j)|^2 V(r_{12})|U(r_2-r_k)|^2 d^3r_{12} \right]. \quad (4.3)$$

The second term should be much smaller than the first because $V(r)$ is heavily screened. To a first approximation we shall ignore it compared to the first. A reasonable exact value of the dominant term could be obtained by using the known form of the sp carbon orbital and evaluating the integral, however, one can obtain a reasonable estimate of the approximate magnitude of the term by considering the electron density in the orbital as constant and occupying a volume of about

half the volume of the atom. The integral is easily done and gives $6e^2/\lambda^2R^3$ where R is the van der Waals radius of the carbon atom. Using the value obtained above for λ and a van der Waals radius of 1.5 Å we obtain a value of 6 eV for the integral. This seems to be a reasonable estimate for one might expect it to be comparable to the energy necessary to add one additional electron to a carbon atom. The energy necessary to form the C^- ion is known as the electron affinity A and is related to the ionization energy I and the electronegativity x of the atom by the relation[8] $(I+A)/5.4=x$ where I and A are expressed in electron volts. The ionization energy of carbon is 11.3 eV and x is 2.5 giving an electron affinity for carbon of 2.2 eV. This is of the same order of magnitude estimated above. Taking the larger value of 6 eV to be safe we find

$$V(0)_{\text{Coulomb}} \approx 1.5 \text{ eV}/G. \quad (4.4)$$

For larger values of Q the first term in (4.3) remains unchanged in the tight-binding approximation while the second term which we have neglected above is reduced by an additional factor of $\cos Qa$ where a is the carbon-carbon spacing. Over the whole range of Q, then we can take the Coulomb repulsion to be of the order of that given by (4.4)

Side-Chain Interaction

The wave function of the side chain must be modified in the same way as the wave function for the electrons of the spine. It is now

$$X_{qn} = \frac{1}{G^{1/2}} \sum_{i=1}^{G} e^{iqR_i} \sum_{j=1}^{4} \beta_{q,j}\psi_n(r - r_{j,i}, \cdots). \quad (4.5)$$

At the sites, j, where side chains are attached β is approximately $e^{iqr_{ji}}/2^{1/2}$, and zero where there are no side chains.

The side chain we have chosen is a resonating hybrid of the two extreme structures shown in the inset to Fig. 3. If the wave functions of these two extreme structures are ψ^+ and ψ^-, respectively, then the ground state, ψ_0 is $(1/\sqrt{2})(\psi_+ + \psi_-)$ and the excited state of interest to us, ψ_1 is the orthogonal hybrid structure $(1/\sqrt{2})(\psi_+ - \psi_-)$. In the matrix element (2.4) we require $\psi_1^*\psi_0$ which is simply $\frac{1}{2}[|\psi_+|^2 - |\psi_-|^2]$.

Using the new wave functions for the spine (4.1) and (4.5), Eq. (2.4) reduces to

$$V(Q)_{\text{side chains}} = \iint \sum_{j=1}^{4} \sum_{n=-1}^{+1} \alpha_{k+Q,j}^*\alpha_{kj}|U_m(r - r_{ji})|^2$$
$$\times V(r)e^{iQn4a}\frac{1}{2}\sum_{l=1}^{4} \beta_{q+Q,l}^*\beta_{q,l}$$
$$\times \{|\psi_+(r_{l,i+n})|^2 - |\psi_-(r_{l,i+n})|^2\}. \quad (4.6)$$

---

[8] L. Pauling, *The Nature of the Chemical Bond* (Cornell University Press, Ithaca, New York, 1960), pp. 257, 95.

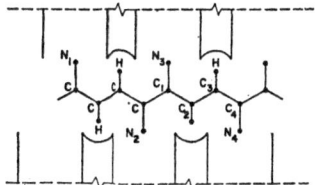

Fig. 5. Enlarged view of atoms of the spine of the proposed polymer.

where we have confined ourselves to a single atomic level m. The different k levels then describe the bands of the type discussed in Sec. III. We recall that j and l refer to each of the four atomic sites within the unit cell; and n has values 0 and ± 1 which refer to the same unit cell, i and the two adjacent cells, respectively.

In the superconducting state two types of matrix elements are important; those which describe scattering between states close to the Fermi surface with momentum transfer $Q \approx 0$ and those for scattering from one side of the Fermi surface to the other. The latter scattering involves a $Q \approx \pi/a$ because there is one electron per atom in the band originating from the fourth sp_3 orbital of the carbon atom. We shall estimate the side-chain interaction matrix element in the neighborhood of these values of the momentum transfer.

Firstly, the interaction between the spine and the iodine ion of the side chain cancels in the matrix element (4.6) because the ion is located at the same point in the two extreme structures of the hybrid. Secondly, in the ψ_- structure the positive charge is located on the nitrogen site which is remote from the spine, consequently, the interaction with this structure is weak. The only important contribution to the matrix element then, is that which is due to the positive charge on the nitrogen close to the spine in the ψ_+ structure, and the electron on each of the sites r_j.

In Fig. 5 we show an enlarged drawing of the spine and part of the side chains. In view of our earlier discussion of screening, it would be consistent to treat the interaction between the positive charge on nitrogen site N_3 (see Fig. 5) and carbon site 1 as virtually unscreened; the interaction with sites 2 and 3 as half screened and sites beyond this as completely screened. Our choice is based on the picture that the migration of negative charge from the vicinity of C_2 and C_3 towards C_1 would partially screen the field produced by the positive charge on N_3 so that at C_4 the interaction would be effectively reduced to zero. We arrive at the figure of $\frac{1}{2}$ for the screening of sites 2 and 3 because half of the atomic orbital adjacent to the nitrogen site on these sites is practically unscreened, while the opposite side is quite strongly screened. There is very little material between the nitrogen site and each of these three carbon sites, so that it seems reasonable to leave the Coulomb interaction with the dielectric constant that of free space. In Table I we tabulate the distances between the nitrogen site and each carbon site. We include also the matrix element for each site, j, in Eq. (4.5) computed for $Q=0$ and $Q=\pi/a$ using the approximate form of the α's and β's given after Eq. (4.1) and (4.5) and the unmodified Coulomb interaction limited in the manner described above.

It is reasonable to neglect the interaction with positive charge on the remote nitrogen site on the ψ_- structure because the distance to the spine is about 14 Å and the interaction is further reduced by the dielectric constant of the material of the side chain. This dielectric constant must be similar to that of benzene which is about 2.2. This gives a total matrix element of 0.1 eV for $Q=0$ which we can neglect compared to the total computed in Table I considering the approximations we have made.

In order to calculate the side-chain induced electron-electron interaction, we must know the energy $\hbar\omega$ for the transitions of the side chains. For an isolated side-chain molecule of 1,1'-diethyl-4,4'-cyanine iodide the absorption maximum for the transition we are considering occurs at 600 mμ giving a value of 2 eV for $\hbar\omega$.[7]

In the polymer the side chains will interact with one another and change the frequency of oscillation to some extent, but as the molecules are quite well separated and the charges are quite well screened from one another, let us take the frequency to be about 2 eV nevertheless. Then the side-chain induced attraction $V \approx \langle 2V(Q)^2 \rangle_{av} \, 'G\hbar\omega$ will be approximately -3.5 eV$/G$ where we have taken the mean of the square of the interaction for $Q=0$ and $Q=\pi/a$ as given in Table I. This is greater than the Coulomb repulsion of 1.5 eV$/G$ estimated in (4.4) so that the net interaction is an attractive one,

$$V \approx -2 \text{ eV}/G. \qquad (4.7)$$

The reason this is so strongly attractive is that we have seen to it that the nitrogen sites lie close to the spine so that the matrix element (4.6) is large and at the same time have chosen a side chain with a fairly low-frequency transition so that $\hbar\omega$ is small.

Some idea of the superconducting transition temperature now can be obtained by estimating the density of states for the electrons in the spine, $N(0)$. This can be done in the following way. The spine itself is very similar to a conjugated polyene chain $\{CH=CH-\}_n$ except that the side chains replace certain of the hydrogens so that one can crudely estimate the density

TABLE I. Carbon-nitrogen distances and matrix elements.

Carbon site	Distance to N_3 (Å)	Matrix element $V(Q)$ (eV) $Q=0$	$Q=\pi/a$
1	1.5	0.76	0.41
2	2.5	0.47	0.11
3	3.0	0.47	0.11
4	(4.5)	0.76	0.41
		Total 2.46	1.04

of states in the spine from a knowledge of it in a polyene chain. The benzene ring is essentially a triene {CH=CH−} tied back on itself in the form of a loop and can be described in terms of a simple band picture.[9,10] The first absorption at 250 mμ is believed to correspond to a transition of a π-electron from a $m=1$ to $m=2$ state which is approximately half the total width of the band. The total width of the band for benzene is thus ≈ 10 eV and should be approximately the same for the polyene chain. If we assume the ϵ versus k curve is parabolic up to the halfway point in the band (up to 5 eV), then the density of states of one spin at the Fermi surface is approximately $\frac{1}{8}G$ states/eV. This is probably a reasonable estimate of the density of states as it corresponds to an effective mass for the electrons in the spine of 0.7 the electron mass.

If there was no bond localization in the spine, then we could use Eq. (2.9) to estimate the superconducting transition temperature using (4.3) and the above density of states. One obtains a temperature $\approx 2200°$K in this case! This extremely high transition temperature can be understood when it is realized that in the chosen structure it is an electronic oscillation which provides the coupling between the electrons rather than the oscillation of the nuclei as in a conventional superconductor. The simple argument of the isotope effect that the transition temperature for a phonon-coupled superconductor is proportional to $1/M^{1/2}$, where M is the isotopic mass of the nuclei indicates that for an electron-coupled superconductor the transition temperature should be a factor of $(M/m)^{1/2}$ (i.e., ≈ 300) times larger. This is, perhaps, too glib an answer for it is necessary to choose the over-all structure so as to obtain a sufficiently strong coupling matrix element (2.6). Our particular model illustrates this in detail.

If there is considerable bond localization in the spine, then our inequality (3.4) shows that the superconducting state can still occur if the semiconductor gap δ is somewhat less than 0.67 eV. If this is satisfied, the transition temperature in this case should still be several hundred °K.

For transition temperatures as high as this the coherence energy of the superconducting state becomes comparable to the chemical binding energy. This energy is approximately

$$W_0 = -2N(0)(\hbar\omega)^2 \exp[-2/(N(0)|V|)]. \quad (4.8)$$

In our example, the coherence energy is about 0.1 eV per unit cell of the chain. This is not very large, however, if one synthesized a polymer in which the density of states is large but $|V|$ is small so as to obtain the same transition temperature, then the coherence energy would become quite large. A coherence energy of as much as 1 eV per unit cell of the chain appears possible

[9] J. C. Slater, *Quantum Theory of Molecules and Solids* (McGraw-Hill Book Company, Inc., New York, 1963), p. 234.
[10] C. R. Noller, *Chemistry of Organic Compounds* (W. B. Saunders Co., Philadelphia, 1960), 2nd ed., p. 665.

in such a polymer. As this energy is comparable to the resonance energy of the benzene ring, one should expect a considerable stabilization of the polymer on this account. It is interesting to note, too, that the destruction of superconductivity at one point in the chain raises the energy by the coherence energy per unit length times the coherence length ζ_0. The coherence length ζ_0 for these molecules should be about 30 Å as it is inversely proportional to T_c and $\zeta_0 \approx 10^4$ Å for conventional superconductors. Consequently, it would require a large amount of energy to destroy the superconductivity locally.

V. DISCUSSION

We believe that while the estimates for the various matrix elements in the above example are crude, they are not unrealistic. This forces upon us the remarkable conclusion that superconductivity could and should occur in structures such as this even at room temperatures. There are many other possible structures similar to the one shown involving a semiconducting chain for the spine and a dye-like molecule for the side chain which would also be superconducting. It is unlikely that our particular choice described above would be the easiest to synthesize or have the optimum superconducting properties, but it illustrates the possibility in a detailed manner.

In these molecules we should expect the usual electrical properties of a metallic superconductor, however, in order to observe such effects contact would have to be made to the ends of the spine. This could be a difficult problem to solve, but may be possible by cross-linking the spines so as to form a three-dimensional net of the filamentary molecules. Because of the large transition temperature, one would expect the critical field for the destruction of superconductivity to be very high compared to that of conventional superconductors. The highly divided filamentary structure of a bulk sample of the polymer should mask any appreciable Meissner effect. Perhaps the most interesting feature of these molecules, however, lies in the phase correlation of the electron pairs throughout the molecule. This phase correlation should impose certain restraints upon the ability of the molecule to react chemically with other such molecules. The reason is that in order to form a covalent bond, the electrons must interfere constructively in the region of positive potential. Consequently, the relative phase of the electrons forming the bond are important. Such an effect has been discussed briefly by Ambegaokar and Baratoff[11] in regard to tunneling between conventional superconductors in the Josephson effect. The superconducting state is unique in that this long-range phase correlation, "off-diagonal-long-range-order" (ODLRO)[6] distinguishes it from the normal or insulating states. This we have shown can occur even

[11] V. Ambegaokar and A. Baratoff, Phys. Rev. Letters 10, 486 (1963).

in our structure which is essentially a one-dimensional chain, which is in striking contrast to a classical one-dimensional interacting chain such as the Ising chain which cannot exhibit long-range order.[12] Because of this ODLRO, which is a property of the molecule as a whole and the above chemical properties which are related to it, these molecules will have the property of reacting as a single entity which is precisely what London[1] was seeking to understand in regard to biologically important macromolecules.

In regard to the possible biological significance of our results, it is appropriate to mention a theorem which was established by Wigner[13] on the probability of a quantum mechanical system reproducing itself. He succeeded in showing that under two reasonable assumptions, this probability is essentially zero. The relevant assumption is that his "collision matrix" S, which generates the final state from the initial state, is assumed to be a random matrix. This assumption may be violated in the superconducting state because of the singular nature of the pair distribution associated with ODLRO. It would be useful to reexamine Wigner's theorem to see whether a *superconducting* quantum system would be capable of reproducing. The curious chemical selectivity mentioned earlier suggests that this may be the case.

ACKNOWLEDGMENTS

I should like to thank Derek Griffiths for many useful discussions and critical comments during the preparation of this work. I should also like to thank William Fairbank, Bascomb Deaver, and Felix Bloch for many stimulating conversations over the past few years related to this problem.

Note added in proof. It is appropriate to clarify a point in regard to the band structure of our chosen model. The partial charges on the nitrogen sites of the side chains produce a periodically varying potential along the spine with a fundamental period of $4a$. This produces a gap at $k = \pi/4a$ but no gap at $\pi/2a$ where the Fermi surface lies. The reason there is no gap at $\pi/2a$ is that in the tight binding approximation, where one can consider an electron as sampling the potential at all points within a particular orbital before moving on, the effective potential of each orbital may be taken as that of the appropriate carbon nucleus. From a calculation similar to that of Table II, one can easily show that there is then no Fourier component of period $2a$ and therefore no gap at the Fermi surface. The only effect which tends to produce a semiconductor gap (i.e., one at the Fermi surface) is the tendency for the double bond to localize at alternate sites. This was considered in Sec. III, where we showed that for a fixed semiconductor gap, the superconducting state should occur if a particular inequality could be satisfied. The band gap produced by bond alternation, however, is not fixed, but depends upon the amplitude of the periodic distortion; consequently, the actual gap and distortion must be determined in a self-consistent manner. In this case our earlier arguments do not apply and whether the superconducting state or the semiconductor state occurs depends upon which has the lower energy. Longuet-Higgins and Salem [Proc. Roy. Soc. (London) **A251**, 172 (1959)] have calculated the stabilization energy for a polyene with alternation of bond lengths and obtain an energy of 0.019 kcal/mole per bond (≈ 0.001 eV/bond) for the semiconductor state. We have calculated it for the superconducting state [Eq. (4.8)] and obtain (≈ 0.025 eV/bond) so that, in our model, the superconducting state should be favored.

Finally, our calculation has shown that a phase transition from the normal to the superconducting state should occur even in our one-dimensional system. This is unusual [see, for example, L. van Hove, Physica **16**, 137 (1950)] and one may question whether our result follows because the BCS theory, upon which it is based, is not sufficiently exact. We have some reason for believing that our conclusions are valid nevertheless, but this point requires further investigation.

[12] J. Ashkin and W. E. Lamb, Jr., Phys. Rev. **61**, 159 (1943).
[13] E. P. Wigner, in *The Logic of Personal Knowledge* (Polanyi Festschrift) (Routledge and Kegan Paul, London, 1961), p. 231.

SCIENTIFIC
AMERICAN

Established 1845

February 1965   Volume 212   Number 2

Superconductivity at Room Temperature

*It has not yet been achieved, but theoretical studies suggest that
it is possible to synthesize organic materials that, like certain
metals at low temperatures, conduct electricity without resistance*

by W. A. Little

Several years ago an experiment was performed at the Massachusetts Institute of Technology that demonstrated the possibility of constructing a perpetual-motion machine. An electric current was induced to flow around a small ring of metal. The ring was then set aside. A year later the current was found to be still circulating in the material of the ring; what is more, it had not diminished by a measurable amount during this period! Although physicists object instinctively to the idea of perpetual motion and refer to such currents euphemistically as "persistent currents," they are obviously extremely persistent currents.

The secret of this extraordinary phenomenon is of course that the metal must be kept very cold—in fact, within a few degrees of absolute zero (−273 degrees centigrade). Below a characteristic "transition temperature" certain metals spontaneously enter what is known as the superconducting state, in which a stream of electrons can flow without encountering any resistance in the form of friction. Since friction is the cause of the failure of all mechanical perpetual-motion machines, its total absence in this case allows the initial current to persist indefinitely without any further input of energy, thereby violating the traditional doctrine of the impossibility of perpetual motion.

Actually the phenomenon of superconductivity is not at all rare. Since its discovery by the Dutch physicist Heike Kamerlingh Onnes more than 50 years ago many different metals and several hundred alloys composed of these metals have been identified as superconductors. As might well be expected, the technological potential of perpetual-motion machines based on the principle of superconductivity is virtually unlimited. Lossless power transmission, enormously powerful electromagnets, more efficient motors, amplifiers, particle accelerators and even computers are just a few of the serious proposals for the exploitation of superconductivity that have been put forward in the past 50 years. The main drawback of all these schemes involves the very low temperatures typically associated with superconductors; the complex and bulky refrigeration equipment required to maintain such metals in the superconducting state makes most of the proposed applications as yet economically unfeasible. The hope that the problem of refrigeration might someday be circumvented by the discovery of superconductors with higher transition temperatures has led to the investigation of a large number of alloy combinations of the known superconducting metals. Although many new superconducting alloys have been found, the outlook for high-temperature metallic superconductors is not bright. The highest transition temperature recorded so far is only 18.2 degrees Kelvin (degrees centigrade above absolute zero), which is still well below the temperature range accessible to simple refrigera-

tion systems. Moreover, this work has yielded a considerable amount of statistical evidence that suggests that it is extremely unlikely that an alloy will ever be found with a transition temperature appreciably higher than about 20 degrees K.

What about the possibility of discovering some other substance—perhaps a nonmetallic one—that would be superconducting at higher temperatures? As a matter of fact it is an especially opportune time to investigate such a possibility in view of the great theoretical advances that have been made in recent years toward understanding the superconducting state. I have been particularly interested in the possibility of synthesizing an organic substance that would mimic the essential properties of a superconducting metal. My calculations have shown that certain organic molecules should be able to exist in the superconducting state at temperatures as high as room temperature (about 300 degrees K.) and perhaps even higher! In order to explain the line of reasoning that led to this prediction I must first discuss some of the theoretical ideas on which it is based.

An understanding of the true nature of superconductivity has proved to be one of the most difficult problems of theoretical physics in this century. A great stride forward was made in 1957 with the publication of a comprehensive microscopic theory by John Bardeen,

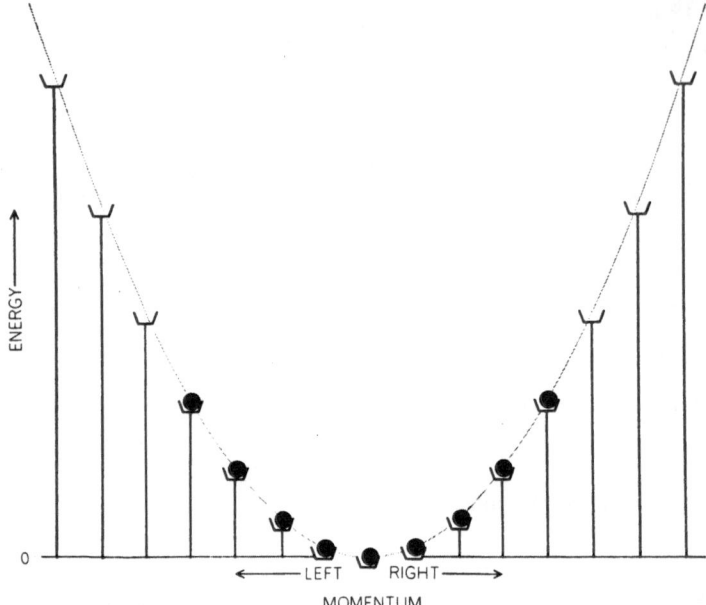

IN AN ORDINARY CONDUCTOR "free" electrons (*colored balls*) can roam in any way consistent with two restrictions: (1) only certain velocities, or energy states, are permitted and (2) only one electron at a time may be in any one of the allowed states. In the most stable energy arrangement all the lower energy states are filled by electrons and all the higher states are empty. No current flows because as many electrons move to left as to right.

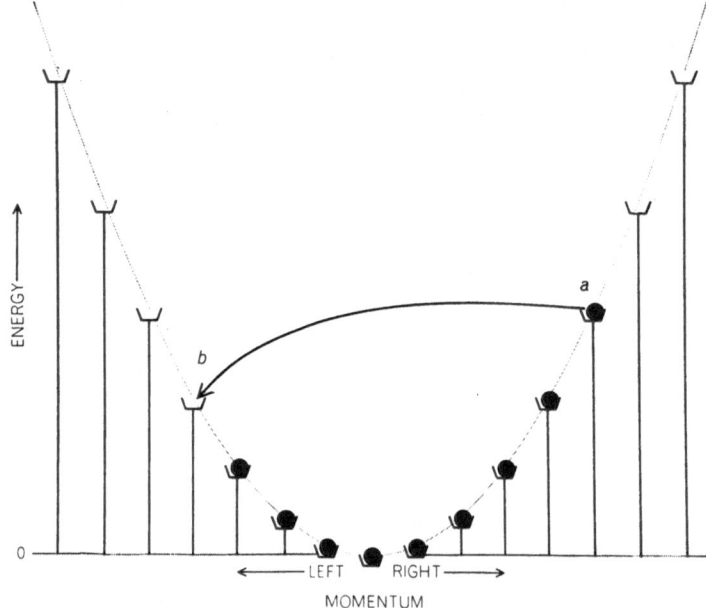

CURRENT FLOWS in an ordinary conductor when all the allowed energy states are displaced in one direction (in this case to right); more electrons now travel to right than to left. Current decays when an electron moving rapidly to right (*a*) collides with some imperfection in the metal and is knocked backward (*colored arrow*); it can then occupy one of the vacant states corresponding to an electron moving somewhat less rapidly to left (*b*).

L. N. Cooper and J. R. Schrieffer, then at the University of Illinois. Their theory, now universally known as the BCS theory, has been successful not only in explaining practically all the experimental data that had accumulated over the past half century but also in predicting a number of new superconducting phenomena.

Like most scientific theories, the BCS theory did not appear out of the blue but was built on a firm theoretical foundation established by earlier investigators. In particular, some of the principal features of the BCS theory were outlined many years before by the theoretical physicist Fritz London, who developed a successful macroscopic theory of superconductivity as early as 1950. In so doing London showed an appreciation of the highly organized nature of the superconducting state and an intuitive grasp of several of the essential criteria a successful microscopic theory would have to fulfill. He recognized that each sample of a given superconductor has a unique character peculiar to itself and that in the superconducting state this character remains unaffected by heat or any other external influence. He was also impressed with the extraordinary stability of the superconducting state, a characteristic that figures prominently in the BCS theory.

It was perhaps through his perception of these rough features of a microscopic theory that London was led to suggest that the phenomenon of superconductivity might be significant in areas of science other than the specialized niche of low-temperature physics. He proposed that the existence of such a state in certain large organic molecules, such as proteins, might help to explain some of the unusual properties of these molecules. Unfortunately London died several years before the advent of the BCS theory and so was unable to develop or test his ideas further in its light. Although his other writings have influenced many workers in the field, little attention appears to have been paid to these suggestions in the decade after his death.

My own interest in the possibility of biological superconductivity was stimulated five years ago while I was working at Stanford University on a rather mundane problem of heat transfer to a metallic superconductor. Like London, I was struck by the great stability of the superconducting state; it occurred to me that if nature wanted to protect the information contained, say, in the genetic code of a species against the ravages of heat and other external influences, the principle of superconductivity would be well suited for the purpose. In view of nature's remarkable record of ingenuity in such matters, I thought it might be useful to determine if the superconducting state could occur in a large organic molecule built along the general lines of the genetic molecule deoxyribonucleic acid (DNA).

One molecule of this general type looked particularly promising in the light of the requirements of the BCS theory. Its structure, incidentally, is quite close to what London appears to have had in mind and suggests that he progressed further along this line of reasoning than his published work reveals. A detailed calculation of the superconducting properties of the hypothetical molecule showed that it should be superconducting at room temperature, and indeed even at temperatures well above room temperature. This bonus, although it is necessary if the phenomenon is to perform any biological function, was quite unexpected. Subsequent investigation has shown that there is a relatively straightforward explanation for this extraordinarily high transition temperature. Before describing how a superconducting current might be transmitted by such a molecule, however, it is necessary to review the mechanisms by which an electric current is transmitted in an ordinary conductor and in a metallic superconductor.

In an ordinary, nonsuperconducting metal each atom loses some of its outer, more loosely bound electrons, which are then free to roam throughout the rest of the metal. The motion of these "free" electrons is not entirely unrestricted: the requirements of quantum mechanics impose the condition that only certain energy states, or velocities, are permitted. Another restriction is imposed on the manner in which the electrons may be arranged in these states. This restriction arises out of the Pauli exclusion principle (named after its discoverer, Wolfgang Pauli), which says that only one electron at a time may be in any one of the allowed states. The electrons are free to arrange themselves in any way consistent with these two restraints. The most stable energy arrangement is one in which all the lower energy states are filled by electrons and all the higher states are empty. For every state that corresponds to an electron moving to the left there is another state of equal energy for an electron moving to the right. Thus in the lowest energy arrangement there are as many electrons moving to the left as to the

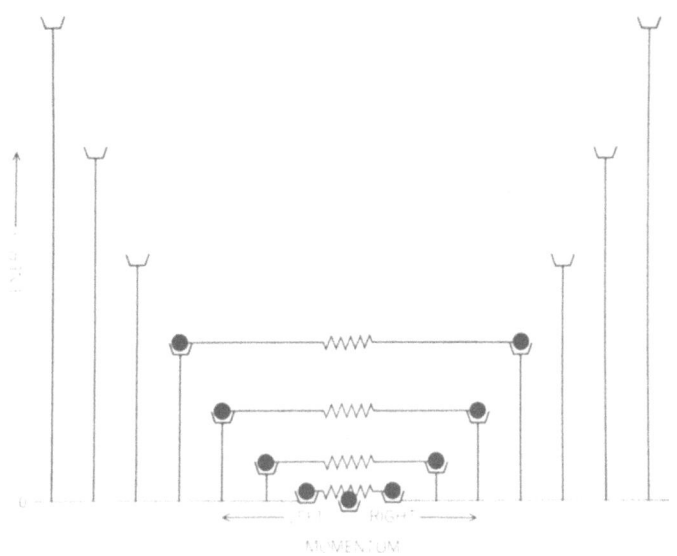

IN A SUPERCONDUCTOR an attractive interaction of "free" electrons binds them together in pairs (see illustration on next page). The paired electrons cannot move freely in the metal, since it turns out that in order to be bound to each other the momentum of the center of mass of each pair must be the same as that of the majority of the other pairs. When no current is flowing in the metal, the momentum of the center of mass for each pair is zero.

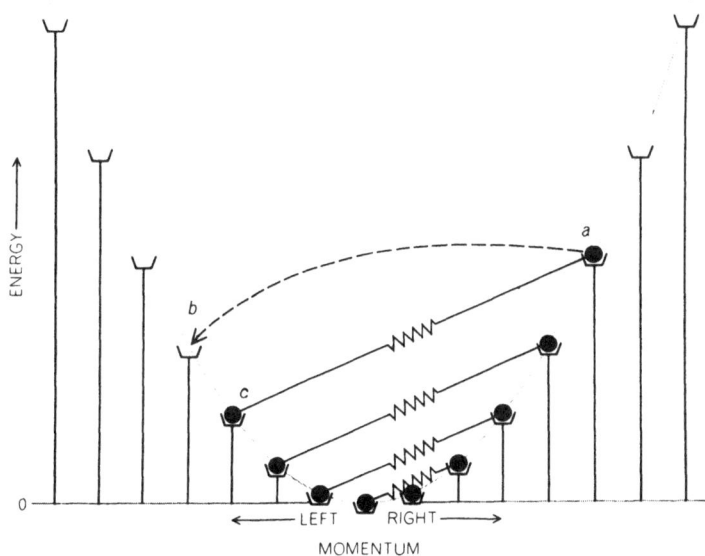

"PERSISTENT CURRENT" FLOWS in a superconductor when more electrons travel to the right than to the left. The current does not decay because if the electron at a were now to jump to b, both it and the electron at c would be left without mates. These two single electrons would not be able to pair up with each other because their center of mass would then have the wrong momentum. Consequently the electron pairs do not as a rule break up.

right and the entire distribution of electrons is symmetric [*see top illustration on page 22*]. In this equilibrium situation the average velocity of the electrons in any one direction is zero and consequently no current flows in the metal.

If a current is now induced in the metal, this is tantamount to saying that all the electrons are forced to move in one direction, say to the right, so to the random velocity of each electron must be added the component of this drift velocity. Electrons moving to the right would therefore have their velocities and consequently their energies increased, whereas electrons moving to the left would have their velocities and energies decreased. On the average this asymmetric distribution of the electrons would have a somewhat higher energy than the original symmetric distribution, owing to the additional kinetic energy of the drifting electrons [*see bottom illustration on page 22*]. The asymmetric distribution does not last long if left to

itself in a normal metal, because if one of the electrons moving rapidly to the right should collide with some imperfection in the metal and be knocked backward, it could then occupy one of the vacant states corresponding to an electron moving less rapidly to the left. The states on the left have a somewhat lower energy because of their lower velocity and so would be preferred by the electron. In this way the asymmetric, current-carrying distribution would rapidly rearrange itself to form the lower-energy, symmetric distribution and current would disappear.

In a superconductor this deterioration of the asymmetric electron distribution does not occur, since there is an attractive interaction of electrons that binds them together in pairs. Each electron in a superconductor has a mate with which it is paired. The paired electrons cannot move freely in the metal; it turns out that in order for them to be bound to each other the momentum of the center of mass of each pair must be the

same as that of the majority of other pairs [*see top illustration on preceding page*]. Now, if the electron moving rapidly to the right collides with an imperfection and is knocked into a state corresponding to the electron moving less rapidly to the left, this electron will have no mate and its old mate will similarly be left single. These two single electrons would not be able to pair up with each other because their center of mass would then have the wrong momentum. Consequently if the cost in energy for breaking up the pair is not offset by a sizable reduction in kinetic energy resulting from the collision, the pair will not break up. The asymmetric distribution will remain and the current will persist [*see bottom illustration on preceding page*]. According to the BCS theory, this is the reason why the current in a superconducting ring can persist indefinitely.

Of course the foregoing argument only partly explains why certain substances suddenly become superconduct-

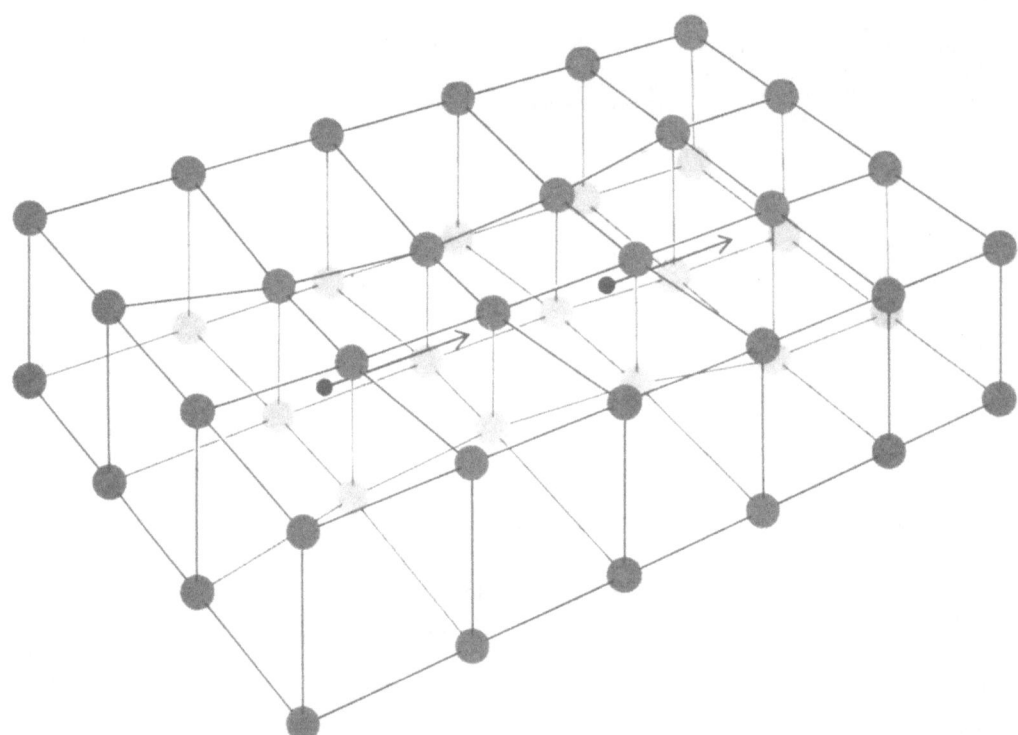

ELECTRON PAIR is formed in a superconducting metal by the attractive mechanism depicted here. As a negatively charged electron moves through a somewhat elastic lattice of positively charged ions it attracts the ions, causing the lattice to "pucker up" in its vicinity. A second electron is naturally attracted to the excess posi-

tive charge created by the higher density of ions in this puckered region of the lattice and is thereby indirectly attracted to the first electron. Since the ions move more slowly than the electrons, the puckered region trails a considerable distance behind the first electron and the second electron can follow at this safe distance.

ing at a specific temperature. Why, for example, should the electrons attract one another when we know that electrons have like charges and thus should repel one another instead? Why should the centers of mass of the pairs be correlated in any way with one another? These are subtler questions that cannot be easily answered straightforwardly, but they can be approached in a somewhat more general way with the aid of an analogy.

Imagine a thin elastic membrane stretched over the head of a drum. On top of the drumhead we put two marbles. If we tilt the drum gently from side to side, how do the marbles move? Obviously the weight of each marble depresses the elastic membrane so that when the two marbles come close to each other, one rolls down into the depression made by the other; to all appearances the two are attracted to each other. If we now tilt the drum to one side, the marbles move about on the surface together as a pair—each moving in the depression made by the other. If the tilting is done more vigorously, however, the agitation can eventually become so violent that the two marbles separate and begin to move about more or less independently of each other.

What has this to do with superconductivity? In a metal the positively charged ions, which remain after the atoms are shorn of their outer electrons, are not rigidly fixed at their sites in the crystal lattice but are able to move elastically about these sites. If one of the "free" electrons moves among the positively charged ions, the ions will be attracted to the negatively charged electron as it passes. This distorts the lattice and causes it to "pucker up" in the vicinity of the electron [see illustration on opposite page]. A second electron is naturally attracted to the excess positive charge created by the higher density of ions in the puckered region of the lattice and is thereby indirectly attracted to the first electron. The situation is closely analogous to the second marble being attracted to the first by the depression in the elastic drumhead; the puckered lattice and the depression in the drumhead play equivalent roles in the two cases.

In a metal the attractive force produced by this mechanism can be such that two electrons can become firmly bound to each other. The binding will only occur, however, if the temperature is sufficiently low, since at higher temperatures the thermal agitation of the electrons will tend to break up the elec-

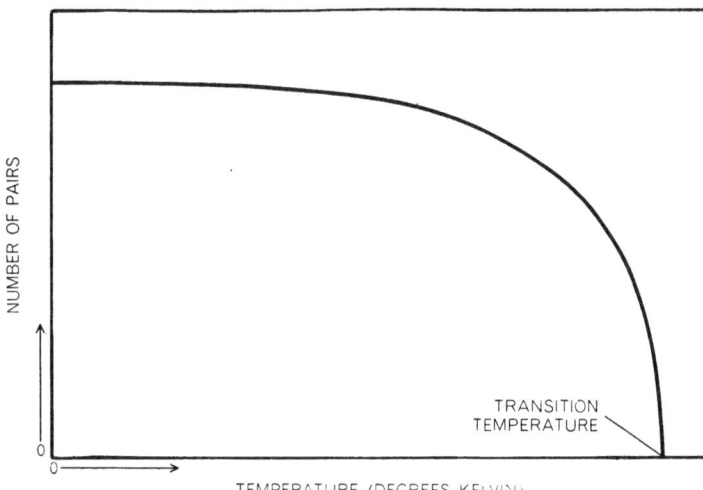

CATASTROPHIC BREAKUP of electron pairs in a superconductor occurs as the superconducting transition temperature is approached from a lower temperature. Although the pairs begin to break up slowly, the unattached electrons resulting from the just broken pairs interfere with the remaining pairs, causing them to break up and accelerating the entire breakup process rapidly. No electron pairs can exist above the transition temperature.

tron pairs in much the same way as the violent tilting of the drumhead can separate a pair of marbles.

It is useful to look more closely at the details of this process. Suppose the first electron is moving down a corridor of positive ions. The range of the electron-ion interaction is quite short; accordingly the electron will give each ion a short, sharp pull as it passes. This impulse starts the ion moving toward the position of the electron, but since the ions move rather more slowly than the electrons, the electron responsible for the motion will be a considerable distance away by the time the ion has moved as far as it can against the elastic restraints of its neighbors. As a result the puckered region of the lattice trails some distance behind the first electron, and the second electron can follow at this safe distance.

Between the two electrons there is enough space for many other electrons to move. What effect do these electrons have on our theoretical model? If their motion is quite chaotic and completely uncorrelated with that of the first electron, they will interfere seriously with the orderly procedure described above. They too would pucker the lattice in their neighborhood and stretch it elsewhere. If the lattice is stretched where it should be puckered and puckered where it should be

stretched by this host of uncorrelated electrons, there will be little left of the puckered region produced by the first electron for the second electron to follow. The scent, so to speak, will rapidly be lost and the pair broken. On the other hand, if the motion of all the other electrons is correlated so that each electron dodges the others and maneuvers in precisely the right way, then each electron can enjoy fully the attraction of its mate and a large number of pairs can coexist with one another.

It turns out that the prerequisite for all the pairs being in harmony with one another in this way is that their centers of mass must all have the same momentum. At low enough temperatures this highly coordinated state of the electrons occurs spontaneously, because the gain in the energy of each pair more than offsets the disadvantages involved in the loss of freedom of the individuals. If the temperature is raised, however, the thermal agitation eventually becomes sufficient for some pairs to break up. The resulting uncorrelated, unattached electrons now become a disruptive nuisance to the electron couples. They interfere with the attractive mechanism and thereby weaken the binding force of the remaining pairs. This in turn causes more pairs to break up. As the temperature is raised still further, the breakup of pairs becomes catastrophic; above a well-defined temperature no pairs can

HYPOTHETICAL SUPERCONDUCTING MOLECULE is built around a "spine" of carbon atoms connected by alternating single and double bonds. Periodically along the spine a side chain consisting of the common dye diethyl-cyanine iodide extends outward. These side-chain molecules are highly polarizable; that is, an electron can move freely from a nitrogen site close to one end of the molecule to another nitrogen site close to the other end. A colored *N* designates the nitrogen atom that contains the resonating electron in the two possible conditions of polarization. Electrons can also move freely along the spine itself.

exist at all [see illustration on preceding page].

The superconducting state is obviously distinguished by a high degree of internal organization. It can exist only below the temperature at which the breakup of the electron pairs becomes catastrophic. The temperature at which this occurs is the superconducting transition temperature.

It is not difficult to deduce from the preceding argument the criteria that would have to be fulfilled for an organic molecule to exist in the superconducting state. The molecule would have to be provided with roughly the same essential ingredients that are found in the superconducting metal. A medium is required in which the electrons can move, and a somewhat elastic charged structure is needed to play the role of the ion lattice.

Imagine a long molecule built of a chain of carbon atoms that form what I shall call a "spine." On each side of the spine molecular side chains extend outward rather like the ribs of the human rib cage. As I have mentioned, this structure was suggested by the genetic molecule DNA, the carbon atoms of the spine replacing the sugar-phosphate sequence of DNA and the side chains replacing the bases. If the carbon chain is

conjugated, that is, if it has alternating single and double bonds along the chain, it will behave much like a metal, with the electrons moving freely from one end of the spine to the other. For the side chains a molecule such as diethyl-cyanine iodide, a dye commonly used to sensitize photographic emulsions, would appear to be suitable. This is a highly polarizable molecule in which an electron can move freely from a site close to one end of the molecule to a site close to the other end [see illustration above]. In an electric field the charge readily shifts from one end to the other and the molecule thus becomes polarized in the field.

Consider now an electron moving along the spine of such a molecule. As the electron passes each side chain its electric field polarizes the side-chain molecule and induces a positive charge at the end nearer the spine. Because of the high speed of the electron in the spine the region of maximum induced positive charge in the side chains trails some distance behind this electron. A second electron is attracted to the region of positive charge and is thereby indirectly attracted to the first electron. This is exactly the same argument we used in describing the superconducting metal and naturally leads to the same phenomenon. When the detailed calcu-

lations of the BCS theory are carried out for the organic molecule, however, one finds that the theoretical transition temperature is enormously high—typically around 2,000 degrees K.! This figure is of course much higher than that of any known superconductor and, if it is to be trusted, must be based on some good physical reason. It turns out that it is.

Let us return to the description of the attractive mechanism between free electrons in a conventional superconductor. As an electron passes the ion it gives it a short, sharp impulse and then is gone. The impulse transfers to each ion a certain amount of kinetic energy, which starts the ion moving. The ion continues to move until the elastic restraints of its neighbors stop it. At this point the kinetic energy of the ion is completely converted into potential energy. It is elementary to show that the maximum distance the ion can be displaced is inversely proportional to the square root of the mass of the ion. If the ions are heavy, the displacement is small; if the ions are light, the displacement is large. The larger the displacement is, the larger is the distortion of the lattice and consequently the larger the magnitude of the excess positive charge in the puckered region. Since the second electron is attracted to this region and is thereby bound to the first electron, we should expect the strength of this binding also to depend on the mass of the ions of the lattice. By the same token, the transition temperature, which is determined by the binding energy, should be inversely proportional to the square root of the ionic mass. This is in fact the case. Experiments carried out on samples composed of different isotopes of a given superconducting element have shown that in most cases the transition temperature does depend on the isotopic mass in just this way. This correspondence is known as the isotope effect; its discovery in 1950 provided an important clue to the understanding of superconductivity.

The isotope effect also plays an important role in explaining the tremendously high transition temperatures of our hypothetical superconducting molecules. In these molecules we have replaced the ions of a metal with polarizable side-chain molecules. Under the influence of the electric field of the electron in the spine, the side chains themselves do not move, but a single electron *within* each side chain does move and this produces the polarization. Instead of being an ion, as in the case of the metal, the moving entity is now an electron with a mass only a hundred-thousandth of a typical ionic mass. According to the requirements of the isotope effect, the transition temperature should be on the order of the square root of 100,000 times larger than that for the average metallic superconductor (that is, 316 times 6 degrees K., or roughly 2,000 degrees K.). This figure is almost exactly the same as the one arrived at by our earlier, independent calculation. Obviously the high transition temperature is a result of an attractive interaction that is mediated by an extremely light particle—an electron rather than an ion.

In order for the motion of the ions and the electrons to be in any way analogous it also seems necessary from the theoretical point of view for the electrons to move in a chainlike spine. This is probably the reason why no metal falls into this category of superconductor and why no high-temperature superconductors have been found as yet.

Once we accept the possibility that superconductivity could occur at room temperature in a molecule of this type, a whole new world of science and technology opens before us. Indeed, speculation as to the uses to which such a superconductor could be put reads more like science fiction than any serious scientific proposal. Of course, the chemical problems and eventually, perhaps, the technological problems in the synthesis and production of such materials are immense. After all, we are asking for the synthesis of a molecule to precise, almost engineering specifications, a task that has never before been demanded of organic chemistry. Nevertheless, many chemists feel that this can be done and that given a reasonable amount of time some such molecules undoubtedly will be synthesized. It is to be hoped that when that day comes, our extension of the BCS theory will pass the acid test.

Suppose for the sake of argument we are presented with a plastic material that is superconducting at room temperature. How could we use it? The obvious applications mentioned at the beginning of this article immediately spring to mind, but even more exciting prospects arise from a superconductor's diamagnetism, or impermeability to a magnetic field. Because of the highly coordinated motion of the electrons a magnetic field cannot penetrate the interior of a superconductor. This property can be demonstrated by placing a bar magnet above a sheet of superconducting metal. The magnet floats freely above the sheet, supported entirely by its magnetic field. The field is unable to penetrate the superconductor and so provides a cushion on which the magnet rests. It is easy to imagine hovercraft of the future utilizing this principle to carry passengers and cargo above roadways of superconducting sheet, moving like flying carpets without friction and without material wear or tear. We can even imagine riding on magnetic skis down superconducting slopes and ski jumps—many fantastic things would become possible.

Have we anything of interest to expect from the biological point of view? If it proves possible to synthesize an artificial superconducting molecule, it seems to me that nature would surely have discovered the fact ages ago. Thus we would expect to find molecules of this type playing some unique role in nature, but we can only speculate as to what this role may be. The highly coordinated motion of the electrons within our hypothetical molecule couples the different parts of the molecule together in an extremely intimate way. As a result reactions in one part of the molecule can influence the reactivity of other groups in any part of the entire molecule, however remote. Could this long-range influence explain some of the intricacies of biological molecules? In our molecule the particular value of the common momentum of the centers of mass of the pairs has a very interesting property: it endows the molecule with a unique, preferred three-dimensional folded structure. Associated with each possible value of this momentum there is a unique, intricate shape for the molecule as a whole. Could these structural requirements have anything to do with the large-scale organization of living systems? We cannot be sure at this stage, but the implications of the idea are intriguing. According to our model there is a highly specific attraction between two molecules whose electron pairs have the same momentum but no such attraction between molecules with different momenta for the pairs. Has this anything to do with the extraordinary selectivity and specificity of certain biochemical reactions? Again we do not know, but the idea is suggestive. When one reflects on all these possibilities, the age-old dream of the perpetual motion of mere mechanical devices appears drab and colorless in comparison.

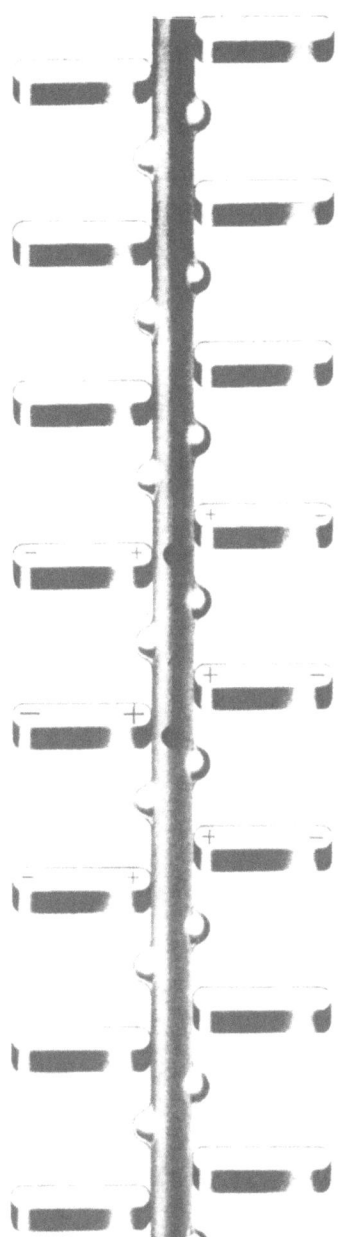

ELECTRON PAIRS are conducted along the spine of a hypothetical superconducting molecule by an attractive mechanism similar to that in a superconducting metal. As an electron passes each side chain its electric field polarizes the side-chain molecule and induces a positive charge at the end nearer the spine. A second electron is attracted to this region of positive charge and is thereby indirectly attracted to the first electron.

SPECIFIC HEAT OF He³ AND He⁴ IN THE NEIGHBORHOOD OF THEIR CRITICAL POINTS

M. R. Moldover and W. A. Little*

Physics Department, Stanford University, Stanford, California

(Received 16 June 1965)

Recently Bagatskiĭ, Voronel', and Gusak[1] showed that the specific heat at constant volume of argon exhibited what appears to be a logarithmic singularity at the critical temperature (T_C) for measurements taken at a density near the critical density. This singular behavior is in sharp contrast to the predictions of the traditional view of this phase transition by Landau and Lifshitz.[2] However, the behavior is precisely that to be expected for the socalled "lattice-gas" model for the liquid-gas transition. Lee and Yang[3] have shown that the partition function of a classical gas of particles moving on a discrete lattice with a repulsive force preventing double occupancy of any site and a nearest-neighbor attraction can be mapped precisely onto the partition function of an Ising model of a spin system in an external magnetic field. The specific heat for this Ising model in zero field exhibits a logarithmic singularity at the Curie point. The specific heat for the corresponding lattice gas on the critical isochore exhibits a logarithmic singularity at the critical point. The measurements on argon then indicate that for a <u>real</u> gas the specific heat behaves in a manner similar to that of a lattice gas. We have investigated this point further by studying the specific heat at constant volume (C_v) of both He³ and He⁴ at densities close to the critical density. We have done this for two main reasons. Firstly, to see whether the behavior observed for argon is also observed for helium, for which quantum effects should be important, and secondly, to investigate the detailed nature of the singularity in the pressure-density plane, not only on the critical isochore but also in its immediate neighborhood. Yang and Yang[4] have conjectured that the quantum effects would reduce the magnitude of the singular contribution to the specific heat in helium. Our results confirm this view.

The calorimeter was built of two OFHC copper parts. The helium was contained in the lower part in 50 slots each 0.3 cm deep and 0.01 cm wide. The large area of these slots facilitated good thermal contact between the helium and the calorimeter. The upper part was a lid, sealed to the lower section with an indium O ring. The lower part was wound with a Constantan heater and had a carbon resistor cemented and clamped to it. The calorimeter was supported on nylon threads in an evacuated chamber. Thermal contact to the bath was made with a mechanical heat switch. A stainless steel capillary 5 in. in length and 0.006-in. i.d. led from the calorimeter to a needle valve. The dead volume of the capillary was about $\frac{1}{2}\%$ of the total volume of the calorimeter. Helium was admitted to the calorimeter via a Toeppler pump which was used to measure the volume of the gas to an accuracy of about 0.2%.

A (1/10)-W Ohmite carbon resistor of nominal resistance 560 Ω was used as a secondary thermometer. Its resistance was measured with a 100-cps bridge using a lock-in amplifier. Temperature changes of about 2×10^{-6} °K at 5.2°K and 1×10^{-6} °K at 3.3°K could be observed. At the end of a run, exchange gas was admitted to the vacuum space and the resistor was calibrated against the vapor pressure of the He⁴ bath from 2.9 to 5.15°K. A smooth resistance temperature relation was fitted, hence relative specific-heat measurements near the critical points are not sensitive to errors either in the vapor-pressure–temperature scale or in the calibration. Stray heat input to the calorimeter was measured before and after each data point was taken. The approach to temperature equilibrium of the calorimeter was observed after each heating interval. As T_c was approached, the equilibrium time increased from a few seconds to several minutes and became one of the limiting factors in this measurement.

For each run T_c was tentatively defined as the temperature at which C_v fell most abruptly. The final value of T_c was then chosen to give the best fit to

$$\frac{C_v}{R} = -a \ln\left(\frac{T_c - T}{T_c}\right) + b, \tag{1}$$

over the range 10^{-3} °K $< (T_c - T) < 10^{-1}$ °K. The adjustment involved was less than 3×10^{-4} °K.

Figure 1 is a general view of the specific

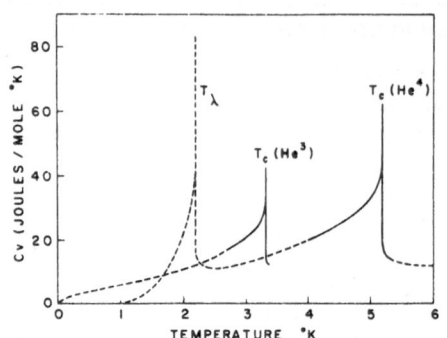

FIG. 1. C_v of He³ and He⁴ at their respective critical densities is plotted as a function of temperature. The solid curve is the present work.

heats at the critical volume of He³ and He⁴ at low temperatures.[5] We have indicated the temperature range covered by these measurements. In Fig. 2 we have plotted on both a linear and logarithmic scale measurements of C_v of He⁴ at a density within 0.5% of the critical density.

In the two-phase region below T_C, C_v clearly exhibits within 0.2°K of T_c, logarithmic behavior similar to that observed in argon[1] and oxygen.[6]

In the one-phase region above T_c two alternatives have been suggested. $C_v \propto \ln(T-T_c)$ has been suggested by an exact calculation on a two-dimensional Ising model,[7] by C_v measurements on argon[1] and oxygen,[5] and by velocity-of-sound measurements on helium four.[8] $C_v \propto (T-T_c)^{-\alpha}$ for some small positive α (such as $\frac{1}{8}$) has been suggested by Fisher[9] on the basis of an approximate calculation on a three-dimensional Ising model and his reanalysis of the argon and oxygen data. We do not believe the present data permit one to distinguish between these two alternatives.

In Fig. 3 similar plots are given for He³ for a density of $0.985\rho_c$. Logarithmic behavior is again observed for $T < T_c$. The data for argon, oxygen, and both isotopes of helium expressed in the dimensionless form (1) are summarized in Table I. We see that the coefficient a decreases from argon to He⁴ to He³ in agree-

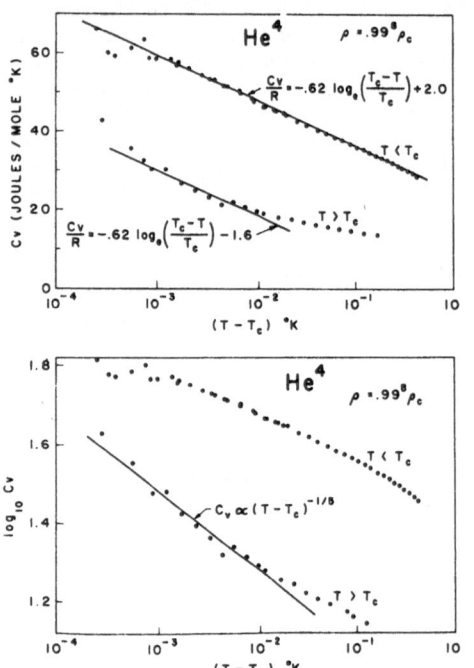

FIG. 2. C_v of He⁴ is plotted against $|T-T_c|$ at a density of $0.99^8\rho_c$.

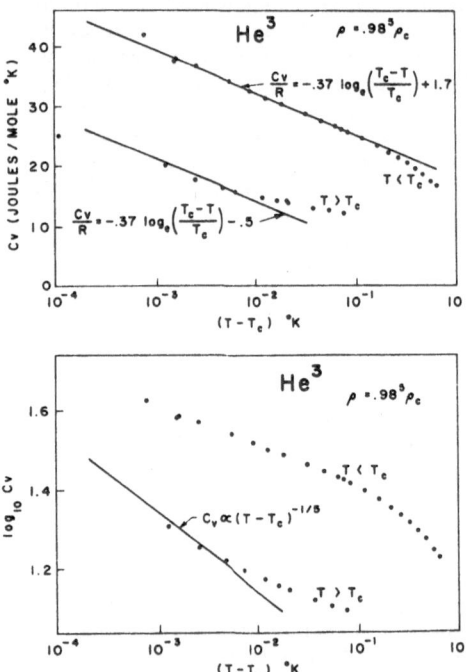

FIG. 3. C_v of He³ is plotted against $|T-T_c|$ at a density of $0.98\rho_c$.

Table I. Coefficients a and b of Eq. (1) for Ar, O_2, He^4, and He^3.

Element	a	b $(T < T_c)$
Ar	1.8	8
O_2	2.4	10
He^4	0.62	2.0
He^3	0.37	1.7

ment with the conjecture of Yang and Yang.[4] We note that the coefficient 0.62 for He^4 is very close to that observed by Buckingham and Fairbank[5] for the specific heat along the saturated vapor curve at the λ point of He^4 (0.64) and may have particular significance in explaining the quantum transition.

Preliminary measurements of C_v for He^4 at densities not as close to ρ_c as the measurements presented above indicate that in the two-phase region $d^2\mu/dT^2$, where μ is the chemical potential, does not approach zero as T approaches T_c. This differs from the result of the two-dimensional lattice gas and supports the prediction of Yang and Yang.[4] The behavior of C_v in the one-phase region is a sensitive and complicated function of density. It will be discussed elsewhere.

We wish to thank Dr. Derek Griffiths, Dr. R. S. Safrata, Dr. C. F. Kellers, and Dr. M. H. Edwards for many useful discussions, and Dr. Derek Griffiths and Donald Moldover for much help in analyzing data. We wish to thank Professor C. N. Yang for his suggestions, interest, and advice. Some of these data were presented previously.[10]

*Work supported in part by the National Science Foundation and the Office of Naval Research.

[1]M. I. Bagatskiĭ, A. V. Voronel', and V. G. Gusak, Zh. Eksperim. i Teor. Fiz. 43, 728 (1962) [translation: Soviet Phys.−JETP 16, 517 (1963)].

[2]L. D. Landau and E. M. Lifshitz, Statistical Physics (Addison-Wesley Publishing Company, Inc., Reading, Massachusetts, 1958).

[3]T. D. Lee and C. N. Yang, Phys. Rev. 87, 410 (1952).

[4]C. N. Yang and C. P. Yang, Phys. Rev. Letters 13, 303 (1964).

[5]The dashed curve for He^4 was calculated from data of M. J. Buckingham and W. M. Fairbank, Progress in Low Temperature Physics (North-Holland Publishing Company, Amsterdam, 1961), Vol. III, p. 86; R. W. Hill and O. V. Lounasmaa, Phil. Mag. 2, 145 (1957); O. V. Lounasmaa and E. Kojo, Physica 36, 3 (1959); H. C. Kramers, J. D. Wasscher, and C. J. Gorter, Physica 18, 329 (1952); M. H. Edwards, Can. J. Phys. 36, 884 (1958); H. van Dijk, M. Durieux, J. R. Clement, and J. K. Logan, Nat. Bur. Std. (U. S.), J. Res. 64A, 1 (1960). The dashed curve for He^3 was calculated from data of Henry Laquer, Stephen G. Sydoriak, and Thomas R. Roberts, Phys. Rev. 113, 417 (1959); R. H. Sherman, S. G. Sydoriak, and T. R. Roberts, Nat. Bur. Std. (U. S.), J. Res. 68A, 579 (1964); and T. R. Roberts and S. G. Sydoriak, Phys. Rev. 98, 1672 (1955).

[6]A. V. Voronel', Yu R. Chashkin, V. A. Popov, and V. G. Simkin, Zh. Eksperim. i Teor. Fiz. 45, 828 (1963) [translation: Soviet Phys.−JETP 18, 568 (1964)].

[7]L. Onsager, Phys. Rev. 65, 117 (1944).

[8]C. E. Chase, R. C. Williamson, and Laszlo Tisza, Phys. Rev. Letters 13, 467 (1964).

[9]Michael E. Fisher, Phys. Rev. 136, A1599 (1964).

[10]Proceedings of the Ninth International Conference on Low Temperature Physics, Columbus, Ohio, 1964 (to be published); Proceedings of the Conference on Phenomena in the Neighborhood of Critical Points, National Bureau of Standards, Washington, D. C., 1965 (to be published).

M501 3-3

Reprinted from THE PHYSICAL REVIEW, Vol. 156, No. 2, 396–403, 10 April 1967
Printed in U. S. A.

Decay of Persistent Currents in Small Superconductors*

WILLIAM A. LITTLE

Department of Physics, Stanford University, Stanford, California

(Received 27 May 1966; revised manuscript received 1 December 1966)

The thermodynamic fluctuations of the order parameter in a superconductor are shown to be able to cause the decay of a "persistent" current in a ring-shaped conductor. Calculations have been made of the lifetimes of these currents, which indicate that in very thin wires this decay should be detectable. We also show that a true phase transition, distinguished by an infinitely sharp change of resistivity, is possible only in an infinite three-dimensional specimen. In one- and two-dimensional samples, on the other hand, no *infinitely* sharp change of resistivity occurs but, instead, the resistance drops smoothly and rapidly towards zero as $T \to 0°K$.

O NE of the outstanding problems of superconductivity is that of explaining the lifetime of persistent currents in ring specimens. In this paper we will show that it is possible to set an upper limit on this lifetime by considering the effects of thermodynamic fluctuations in the specimen. These fluctuations provide a means by which the persistent current can decay. Decay by this means should be observable in small specimens where the thermodynamic fluctuations can be appreciable. In large specimens, on the other hand, the lifetimes are predicted to be enormously large, in agreement with the experimental evidence.

This work was motivated by the consideration of Ferrell[1] and Rice[2] of the role played by the dimensionality of the specimen on the existence of superconductivity or off-diagonal–long-range order[3] (ODLRO). They have shown that in one-dimensional specimens, in particular, fluctuations in the density or of the superconducting order parameter make it impossible for ODLRO to occur in these specimens. One might then be led to conclude that superconductivity would be impossible in a one-dimensional sample, that is, one in which the transverse dimensions are small compared to the length. In particular, one might conclude that superconductivity would be impossible in a linear macromolecule of the type discussed earlier by the author.[4] This does not necessarily follow, for we do not know at present whether ODLRO is a sufficient *and necessary* condition for the existence of superconductivity. In order to investigate whether superconductivity can be ruled out in these systems, we have extended the arguments of Ferrell and Rice and used them to determine directly a limit on the lifetime of a "persistent" current in finite ring-shaped superconducting samples. We obtain the result that for a system in which the range of the interaction force is finite, a phase transition defined by an infinitely sharp change of conductivity can occur only in a sample which is infinite in at least three dimensions. We are particularly interested in samples which are infinite in

only one of the three dimensions. In these, the average electrical resistivity at finite temperatures never drops to a value which is absolutely zero. However, in most cases, for temperatures appreciably less than the bulk T_c, the resistance drops to an exceedingly small fraction of the normal resistance. We find then that while we have neither a true phase transition nor the existence of ODLRO because of the fluctuations in a one-dimensional system, we can still have a state of greatly enhanced conductivity at low temperatures. We must stress, however, that in our argument we follow Ferrell and Rice in assuming the existence of an order parameter locally. There is some reason to believe that for specimens so narrow that the lateral dimensions are less than the Fermi wavelength, it may be impossible for an order parameter to occur here at all.[5] This is a different problem from that of the fluctuations and we make no attempt to examine this here, except to discuss what bearing our results will have on this problem. Also, in our arguments we have confined ourselves to the problem of thermodynamic fluctuations alone: i.e., fluctuations in which a small part of the specimen deviates, for example, in its temperature or density from that of the rest of the specimen, but within each small part equilibrium at this different temperature or density is maintained at all times. We have not considered fluctuations to states which are not describable by such local values of the intensive parameters. Because of this, our results give only an upper limit on the lifetime of the persistent current. It is for this reason that an experimental investigation of these lifetimes would be particularly valuable, for it would show how important these microscopic fluctuations were in determining the properties of the superconducting state.

THERMODYNAMIC FLUCTUATIONS

It has been shown by Gor'kov[6] that the Ginzburg-Landau theory[7] of superconductivity is equivalent to

* Work supported by the U. S. Office of Naval Research.
[1] R. A. Ferrell, Phys. Rev. Letters 13, 330 (1964).
[2] T. M. Rice, Phys. Rev. 140, A1889 (1965).
[3] C. N. Yang, Rev. Mod. Phys. 34, 694 (1962).
[4] W. A. Little, Phys. Rev. 134, A1416 (1964).

[5] S. Engelsberg and B. B. Varga, Phys. Rev. 136, A1582 (1964); D. C. Mattis and E. H. Lieb, J. Math. Phys. 6, 304 (1965).
[6] L. P. Gor'kov, Zh. Eksperim. i Teor. Fiz. 34, 735 (1958) [English transl.: Soviet Phys.—JETP 9, 1364 (1959)].
[7] V. L. Ginzburg and L. D. Landau, Zh. Eksperim. i Teor. Fiz. 20, 1064 (1950).

156

the BCS[8] theory, at least in the vicinity of the transition temperature. These Ginzburg-Landau equations provide a simple means for incorporating spatial and time varying fluctuations in the condensate which would be cumbersome in the more microscopic forms of the BCS theory. Recently it has been shown by Werthamer and Tewordt[9] that the Ginzburg-Landau theory has a wider range of validity for materials with a short mean free path. Furthermore, there is an impressive amount of experimental evidence to show that it gives an excellent account, both qualitatively and quantitatively, of the known phenomena of superconductivity.[10,11] In this paper we shall be concerned primarily with superconductivity in specimens with at least one dimension extremely small. In this domain, the Ginzburg-Landau equations have been particularly successful in explaining old phenomena and predicting new.[12,13] For the above reasons, we believe that these equations form a valid basis for our arguments except, perhaps, for samples so small that the lateral dimensions are small compared to the Fermi wavelength. The resultant lamination of the Fermi sphere introduces special considerations which in this case may prevent the existence of an order parameter.

For our arguments, the existence of an order parameter plays an essential role. However, Ferrell in his paper[1] has argued that the Gor'kov function $F(x)$, which is related to the Ginzburg-Landau order parameter, vanishes in a one-dimensional system. This aspect of his argument is not correct. The error lies in his incorrect replacement of the definite integral in his equation (3) by the indefinite integral. The definite integral has as its limits the phase of the order parameter at points x and x', respectively. This phase difference is, in principle, subject to measurement. The use of the indefinite integral, however, replaces the appropriate phase difference with the absolute value of the phase. This is tantamount to measuring the phase at x with respect to some arbitrary external standard, and this is not subject to physical measurement. Ferrell's conclusion, however, that ODLRO should not occur in one dimension survives, nevertheless, because $G(x,x')$, which occurs in the criterion for ODLRO, involves the phase difference between x and x' rather than any absolute phase.

It appears, then, that one is on safe ground in assuming the existence of an order parameter at least locally subject to the above restrictions and that the free energy depends upon it through the Ginzburg-Landau equations. We wish to show, then, first that

the thermodynamic fluctuations which have been shown by Rice to be capable of destroying ODLRO in a one-dimensional system will not destroy flux quantization or a persistent current in a closed loop, unless a fluctuation occurs which is of such an amplitude that the order parameter is driven to zero for some section of the loop.

Consider the Ginzburg-Landau equations:

$$F(\psi) = \int [a|\psi(r)|^2 + b|\psi(r)|^4 + c|\nabla\psi(r)|^2]d^3r, \quad (1)$$

where the order parameter $\psi(r)$ is a complex function with real amplitude $\Delta(r)$ and phase $\phi(r)$, and where

$$\psi(r) = \Delta(r) \exp i\phi(r). \quad (2)$$

In the usual treatment, one ignores fluctuations and determines the equilibrium value of $\psi(r)$ by minimizing (1) with respect to $\Delta(r)$ and $\phi(r)$. However, other functional forms of $\Delta(r)$ and $\phi(r)$ are possible and will occur with a probability $e^{-\beta F(\psi)}$, where $F(\psi)$ is the computed value of the free energy in (1) for the given functional forms of $\Delta(r)$ and $\phi(r)$. The actual order parameter will fluctuate among these possibilities. Rice took these into account for a one-dimensional system and showed that the appropriately weighted average of the function $\langle\langle\psi(r)\psi^*(r')\rangle\rangle$ over all possible forms of the order parameter was such that

$$\lim_{|r-r'|\to\infty} \langle\langle\psi(r)\psi^*(r')\rangle\rangle \to 0. \quad (3)$$

This is the condition for the absence of ODLRO in the system,[3] and consequently we see that ODLRO cannot exist in such a one-dimensional system. Let us extend this argument by considering a sample in the form of a wire of diameter d and length l joined back on itself to form a closed loop. We will consider initially the case where $d < \xi_t$, so that variations of $\psi(r)$ across the wire can be neglected so that we may use the one-dimensional form of the Ginzburg-Landau equations. Later we will discuss the limitation of this approach. Byers and Yang[14] have shown that if the free energy of such a loop varies with the magnetic flux Φ through it, then persistent currents will flow and it will exhibit the phenomenon of flux quantization. Consequently, we ask whether or not the free energy of the loop as given by (1) varies with the magnetic flux through it.

For such a loop geometry, the boundary condition on $\psi(r)$ is that it should be single valued. This, therefore, imposes upon the phase the condition that

$$\phi(l+r) - \phi(r) = 2\pi n, \quad (4)$$

where n is an integer. Because of this, each possible form of the order parameter $\psi(r)$ can be classified according to the integer describing the phase change

[8] J. Bardeen, L. N. Cooper, and J. R. Schrieffer, Phys. Rev. 108, 1175 (1957).
[9] N. R. Werthamer, Phys. Rev. 132, 663 (1963); L. Tewordt, ibid. 137, A1745 (1965).
[10] A. A. Abrikosov, Zh. Eksperim. i Teor. Fiz. 32, 1442 (1957) [English transl.: Soviet Phys.—JETP 5, 1174 (1957)].
[11] T. G. Berlincourt and R. R. Hake, Phys. Rev. 131, 140 (1963).
[12] M. Tinkham, Phys. Rev. 129, 2413 (1963).
[13] L. Meyers and W. A. Little, Phys. Rev. Letters 13, 325 (1964).

[14] N. Byers and C. N. Yang, Phys. Rev. Letters 7, 46 (1961). See also Ref. 4 above.

round the loop in Eq. (4). Let us consider that we group these ψ's in subensembles of given n.

We must include the contribution of the vector potential $\mathbf{A}(r)$ into the Ginzburg-Landau equations in order to take the contribution of the flux into account. As indicated above, we will take initially the one-dimensional form of these equations:

$$F(\psi) = \int \psi^*(x) \left\{ a + b|\psi(x)|^2 + c\left[i\frac{\partial}{\partial x} - \mathbf{A}(x)\right]^2\right\}$$
$$\times \psi(x)dx. \quad (5)$$

For any arbitrary form for the order parameter belonging to a given subensemble n, we find that

$$F(\psi_n) = a\int \Delta^2(x)dx + b\int \Delta^4(x)dx + \frac{4\pi^2c(n+\alpha)^2}{I}$$

$$+ \int \beta^2(x)\Delta^2(x)dx, \quad (6)$$

where $I = \int_0^l [dx/\Delta^2(x)]$, $\beta(x)$ is a real function independent of the flux but dependent upon $\psi_n(x)$, and α is $\Phi/(hc/2e)$. We see then that for all order parameters which lie in a given subensemble, the free energy will depend on the flux unless the integral I is infinite. For a finite ring this integral can be infinite only if $|\Delta(x)|$ is zero in at least one place. As a result of this, the expectation value of the free energy over the subensemble of states of given n, $\langle\langle F(\psi_n)\rangle\rangle$, must vary with the flux. It is convenient to think of a single system moving in time through the various possible phase points representing the various fluctuations, rather than an ensemble of the systems with a given number in each particular region of phase space. In this view, then, we see that so long as the system fluctuates among the states of just this subensemble the free energy will be flux-dependent, and consequently during this period a persistent current will not decay. It will fluctuate in magnitude depending upon the instantaneous value of I, but its fluctuations will be centered on a value determined by n. Likewise, the trapped flux will fluctuate about a value n times the appropriate flux quantum, and flux quantization will be maintained.

In order to understand the decay of a persistent current, then, we must determine how the system can make a fluctuation from one subensemble n to another n'. We will argue that this is only possible if a fluctuation occurs which drives the order parameter $|\Delta(x)|$ to zero at least at one point in the loop.

FLUCTUATIONS BETWEEN SUBENSEMBLES

We recall that the order parameter $\psi(x)$ describes the behavior of a large number of particles. Changes in this order parameter can occur, for example, because of an influx of heat to the region near x or of an influx

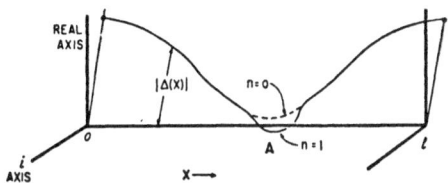

FIG. 1. The order parameter $\psi(x)$ which is complex is drawn as a function of position. Two possible configurations are shown, one for an order parameter in the subensemble $n=0$ and the other for $n=1$. Near A, $\psi_1(x)$ makes an excursion round the Argand diagram while $\psi_0(x)$ does not.

or efflux of particles to or from this region. It seems reasonable to assume that changes in $\psi(x)$ must occur continuously because it involves many particles. Each particle, on the average, must gain a little energy for the over-all energy to increase. It is unreasonable to expect the large number of particles described by the order parameter to gain an increment of energy discontinuously. However, we could expect them to gain energy continuously, and thus the total energy density would change continuously through all neighboring values of the energy density. We would expect, for the same reason, that changes in the density and in the entropy density would also occur continuously. On the basis of this argument, we contend that all fluctuations occur such that the order parameter changes continuously or, in other words, through configurations which lie infinitely close to one another. On this assumption we can determine how a fluctuation can carry the system from one subensemble, n, to another, n'.

Consider the illustration (Fig. 1). This depicts the variation of the amplitude $\Delta(x)$ and phase $\phi(x)$ for two possible configurations of the order parameter, $\psi_0(x)$ and $\psi_1(x)$, lying in the subensembles $n=0$ and $n=1$, respectively. The total change of phase from x to $x+l$ for ψ_0 is zero, while for ψ_1 it is 2π. We have taken the flux Φ to be zero. We have chosen the amplitude and phase of the two configurations to be identical except in the region near A. In this region, $\psi_1(x)$ makes an excursion round the Argand diagram to rejoin $\psi_0(x)$ a little beyond A, while $\psi_0(x)$ makes no such excursion. In this region, $\psi_0(x)$ and $\psi_1(x)$ have the same amplitude but, of course, differ in phase. It is clear from this figure that it is only possible for $\psi_0(x)$ and $\psi_1(x)$ to lie infinitely close to one another everywhere if they both drop to zero at the same point. At this point $\psi_1(x)$'s excursion round the Argand diagram could coincide with $\psi_0(x)$ if both had zero amplitude. Our assumption that fluctuations progress only through configurations which lie infinitely close to one another then leads us to the conclusion that a fluctuation from one subensemble n to another n' can occur only if $|\Delta(x)|$ fluctuates to zero at some point in the loop. Our picture of the decay of a "persistent" current then proceeds as follows: Let us suppose that at $t=0$ the order parameter of the loop lies in a subensemble n.

Now fluctuations will occur among the states of this subensemble as discussed earlier. Eventually a fluctuation will occur with such an amplitude that $\Delta(x)$ is driven to zero in some part of the loop. The system may then proceed into any other subensemble n'. It will then fluctuate in this new subensemble until again $\Delta(x)$ is driven to zero, at which time it may progress to some other subensemble. On the average, it will gradually find its way to that value of n' which minimizes the free energy and will then fluctuate in and around this neighborhood. Within each subensemble, the system behaves like an ordinary superconductor exhibiting flux quantization and a "persistent" current which fluctuates about a mean value. Where $\Delta(x)$ fluctuates to zero, this part of the loop reverts temporarily to the normal state exhibiting its normal electrical resistance. For the rest of the time the loop exhibits zero resistance. We can therefore calculate the time average of the resistance of the loop by determining what fraction of the time some part of the loop is normal and thus has its normal resistance.

We proceed to calculate this in the following way. First, we note that the order parameter $\psi(x)$ in some small part of the loop may be considered to be a function of the local value of the density (of electrons) ρ and of the local temperature T. Fluctuations in either T or ρ can cause the order parameter to drop to zero in some regions of the loop. For small fluctuations, standard fluctuation theory[15] shows us that the probability of a fluctuation occurring of amplitude ΔT or ΔV, where ΔV is the change of volume V containing N particles, is

$$\omega = \exp\left\{ -\left(\frac{C_v}{2kT_0^2}\right)(\Delta T)^2 + \frac{1}{2kT_0}\left(\frac{\partial p}{\partial V}\right)_{T,N}(\Delta V)^2 \right\}. \quad (7)$$

Here C_v is the heat capacity of the small volume which fluctuates in temperature by ΔT. For a given mean temperature T_0, we can readily calculate the probability that a fluctuation can occur such that ΔT is sufficient to raise a small part of the body above the temperature T_c at which the order parameter goes to zero. This is the meaning of T_c for our further calculations. For fluctuations in density or ΔV, we can obtain some idea of the magnitude ΔV must reach for the order parameter to go to zero. This we can do by assuming

that the BCS approximate expression,

$$kT_c = 1.14\hbar\omega \exp[-1/N(0)V] \quad (8)$$

is valid in this small part of the body. Here $N(0)$ and V depend upon the density, and thus density fluctuations at a given temperature T_0 can cause T_c to fall below T_0. An increase of electron density will increase $N(0)$ but decrease V because of increased screening. Thus we can get an upper bound on the variation of T_c with density by assuming that V does not change with density. A simple analysis for a free-electron gas and the use of (8) shows that the rms fluctuation of the temperature in (7) is greater than the rms fluctuations of T_c due to density fluctuations by a factor of the order of $\epsilon_f/16kT_c$. Taking the Coulomb repulsion into account gives a factor which is even larger than this; consequently, it is an excellent approximation to ignore the fluctuations of the density and consider only fluctuations in the temperature.

We have to consider fluctuations which are not small, in which case Eq. (7) is an inadequate approximation. In this case let us imagine that the filament of superconductor is immersed in an environment at some mean temperature T_0 as illustrated in Fig. 2. This seems to be the most realistic situation of experimental interest. Quite generally,[15] then, the probability ω of some small part of the filament fluctuating to some temperature T is

$$\omega = \exp\left\{ -\frac{u(T)-u(T_0)-T_0[s(T)-s(T_0)]}{kT_0}\Omega \right\}, \quad (9)$$

where $u(T)$ is the internal energy per unit volume of the filament at temperature T and $s(T)$ is the corresponding entropy. Ω is the volume of the small part of the filament.

The order parameter cannot change very rapidly in a distance much smaller than the mean-free-path reduced coherence length ξ_l without raising the free energy appreciably; consequently, the smallest volume whose temperature fluctuations we need consider is a volume having the diameter of the filament and a length ξ_l.

The *time average* of the resistance of any part of the filament is then

$$R = R_n\left[\int_{T_c}^{\infty} dT \exp\left\{ -\frac{u(T)-u(T_0)-T_0[s(T)-s(T_0)]}{kT_0}\Omega \right\} \middle/ \int_0^{\infty} dT \exp\left\{ -\frac{u(T)-u(T_0)-T_0[s(T)-s(T_0)]}{kT_0}\Omega \right\} \right]. \quad (10)$$

[15] L. D. Landau and E. M. Lifshitz, *Statistical Physics* (Addison-Wesley Publishing Company, Inc., Reading, Massachusetts, 1958).

In the numerator, the integral should run from 0 to ∞; however, for $T < T_c$, $R = 0$, so this part makes no contribution. For $T > T_c$ we assume that $R = R_n$ and thus can be factored from the integral, and we get the expression above.

We can get a good idea of the variation of R with T by considering samples such that the small volume which can fluctuate in T is large enough so that $u(T)$ and $s(T)$ for this volume can be approximated to the values of the bulk material. In this case, for T less than T_c we know empirically that $u(T)$ and $s(T)$ are given approximately by[16]

$$u_s(T) = (3\gamma/4T_c^2)(T^4 - T_c^4)$$

and

$$s_s(T) = T^3/T_c^2. \tag{11}$$

Here γT is the electronic specific heat per unit volume for the normal state.

For $T > T_c$ we have

$$u_n(T) = \tfrac{1}{2}\gamma(T^2 - T_c^2),$$
$$s_n(T) = \gamma T. \tag{12}$$

The general behavior of the variation of the resistance with temperature is illustrated in Fig. 3. The width of the transition (from $0.9R_n$ to $0.1R_n$) is of the order of $2(2kT_c/\Omega\gamma)^{1/2}$, i.e., twice the root-mean-square (rms) fluctuations of the local temperature. It is clear from this that unless the volume Ω which undergoes the temperature fluctuations is infinite, the resistance will not drop discontinuously to zero at T_c. Thus for a finite coherence length ξ_l and $d < \xi_l$ a discontinuous change in resistance cannot occur at T_c.

For finite samples we have evaluated the expression (10) approximately, using (11) and (12). In the denominator, the approximate expression (7) has been used in evaluating the integral because the major contribution to this integral comes from small fluc-

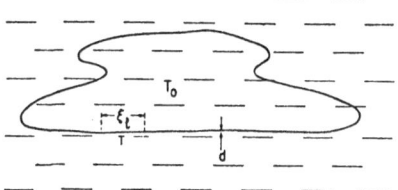

FIG. 2. Illustration of a long thin superconductor of diameter d connected in the form of a closed loop and immersed in a bath at a temperature T_0. Fluctuations of the local temperature T of a small section of length $\approx \xi_l$ can occur and lead to decay of a "persistent" current.

[16] See for example, M. Tinkham, in *Low Temperature Physics*, edited by C. DeWitt, B. Dreyfus, and P. G. de Gennes (Gordon and Breach Science Publishers, Inc., New York, 1962), p. 151.

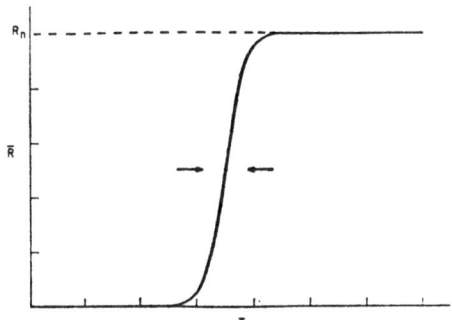

FIG. 3. General form of the mean resistance of a finite superconducting sample as a function of the temperature. The width indicated by the arrows is approximately $2(2kT_c/\Omega\gamma)^{1/2}$.

tuations for which (7) is valid. In the numerator we have dropped terms of order g^2 and higher, where $g = \{kT_0/\gamma\Omega(T_c - T_0)^2\}^{1/2}$, and consequently our expression is not valid for T_0 very close to T_c. The behavior here is continuous, however, and one could, with some labor, obtain the exact details if they were needed. The results we obtain for $T_0 < T_c$ and $T_0 > T_c$ are, respectively,

$$R = R_n\left\{\exp\left[-\frac{[f_n(T_0) - f_s(T_0)]\Omega}{kT_0}\right]\right\}\left(\frac{\gamma T_0\Omega}{\pi kT_c^2}\right)$$
$$\times\left(\frac{kT_0}{\pi\Omega(T_c - T_0)}\right)\exp\left[-\frac{\gamma\Omega}{2kT_0}(T_c - T_0)^2\right], \tag{13}$$

where $f_n(T_0)$ is the free energy per unit volume of the normal state, $f_s(T_0)$ is the corresponding free energy for the superconducting state, and

$$R = R_n\left\{1 - \frac{2kT_0}{\pi(T_0 - T_c)^2\Omega\gamma}\exp\left[-\frac{\gamma\Omega}{2kT_0}(T_0 - T_c)^2\right]\right\} \tag{14}$$

for $T_0 > T_c$.

These can be represented by a single parameter $S \equiv \gamma\Omega T_c/k$, and the reduced temperature $t \equiv T_0/T_c$. For $T < T_c$, we show the behavior in Fig. 4.

From the above we see that in spite of the absence of ODLRO in such a one-dimensional system, a significant reduction in the normal resistance will occur for temperatures appreciably below the bulk T_c value.

One can appreciate better the rapid change of resistance with temperature by considering the decay time for a current in a wire in the form of a circle of diameter 1 cm and wire thickness d. In Table I we give the lifetime computed from the ratio of L/R as a function of temperature for two wire diameters, 100 and 360 Å ($S \approx 20$ and 1000, respectively). Here we assume $\xi_l \approx d$ and $T_c = 3.7°$K.

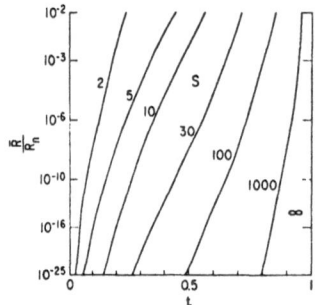

FIG. 4. Plot of the average resistance of a thin superconducting wire as a function of reduced temperature $t = T/T_c$ and parameter $S \equiv \gamma \Omega T_c/k$, where Ω is of the order of $(\pi d^2/4)\xi_t$ and d is the wire diameter.

For wires of diameters appreciably larger than this, it is clear that the transition from the relaxation time characteristic of the normal state to that of an immeasurably long relaxation time would occur in a minute temperature interval and normally would be masked by small differences of the transition temperature due to impurities and strains. Only in samples with

TABLE I. Computed lifetimes for two Sn wires, one of 100 Å diameter and the other of 360 Å.

Lifetime	100 Å wire (°K)	360 Å wire (°K)
$\approx 10^{-14}$ sec	3.7	3.7
10^{-7} sec	2.0	3.4
10^{-4} sec	1.5	3.26
1 sec	1.25	3.10
10 days	1.0	3.06
10^8 years	0.75	2.90

dimensions comparable to those discussed above would it be possible to observe directly the decay of the persistent current due to the thermodynamic fluctuations. Measurements on samples of this size would be particularly valuable in determining the role other nonthermodynamic fluctuations play in determining the decay lifetime.

DISCUSSION

By considering just the thermodynamic fluctuation of the temperature of a thin superconducting specimen, we have shown that the lifetime for the decay of a persistent current is not infinite. However, for wires larger than a few hundred angstroms in diameter the lifetime for the decay attains an immeasurably large value within a small fraction of a degree below the bulk transition temperature.

The extension of these ideas to samples of macroscopic size in the second and third dimensions is not trivial. In the first place, the classification of the order parameters into subensembles of definite n, strictly speaking, fails for any finite width of the sample. For,

in principle, a variation of $\psi(r)$ across the sample can make the phase change of the order parameter along a path on one side of the sample differ from that along the other, so that a vortex line is enclosed by the specimen. For the case we have considered, however, where $d < \xi_t$, the free energy for such a vortex configuration is so high away from the immediate neighborhood T_c that the system spends a sufficiently small time in it that our classification according to a definite n is a workable approximation. For samples in which one of the transverse dimensions is greater than ξ_t, we must include the possibility of the existence of such vortex configurations and the interactions these would have with an external field and a conduction current. If the magnetic field is large enough, then an equilibrium array of these vortices can exist and one obtains the Abrikosov state characteristic of the type-II superconductor.

It is worthwhile noting that the decay mechanism we have considered amounts to the entry of one flux unit (not $hc/2e$, but this value corrected to the dimensions of the sample[17]) in the average flux during a transition between subensembles. This is superficially similar to the mechanism which generates the flux flow resistance of type-II superconductors. The dissipative mechanism, i.e., the movement of flux through the specimen, is the same in both cases; however, the important difference lies in the role played by the magnetic field. In the type-II superconductor, a magnetic field greater than H_{c1} is necessary in order to maintain the sample in the Abrikosov state with an *equilibrium* distribution of vortices throughout the sample. The movement of this array due to the Lorentz force interaction with the conduction current gives rise to the dissipation of the current. The magnitude of this dissipation is determined by the strength of the pinning sites of the vortex array. This problem has been treated by several authors.[18] The situation which we have examined is that were the magnetic field is so small, $H \ll H_{c1}$, that there is, firstly, no equilibrium vortex array and, secondly, the magnetic field interactions, can be neglected. In this case, it is not the strength of the pinning sites of the vortices which determines the dissipation but rather the probability for the creation of a vortex. In the very thin sample we have considered, i.e., $d < \xi_t$, the existence of a vortex in the wire would require a higher free energy than that which is necessary to drive the order parameter to zero across a section of the wire. For this reason the preferred decay mechanism is by the latter means. For samples with transverse dimensions comparable to the penetration depth or larger, the free energy required for the creation of a vortex will be lower than that needed to drive the order parameter to zero across a section of the wire. In

[17] J. Bardeen, Phys. Rev. Letters **7**, 162 (1961).
[18] See for example, P. W. Anderson and Y. B. Kim, Rev. Mod. Phys. **36**, 39 (1964).

this case the current can still decay but the mechanism is more closely akin to the flux flow resistance found for type-II superconductors.

In a Josephson junction the penetration depth is anomalously large and the order parameter anomalously small; consequently, the most probable way for flux to cross the junction, assuming that the junction is short compared to the penetration depth, is by a fluctuation in which the order parameter fluctuates to zero as discussed here rather than by the migration of a flux line across the junction as in a type-II superconductor.

It is useful to note that in the absence of an external magnetic field the energy per unit length of a vortex is of the order of $\Phi_0^2/4\pi^2\lambda^2$, with approximately equal contributions coming from the field energy and from the current. Here $\Phi_0 = hc/2e$, and λ is the penetration depth. Even for quite thin films at temperatures a little below T_c this energy is appreciably larger than kT_c so that the equilibrium density of these thermally excited vortices will be extremely small. For example, consider a film of Sn, 100 Å thick, the mean free path, $l \approx 100$ Å, $\xi_0 = 10^4$ Å, $\lambda_L(0) \approx 500$ Å at $t = T/T_c = 0.99$. We have[19] $\lambda \approx 0.62\lambda_L(0)[\xi_0/l(1-t)]^{1/2}$, giving $\lambda \approx 3.1 \times 10^4$ Å. The total energy of this vortex line is therefore of the order of $18kT_c$, and the probability of finding a vortex in this region at this temperature is $\approx e^{-18}$. We observe, however, that for any finite film thickness the energy of such vortices, while large, cannot be infinite, so that near enough to T_c an appreciable density of these will occur. Similarly, the pinning energy cannot be infinite for a finite film. The resistive transition to the superconducting state must therefore be continuous because of the gradual decrease in the number of these vortices per unit area as the temperature is lowered below T_c. As the film thickness is made larger, the total energy of a vortex line through the film increases, and consequently we should expect a sharper transition to the superconducting state to occur. Only for an *infinite* *three*-dimensional sample can one expect a discontinuous transition. Such a discontinuity is characteristic of a phase transition. One may conjecture then that no true phase transition should occur in either the one- or two-dimensional cases.

In the presence of a magnetic field, the problem becomes considerably more complicated in detail, although it is easy to see that such a field lowers the energy of those vortices with the appropriate orientation with respect to the field until at H_{c1} an equilibrium array becomes possible. Below H_{c1}, the number of thermally excited vortices will then be field-dependent and likewise the resistance.

One can also apply these considerations to the hypothetical superconducting macromolecule suggested earlier.[4] If an order parameter can exist in such a linear

---

[19] B. B. Goodman, Rev. Mod. Phys. **36**, 12 (1964).

molecule and we accept the estimates[4] of $T_c \approx 2000°K$, a coherence length of 30 Å, and the density of states at the Fermi surface used in that calculation, we obtain a value of $S \approx 2.5$. (In this system where the lateral degrees of freedom of the electrons are frozen out, we calculate the electronic specific heat of the molecule per unit length, and Ω is replaced by the coherence length ξ_l.) Referring to Fig. 4, we see that below room temperature ($t < 0.15$) the average resistance will have fallen to a very small fraction of its normal resistance. A much more rapid drop in resistance with temperature can be gotten by cross-linking the individual polymer filaments so that they form a three-dimensional net. By so doing one raises the energy required for a flux line to penetrate through the mass of filaments. Also, if the polymer filament is imbedded in some other material or fluid instead of vacuum, the fluctuations of the temperature of the filament will be tied to the fluctuations of some part of its immediate environment. This will effectively increase the volume which can undergo the thermal fluctuations and again increase the effective value of S.

We see then that the thermodynamic fluctuations are not sufficient in themselves to rule out the possibility of a state of greatly enhanced conductivity occurring at the low temperatures in a linear macromolecule. This point can only be settled by examining whether or not an order parameter can exist locally in such a system. In this regard we may note that the exactly soluble one-dimensional model of Mattis and Lieb[5] shows no evidence of the superconducting state for any form of the interaction potentials. This must not be construed as evidence that such an order parameter cannot exist here, for we have shown that in any one-dimensional system such as that of Mattis and Lieb, an order parameter, if it exists, must fluctuate so that it moves through all the different subensembles n discussed earlier. The exact solution to the thermodynamic properties of this system, therefore, must be an average over all these subensembles of different n. It is not difficult to see from Eq. (6) that such an average washes out all the flux dependence in the free energy and thus yields a state which in equilibrium has no superconducting properties. The true equilibrium state in this case is not superconducting but, as we have seen in our examples above, the approach to equilibrium can be so slow at low temperatures that a *nonequilibrium state* can persist for a sufficient length of time to give a greatly enhanced conductivity.

CONCLUSION

We have shown that the fluctuations of both the amplitude and the phase of the Ginzburg-Landau order parameter do not destroy a persistent current in a "one-dimensional" superconducting loop unless a fluctuation occurs which drives the amplitude of the order parameter to zero at all points on a surface which severs the loop. We calculate the probability of this occurring by using standard fluctuation theory. From this we are able to calculate the time-average resistance of the samples, and we find that while no infinitely sharp change of resistance occurs at any temperature, nevertheless, the resistance falls significantly below the normal resistance of the specimen as the temperature is lowered appreciably below the bulk T_c. A true phase transition to the superconducting state appears to be possible only in an infinite three-dimensional sample. In one dimension, if the range of the interaction force is finite, no phase transition is possible. The resistance of the one-dimensional system does approach zero, however, as $T \rightarrow 0°K$.

ACKNOWLEDGMENTS

We wish to thank Felix Bloch and Alexander Fetter for helpful discussion and criticism.

From: HIGHLY CONDUCTING ONE-DIMENSIONAL SOLIDS
Edited by Jozef T. Devreese, Roger P. Evrard,
and Victor E. van Doren
(Plenum Publishing Corporation, 1979)

7

The Prospects of Excitonic Superconductivity

H. Gutfreund and W. A. Little

1. Introduction

The present era of intensive research in the field of one-dimensional conductors that started with the work of the Penn group[1] on TTF-TCNQ was preceded by the discussion of the possibility of superconductivity in one-dimensional organic materials which was suggested by one of us[2] in 1964. This suggestion was based on a new mechanism of superconductivity, the so-called exciton mechanism. The term "exciton" applies here broadly to any electronic excitation. In this new mechanism the effective attraction between electrons at the Fermi surface is induced by the exchange of excitons, rather than by phonons as is believed to be the case in all presently known superconductors. The specific one-dimensional structure proposed for the realization of the exciton mechanism consisted of a spine of conducting electrons with organic dye molecules chemically bound to this spine at regular intervals. Under favorable circumstances one may expect that the excitons propagating along the array of the dye molecules will induce an effective attraction between the electrons on the spine. The exciton mechanism is in principle not restricted to one-dimensional structures. Ginzburg[3] has discussed the possibility of excitonic superconductivity in two-dimensional systems consisting of thin metallic films sandwiched between, or coated on one side, by layers of a highly polarizable material. Allender et al.[4] discussed along the lines of Ginzburg's ideas a specific system consisting of a thin metallic layer coated by a

H. Gutfreund • Racah Institute of Physics, The Hebrew University, Jerusalem, Israel.
W. A. Little • Department of Physics, Stanford University, Stanford, California 94305.

semiconductor with a high dielectric constant. Their work was followed by an unsuccessful attempt to find superconductivity with an enhanced transition temperature in such a system.[5] The common feature of the proposed models, based on one- and two-dimensional structures, is the distinction between the electrons that are expected to form Cooper pairs and those that participate in the virtually excited excitons. These two types of electrons are confined to two spatially separated regions in close contact with each other.

One of the main attractions of the proposed exciton mechanism is the apparent possibility of higher transition temperatures. This follows from the BCS formula for T_c

$$k_B T_c = \hbar \omega_D \exp\left(\frac{-1}{\lambda - \mu^*}\right) \qquad (1)$$

where λ characterizes the strength of the attractive interaction necessary for the creation of Cooper pairs and μ^* measures the repulsive Coulomb interaction. In the case of the phonon mechanism $\hbar \omega_D$ is approximately the Debye temperature, which is of the order of several hundred degrees. For the exciton mechanism, on the other hand, $\hbar \omega$ is expected to be a typical electronic excitation temperature of the order of 10^4–10^5 °K, thus leading to much higher values of T_c. This argument is oversimplified, and the conclusion is in general not correct but depends upon the details of the system. While it appears that under special conditions the exciton mechanism may lead to higher values of T_c, the high temperature of an electronic excitation is by itself not sufficient for that. This will be discussed later in greater detail.

The proposed exciton mechanism raises many questions that have to be examined with great care, for one is attempting to apply the BCS theory in a predictive manner outside of the area where it has achieved its most spectacular successes. First, one has to face the general problems that arise when one replaces phonons by excitons, then one encounters a host of other problems that are associated with the restricted geometries of the proposed models. In addition one should bear in mind that there is still no experimental evidence for the existence of the excitonic mechanism of superconductivity and thus no empirical input to guide the theory. On the positive side, we feel that the progress in the understanding of superconductivity and of the above problems together with the accumulated experience gained from the theoretical and experimental research on quasi-one-dimensional metals makes the time right now for a critical reevaluation of the whole idea of excitonic superconductivity and an examination of the arguments and counterarguments used in the controversy that this idea has provoked. This is the purpose of the present review article.

The subject has a rich history, and many of its facets have been discussed in review papers over the past decade. A summary of the early work and an introduction to the basic problems is contained in the *Proceedings of the International Conference on Organic Superconductors*[6] and in excellent reviews by Keldysh[7] and Ginzburg.[8,9] Contributions to the understanding of several of the key theoretical problems are contained in later reviews by Ginzburg,[10] Ginzburg and Kirzhnits,[11] and Bulaevsky *et al.*[12] This early work of Ginzburg and his co-workers made a particularly illuminating contribution to the understanding of superconductivity as a whole by establishing the connection between the most complete theoretical treatments and the simple form of the BCS equation for T_c [Equation (1)]. With suitable redefinition of the parameters of that equation the expression represents correctly the essence of the complete theory.

More recently there has been a review by Andre *et al.*[13] This contains a summary of recent theoretical developments using renormalization group methods and exact field-theoretical results, which have contributed significantly to our understanding of idealized one-dimensional systems. This review also contains a discussion of many of the materials of a filamentary or one-dimensional nature whose properties are relevant to the model discussed here and an extensive list of references. A comprehensive review of synthetic endeavors directed toward the preparation of a one-dimensional organic superconductor is given by Yagubskii and Khidekel.[14] The broader question of conductivity in complexes of TCNQ and the platinum square-planar compounds is contained in a review by Shchegolev,[15] while Zeller[16] had given an overview of the present understanding of the mixed-valence square planar system. A comprehensive source of information on other mixed-valence inorganic complexes is contained in a recent review by Miller and Epstein.[17] A feel for developments in one-dimensional metals and some background on excitonic superconductivity can be got from the proceedings of the two NATO ASI meeting edited by Keller.[18,19]

In this review we have attempted to present the material at two levels. First, much of the initial section is presented at an introductory level for the newcomer to the field who is not familiar with all the techniques of many-body-theory. Second, we discuss the most recent advances at a more sophisticated level where we assume the reader is familiar with that theory However, in discussing the latter we have not given the usual formal derivations of the results discussed but instead have presented the essence of the underlying physics. By so doing we believe it will be useful to both groups of readers.

We begin with a nontechnical discussion of the nature of superconductivity emphasizing the basic ideas of the BCS theory and pointing out the limitations on T_c imposed by the ordinary phonon mechanism. We then introduce the exciton mechanism, which we hope is not subject to these limitations, but at the same time we point out the problems posed by this mechanism. The next two sections are devoted to the discussion of these problems. First we discuss some general questions which are due to the difference between phonons and excitons. These include the role of exchange, vertex corrections, and the difference in local corrections or umklapp processes. Following that, we consider the problems introduced by the one-dimensional structure of the proposed system. The emphasis in this review is on a one-dimensional model, and this emphasis is based on our belief that the one-dimensional structure has the best prospects for a successful realization of the exciton mechanism. The reasons for this opinion will be presented in the course of this article. The answers to many of the questions raised in these two sections depend on the specific features of the particular model considered. In Section 5 we discuss in detail one such model. It consists of a linear chain of partially oxidized platinum atoms closely similar to the structure of the much-studied KCP system[15] $[K_2Pt(CN)_4Br_{0.3} \cdot 3H_2O]$, but with cyanoligands replaced by polarizable cyanine-dye-like ligands. For this model we perform a detailed calculation of the various interaction parameters which determine the superconducting phase transition. The model proposed raises important questions of synthetic and structural chemistry. Among these are: Will such a structure be stable? Will the monomers stack? Can the Pt spine be oxidized without destroying the ligands? These and other questions of chemical nature will be discussed in a later section.

It should be stressed that there must surely be other systems based or other metals such as Ni, Ir, Os, etc. with similar or with other types of polarizable ligands that might exist as crystalline stacks or as true polymers and that could also satisfy our theoretical criteria for superconductivity. It is not claimed that the model discussed in Section 5 is unique but rather that it is a prototype of the type of system in which excitonic superconductivity might occur. The common features that all of these systems would need to have are the following: (a) a dense packing of the polarizable substituents, (b) a polarizability of the ligands comparable to that of a cyanine dye, (c) a strongly concentrated conduction-electron density in the center of the plane of the ligand system, and (d) the possibility for these electrons to move more or less freely along the axis of the stack perpendicular to the plane of the ligands. The value of the specific model discussed in Section 5, then, is that it provides a guide as to *how* dense the substituents must be packed, *how* polarizable the ligand must be, *how* concentrated the electron density, etc. If comparable values for these factors can be achieved in other systems then they too would be expected to satisfy the criteria for superconductivity. Within these constraints, demanding though they may be, the synthetic chemist is free to innovate.

The conclusion of the discussion and calculations described in this article is that the exciton mechanism does not violate any physical principle and its existence or nonexistence will depend on the specific properties of the system under consideration. We shall specify the requirements for the realization of this mechanism and show that they are exacting and not easily met. This may explain why it is so hard to "construct" an excitonic superconductor and why a lot more effort in that direction is needed before one can hope to test a real system that meets the conditions of the proposed model.

2. The Nature of Superconductivity

2.1. Background

The superconducting state occurs in a very large number of metals and alloys. As one cools such a metal through a temperature that is characteristic for the material, one finds that the electrical conductivity changes abruptly to an infinitely large value. The temperature at which this occurs is known as the critical temperature T_c, and typically this lies at a temperature within a few degrees of absolute zero. The distribution of the transition temperatures of several hundred of the known superconducting alloys is illustrated in Figure 1.[20] At present the highest known superconducting transition temperature occurs in the alloy Nb_3Ge at 23.2°K. Most others lie below the boiling point of liquid hydrogen (20.4°K) in the liquid-helium range (≈ 4.2°K).

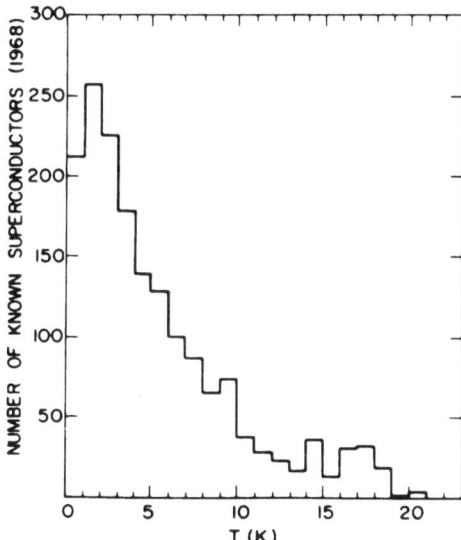

Figure 1. Plot of the number of known superconducting compounds as a function of their transition temperature (1968). A small number of compounds are now known up to 23.2°K.

Our understanding of superconductivity is based on the highly successful theory developed by Bardeen, Cooper, and Schrieffer[21] in 1957—the BCS theory. It has successfully explained and predicted a vast number of effects and phenomena in the field. The original theory has been expanded upon, reformulated, and developed enormously over the years, but the essence of the theory has remained unchanged and thus one still refers to the present theory as the "BCS theory." The one area where the theory has not played a significant role is in the prediction of new superconducting compounds or compounds of high T_c. Here the experimentalists have continued to play the key role. This has often been cited as a failure of the theory. However this is not a fair criticism of the *theory of superconductivity* itself, for it is our inability to predict the normal-state properties of new materials that limits the predictability of the superconducting properties. Where the normal-state properties can be predicted, values of T_c for the superconducting state can be predicted correctly to within a few percent. Our discussion of excitonic superconductivity will be based on the BCS theory but certain features of the exciton mechanism introduce new types of problems, the understanding of which is far less securely based than those of conventional superconductivity. Thus, not only are we limited by our lack of knowledge of the properties of the normal state of the materials one hopes to synthesize, but in addition these theoretical problems make calculations of T_c even less certain. For this reason our calculation of T_c in Section 5 is not to be taken as of great numerical significance but rather as an estimate of the order of magnitude of T_c and a guide for the design of new materials.

To understand the concept of excitonic superconductivity one must appreciate the essential elements of the BCS theory. It is based on two key features. First it requires the conduction electrons to form bound pairs, and second these pairs must form a Bose–Einstein-like condensate. It is illuminating to use a simple mechanistic model to explain how the pairing interaction comes about and how it is related to the existence of the condensate. On the basis of this one can give a crude plausibility argument to explain the limitations on T_c, to gain some insight into the role played by the Coulomb repulsion and the possibility of other mechanisms, such as the exciton mechanism, leading to high-T_c materials.

2.2. Phonon Mechanism

Consider a small portion of a metal with the positive ions located at more or less regular sites within the crystalline lattice and the conduction electrons moving among them (Figure 2). It should be recalled that the Coulomb interaction between the electron and the ion, and between the

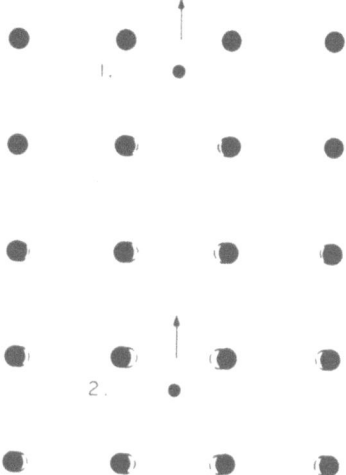

Figure 2. Schematic representation of the electron–electron attraction resulting from the polarizability of the lattice. The large circles represent the positive ions and the small circles the electrons. As #1 electron moves through the lattice the ions move towards its instantaneous position. A second electron can then move into the distorted part of the lattice left behind and benefit from the excess positive charge density of this region. An effective attraction between the electrons thus results from the momentary lattice distortion.

electron and another electron, is screened by the collective motion of the other electrons, so the effective interaction is one of short range, $(e^2/r)\,e^{-\kappa r}$. As one electron moves through the lattice the short-range screened interaction will result in the electron giving each ion an impulsive force as it passes by. The ions will begin to move towards the instantaneous position of the electron. However, the electron, which moves at the Fermi velocity, moves much more rapidly than do the ions, so that the ions will reach their maximally distorted position when the electron is some distance away. Thus some distance behind this electron there will be a puckered region of the lattice containing a slight excess density of positive ions. A second electron will be attracted to this region and thus, indirectly, will be attracted to the first electron. This indirect attraction is what is known as the phonon-mediated electron–electron interaction. It derives its name from the fact that the distorted lattice may be described in terms of the phonon coordinates.

This model is too simple as it stands, for one can easily show that a large number of electrons will share the same region of the lattice with the "first" two electrons described above. If the motion of these other electrons is random and uncorrelated with respect to the first two, then they will create random fluctuating forces which will interfere with the interaction between the first two electrons. On the other hand, if their motion is coordinated with that of the other two, then the interaction will be constructive, the interaction between the first two electrons will remain, and they in turn will not interfere with the motion of the other coordinated electrons. This harmonious coordination can be achieved if the pairs form a Bose–Einstein-like condensate in which the center of mass and internal states of each pair of the condensate are identical to those of all others. At low temperatures it is energetically favorable to form this coherent, coordinated state in the metal. This is the superconducting state. At higher temperatures thermal agitation causes the disruption of the pairs. The disrupted pairs now make no further coherent contribution to the inter-action and the effective binding energy of the other pairs is reduced. The weakened binding then makes the pairs more vulnerable to thermal dis-sociation, with the result that above a critical temperature T_c the condensate disappears and the system reverts to the normal state.

The BCS theory shows us that within the approximations of a simple model where we neglect the Coulomb interaction the value of this tran-sition temperature is expressible in the form

$$k_B T_c = \hbar \omega_D \exp\left(\frac{-1}{\lambda}\right) \tag{2}$$

where $\hbar \omega_D$ is a characteristic energy of vibration of the ions of the lattice, and $\lambda = N(0)V$ where $N(0)$ is the density of states at the Fermi surface and V is the effective electron–electron attraction. The factor $\hbar \omega_D$ in (2) comes from the width of the energy region within which the electrons can make a coherent contribution to the pair binding energy. It is thus a measure of the number of electrons which make such a contribution.

Also the expression (2) is valid only for weak coupling of the phonon–electron system. With stronger coupling, ω itself is affected by the inter-action with the electron system and λ is also renormalized by this inter-action. Indeed, as will become more apparent in our later discussion, the parameters ω, N, and V are not independent of one another, but each influences the other in a complicated manner. However, Equation (2) with the parameters suitably defined does give one a useful expression for the critical temperature. It may also be used to illustrate why there is a natural limit on T_c.

2.3. Limitation on T_c

An understanding of the limitation on the value of T_c can be obtained by using our simple mechanistic model and the above expression for T_c. Let us suppose that each ion is attached to its lattice position by a simple spring of spring constant k. Then the phonon spectrum will be the single Einstein frequency $\omega = (k/M)^{1/2}$, where M is the mass of the ion.

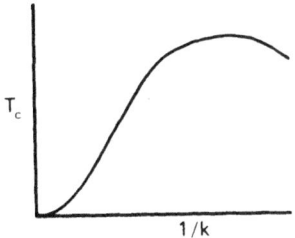

Figure 3. Variation of T_c as a function of effective spring constant k which describes the elasticity of the lattice.

Let us suppose that we can vary the strength of the spring constant k and then consider the value of T_c as a function of $1/k$. The point of this argument is that one can consider small variations of composition and structure of a series of alloys as variations in the stiffness of the lattice with only a modest variation of the mass of the ions. So our simple model can give us insight into the variations that can occur in T_c for such a series of alloys. If the spring is extremely stiff (k large) then the electron moving past the ion will be unable to displace it and, thus, there will be no phonon-mediated attractive interaction. Consequently λ in (2) will be zero and, even though $\hbar\omega$ would be large [$\omega = (k/M)^{1/2}$], T_c will vanish because of the dominance of the exponential term. As we make the spring less stiff by reducing k, a small attraction will begin to appear and T_c will become finite and will rapidly increase with the decreasing negative exponent. This tendency is illustrated in Figure 3. If we continue to make k smaller, then eventually the variation of the exponential term will flatten out and the reduction in the pre-exponential factor $\omega = (k/M)^{1/2}$ will begin to contribute. Ultimately the decrease of this factor will cause T_c to reach a maximum and then to decrease. Thus we see even within this simple model one should expect a natural limit to occur in the value of T_c.

One other factor needs to be taken into account to understand more fully the limitations on T_c. As we decrease the value of k then eventually the electron–phonon interaction itself will begin to contribute to the effective restoring force of the lattice. This interaction has the effect of renormalizing the phonon frequencies. One finds that in the simplest model of this effect[22]

$$\omega^2 = \omega_0^2(1 - 2\lambda) \tag{3}$$

where ω_0 is the "bare" phonon frequency, defined as the phonon frequency in the absence of the electron–phonon coupling, and ω is the renormalized frequency resulting from this interaction. The coupling parameter λ is the same as that occurring in (2).

Now as k is reduced the attractive interaction increases and thus λ increases and approaches unity, causing ω^2 to fall and eventually to become negative. This yields an imaginary value for ω, which gives a nonoscillatory solution for the motion of the ions. This signals the onset of an instability resulting in a distortion and change of symmetry of the lattice. (This simple model would require $\lambda \leq \frac{1}{2}$.) Such lattice instabilities impose even stricter limits on the maximum attainable value for T_c. In a more sophisticated treatment of the problem where one includes the effects of the Coulomb field and the effects of renormalization on both the phonons and the electrons and upon the electron–phonon coupling, a more compli-

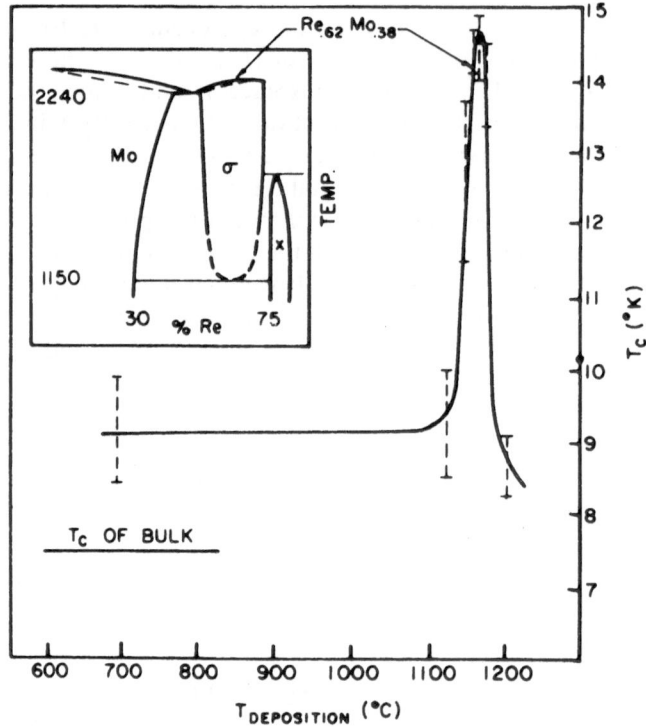

Figure 4. T_c vs. film deposition temperature for $Mo_{0.38}Re_{0.62}$. Insert shows part of the phase diagram. Note the large enhancement of T_c for films prepared at temperatures close to the structural transformation temperature at 1150°C (Reference 23).

cated criterion for the stability of the lattice would result. This problem has not been analyzed completely as yet, but there is ample empirical evidence to show that high transition temperatures are limited by the stability of the lattice.

On the basis of the above argument one might expect T_c to vary strongly as a function of composition in the vicinity of a structural phase boundary. This is well illustrated in Figure 4, where T_c is plotted as a function of the deposition temperature of an alloy in the vicinity of the structural transformation temperature.[23] It explains also why certain alloys such as Nb_3Ge can be prepared with high T_c by suitable treatment which stabilize the incipient lattice instability while without such treatment T_c is substantially lower.[24]

The above argument of a limitation on T_c assumes, of course, that there is some limit on the Debye frequency. This does appear to be the case, for this frequency depends upon such factors as crystal structure, bond strengths, and ionic masses which combine to restrict $\hbar\omega_D \lesssim 10^3 \,°K$. On the other hand a different type of interaction such as the exciton interaction which we will describe shortly would be subject to somewhat different limitations. It is on this basis that hope exists for attaining substantially higher transition temperatures. In principle, at least, a possibility exists for devising a system in which the effective interaction can be maintained at a large value over an energy range substantially greater than the Debye energy $\hbar\omega_D$ which occurs in (2). These are systems in which an electronic or excitonic interaction is used to provide the attractive interaction instead of the phonons, and in these it has been argued that one should be able to obtain substantially higher transition temperatures. Our simple model may be used again to illustrate the physical basis for these arguments.

2.4. Isotope Effect

Consider again Figure 2. As the electron moves past an ion of the lattice it gives one such ion a momentum p. This ion is then set into motion having a kinetic energy $p^2/2M$. After the electron has passed, the ion continues to move until it is limited by the elastic restraints of its neighbors. The original kinetic energy of the ion is then converted into potential energy. If the maximum displacement is X then $\frac{1}{2}kX^2 = \frac{1}{2}p^2/M$.

So for a given momentum transfer p and spring constant k, the displacement is inversely proportional to the square root of the ionic mass. As explained earlier this displacement is responsible for the attractive component of the electron–electron interaction. The larger this displacement the greater this attraction can become. The attraction depends on both the coupling strength and the value of the phonon frequencies ω. The net effect of both these factors is to make the term λ in (2) independent of the ionic mass.

McMillan[25] has shown that the coupling constant λ has the form

$$\lambda = \frac{N(0)J^2}{M\langle\omega^2\rangle} \qquad (4)$$

where J is a matrix element for the phonon–electron interaction averaged over the Fermi surface which is independent of the mass of the ions M, and $\langle\omega^2\rangle$ is an average of the square of the phonon frequency. Recalling that $\omega \approx M^{-1/2}$, as in our Einstein model, we see that λ is independent of M. Because of this, the only mass-dependent term in the expression (2) for T_c is that of the pre-exponential term $\hbar\omega$, which then gives a mass dependence for T_c of $M^{-1/2}$. This result is almost exactly true in the absence of the Coulomb interaction. When the Coulomb interaction is included, deviations from a simple power dependence occur. The reason for this is that the electrons can interact via the Coulomb interaction with all the other electrons regardless of their energy, whereas the interaction via the phonon mechanism is limited to electrons within $\pm\hbar\omega_D$ of one another. It is possible to replace the real Coulomb interaction μ by an effective or pseudointeraction μ^* which has the same energy cutoff $\pm\hbar\omega_D$ as the phonons. When this is done one finds that[26]

$$\mu^* = \mu[1 + \ln(E_F/\hbar\omega_D)]^{-1} \qquad (5)$$

and (2) becomes

$$k_B T_c = \hbar\omega \, \exp\left(\frac{-1}{\lambda - \mu^*}\right) \qquad (6)$$

Now, however, $\lambda - \mu^*$, unlike λ, does depend upon ω through (5) and thus upon the mass M. The contribution from this term within the exponential causes deviations from the simple isotope effect.

The reasons for this deviation can also be seen from our model. Let us assume that one can vary the value of the ionic mass. If one makes it smaller, then the resultant larger ionic displacements $J/M^{1/2}$ contained in (4) will result in a stronger coupling constant for the electron–phonon interaction. However, the ionic motion also becomes more rapid because the frequency ω in our simple Einstein model depends on the mass through the expression $\omega = (k/M)^{1/2}$. Because of this, the most heavily distorted region of the lattice will now lie closer to the "first" electron which is responsible for this distortion than it would for the case of a heavier ionic mass. The second electron is attracted to this region, but some of the enhanced attraction due to the larger lattice distortion will be offset by the greater direct Coulomb repulsion of the nearby "first" electron. Thus the contribution of the Coulomb repulsion causes a deviation from the simple isotope effect, giving a reduced isotopic dependence,[27] $T_c \approx M^{-1/2(1-\varepsilon)}$. We shall see that an appreciation of the factors that lead to the isotope effect can give insight into the possibility of obtaining high T_c through the exciton mechanism.

2.5. Exciton Mechanism

Consider now a system illustrated in Figure 5. Let A be a conductive polymeric chain which we call the "spine" and attached to this spine is a series of polarizable molecular side-chain substituents, B. As an electron moves along the spine it will induce a movement of charge in the side chains. This induced positive charge will reach its maximum value not at the time the electron is adjacent to the side chain but a short time later.

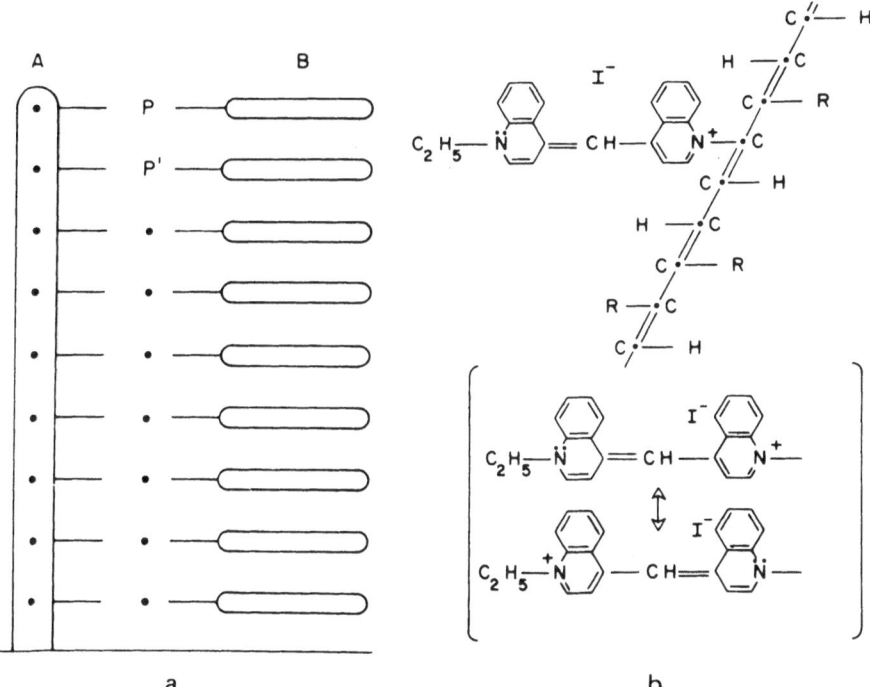

Figure 5. (a) Schematic diagram of an excitonic superconductor. A is the conductive "spine" and B the polarizable substituents to provide the excitonic attraction. (b) Suggested realization of the model (Reference 2).

The reason for this delay lies in the finite frequency of oscillation of the charge in the side chain, just as in the analogous case of phonons the finite frequency of oscillation of the ions give a similar delay. Consequently the region of maximum induced charge will lag some distance behind this electron. A second electron will be attracted to this induced charge and, just as for the phonon case, one might expect pairing to occur between the electrons. However, in this case the interaction is mediated by the virtual electronic excitations of the side chains, which may be described in terms of "exciton" coordinates rather than phonon coordinates. The characteristic energy, $\hbar\omega$, associated with these excitations is much higher and can be of the order of several electron volts rather than a fraction of an electron volt characteristic of phonon frequencies. Thus, provided one can find a system where the exciton-mediated electron–electron interaction is comparable to that of a phonon system *and* the electron density of states is also comparable, then expression (1) would lead one to expect substantially higher values of T_c because of the large pre-exponential factor $\hbar\omega_D$.

Alternately one may arrive at the same conclusion using the arguments of the isotope effect. In the proposed excitonic system the interaction is mediated by the movement of an electron in the side chain rather than by the much heavier ion of a phonon superconductor. Provided the momentum transfer p and effective "spring constant" k can be kept comparable to those in the phonon case, then the transition temperature T_c for such a hypothetical superconductor would be scaled up from that of conventional superconductors by a factor of order $(M_{ion}/M_e)^{1/2}$, which is typically of order 300. So if the above conditions could be met then one would have hope of finding superconductors of substantially higher transition temperature than those known today.

There are good physical reasons for believing that in an excitonic system it should be possible to make the momentum transfer p and the effective "spring constant" k comparable to the phonon case. To see this one notes that the momentum transfer p in the exciton case is related to the distance of the nearest approach of the conduction electrons in the spine to an electron in the side chain. This can be of the order of an angstrom or two, which is somewhat greater but of the same order of magnitude as the distance of nearest approach of a conduction electron and an ion core in a conventional metal. So one would expect the resultant momentum transfer for the two cases to be somewhat similar. Provided the above distances lie within the screening length of the two systems then one can show that the matrix element J of Equation (4) will be approximately proportional to the square of this momentum transfer. Second, an examination of typical electronic and Debye energies shows that their ratio is of the order of 100. This shows that the effective spring constants k $(=M\omega^2)$ for the electronic and vibrational systems are of the same order of magnitude. One should note also that this factor $M\omega^2$ is the denominator of (4) and thus the effective coupling constant for a tightly coupled electron–exciton system could be comparable to that of a phonon system. On the other hand, where the polarizable substituents are separated by several angstroms from the conductive spine we would conclude that λ should become very small. This is borne out in our detailed calculations of Section 5.

This is the dogma of those who believe in the possibility of achieving high-temperature excitonic superconductivity. However, these arguments have been known for over a decade, yet no such high-temperature super-conductor has been found to date. Indeed, no excitonic superconductor of any kind is known. We believe, and will present arguments to back this up, that the reason for this is that only in very special structures can the excitonic interaction reach a magnitude large enough to overcome the Coulomb interaction and that until now none of these structures have been prepared. They may be impossible to make but the indications are that though difficult they should not be impossible to prepare.

The above mechanistic model, while useful as a guide to thinking, must not be taken too literally. For one must remember that in discussing the superconducting ground state one is discussing a *single* quantum-mechanical state. A single such state is a stationary state and cannot describe electrons and ions that are moving, for to do so would require a wave-packet formalism involving several quantum-mechanical states. Nevertheless the physical effects of motion in the classical model show themselves in the ground state as an admixture of virtual configurations. Keeping this caveat in mind, we can use the model in establishing the physical principles.

We shall endeavor to show that the above crude arguments are essentially correct and are supported by the detailed sophisticated analysis, and, further that the electronic and structural criteria that must be met to obtain a sufficiently strong interaction can only be realized in a small set of highly contrived systems. We believe that only by the deliberate synthesis of these systems can one hope to find such excitonic superconductors.

3. Problems of Superconductivity Unique to the Exciton Mechanism

In Section 2 we introduced the idea of the exciton mechanism of superconductivity. We emphasized that this mechanism raises several questions stemming from the qualitative differences between an electron–phonon and an electron–exciton system. In the present section we examine these questions.

3.1. Exchange

The most obvious difference between an electron–phonon and an electron–exciton interaction is that in the first case electrons interact with ions, whereas in the second case they interact with other electrons, namely, with identical particles. This results in exchange terms in the interaction between electrons in the conducting and in the excitonic regions. These exchange terms are of opposite sign to the direct Coulomb terms and may reduce substantially the effective coupling between the electrons and the excitons. In view of the delicate balance between the Coulomb repulsion and the exciton attraction this reduction of the latter may be disastrous. The importance of the exchange terms was first pointed out by one of us.[28] Since the exchange interaction is very short-range while the direct interaction does not change greatly over distances of the order of chemical bond lengths, we find that a spatial separation between the conducting electrons and the excitonic medium of about a bond length optimizes the electron–exciton coupling strength for a given system. Such a separation was required by Little[2] in his proposed one-dimensional structure and at least in principle would be possible in the two-dimensional structures of Ginzburg.[3] In a three-dimensional system the elimination of exchange terms would require a very special structure in which the two kinds of electrons would have to occupy orthogonal orbitals such that the "excitonic electrons" would not be hybridized with any of the states within the conduction band. This is certainly not the case in the model discussed by Geilikman,[29] who suggested the possibility of pairing in the *s* band of a transiton metal induced by the interaction with *d* electrons. Another model in which the exchange interaction may be destructive is the model suggested by Allender *et al.*[4] There the electrons of a thin metallic film spend part of their time in a semiconductor layer where they interact with electron excitations across the semiconductor gap. This interaction determines the electron–exciton coupling strength. Exchange effects were not considered in the estimate of the electron–exciton coupling constant, so that the number $\lambda \simeq 0.5$ quoted for this model is probably significantly overestimated.

3.2. Apparent Limitation on $\lambda - \mu^*$

Another argument that has been raised[30,31] asserts that the electron–exciton coupling constant λ_{ex} is restricted to significantly smaller values than the electron–phonon coupling constant λ_{ph}. This argument is based on the assumption that the net electron–electron interaction may be represented in the form

$$V(q, \omega = 0) = \frac{4\pi e^2}{q^2 \varepsilon(q, \omega = 0)} \tag{7}$$

where $\varepsilon(q, \omega)$ is the dielectric function of the medium at momentum q and frequency ω. It contains the direct Coulomb interaction as well as the effect of all the excitations of the system that contribute to the effective interaction between two conduction electrons. This effective interaction may be separated into a Coulomb part and an indirect part due to the exchange of some virtual excitation. This separation will be demonstrated explicitly later. The product of the averages over the Fermi surface of these two parts and the density of states at the Fermi surface $N(0)$ yield the parameters μ and λ, respectively, namely, $N(0)V(q, 0) = \mu - \lambda$. In a simple model of an electron liquid immersed in a uniform background of positive charge, causality and stability arguments imply[32] that $\varepsilon(q, 0) \geq 0$, namely, $\lambda \leq \mu$. This does not exclude superconductivity in such a model because the relevant parameter is $\lambda - \mu^*$ [see Equation (6)], which may still be positive, but because of the close connection between μ and μ^* [Equation (5)] the above inequality restricts severely the value of $\lambda - \mu^*$. The stability argument is based on the fact that a static modulation of the background charge by $\delta\rho(q)$ would modify the total energy of the system by

$$\delta E = \frac{4\pi e^2}{q^2} \frac{|\delta\rho(q)|^2}{\varepsilon(q, 0)} \tag{8}$$

If $\varepsilon(q, 0) < 0$ for some q, the uniform background would be unstable against such a distortion. This would result in a static charge-density wave and a gap in the single-electron energy spectrum. One should, however, keep in mind that this result is derived in a model in which one neglects the change in the kinetic energy that arises from a change in the zero-point motion of the positive background. In a real material $\varepsilon(q, 0)$ may be negative and thus λ may exceed μ as is the case in most ordinary superconductors.[25] For Pb, for example, $\lambda = 1.3$, $\mu^* = 0.1$, and $\mu = 0.6$. This fact is commonly assigned to local corrections or, synonymously, to umklapp processes. These corrections come from the fact that the local fields at the ions are considerably different from those seen on the average by the electrons. It has been argued[31] that the local field corrections would be very ineffective in producing the same effect for the exciton mechanism, because the exciton, being an electronic excitation, occupies a large part of the unit cell and thus there is not much difference between the average field seen by the electron and that seen by exciton. However, for the systems we have considered[2,33] this argument is not correct. For in these systems the unit cell is very large—much larger than an atomic cell;

and while the exciton system fills a large part of it, the region of interaction between the excitonic system and the electron is concentrated in a very small portion of it. The local corrections are therefore very large and the average field seen by the electron and that seen by the exciton are very different. Indeed, as is obvious from the details of the earliest calculations,[2] the spatial separation between the spine electrons and the excitonic system is essential in order to get a net attractive interaction. Should the conduction electrons be distributed uniformly throughout the unit cell the attractive component can be shown to be negligible.

3.3. Vertex Corrections

The formulation of the theory of superconductivity, both in the weak coupling regime leading to the BCS equation or in the strong coupling case resulting in the Eliashberg equations[34] depend on the validity of Migdal's theorem, which asserts that vertex corrections to the electron–phonon interaction are small and may be neglected. The lowest-order correction to the electron–phonon vertex is shown in Figure 6. For an incoming phonon of a phase velocity ω/q much smaller than the Fermi velocity v_F, this correction is of the order of $(\omega_D/E_F) \approx 10^{-2}$. Most of the phonons involved in conventional superconductivity have momentum $q \approx p_F$ and, therefore, a very small phase velocity. This is the basis of Migdal's theorem. If this argument is carried over blindly to the exciton mechanism we would have to replace ω_D by an electronic frequency and this would result in large vertex corrections and would raise questions as to the applicability of the usual theory of superconductivity to the exciton mechanism.

To see that this does not in fact have serious consequences, one has to consider again the specific properties of the model proposed by us. The basic difference between this model and a conventional superconductor is that in the latter the electron–ion interaction is strongly screened, giving rise to a phonon–electron interaction of a very short range. The Fourier

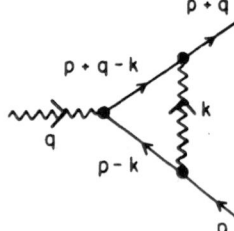

Figure 6. Lowest-order correction to the electron–phonon vertex.

transform of this interaction results in an electron–phonon coupling constant that depends only weakly upon momentum. On the other hand, in the proposed exciton system the bulky polarizable ligands make it impossible for the transition charges in the ligands to come very close to the charges in the spine, and this, together with reduced screening in such filamentary compounds, makes the electron–exciton interaction one of long range. This results in a strongly momentum-dependent electron–exciton coupling constant $|Q(q)|^2$, which falls off sharply at values of $1/b$, where b is of the order of the distance of the nearest terminal group of the ligand system to the spine. The qualitative conclusion is borne out by detailed calculations (Section 5, Figure 21).

It thus follows from the preceding paragraph that the distinctive feature of the model discussed here is that only excitons with small momenta, and hence phase velocities much greater than v_F, are involved in the conjectured superconducting transition. It was shown by Engelsberg and Schrieffer[35] that for phonons with a high phase velocity the vertex correction in Figure 6 is not given by ω_D/E_F but rather is of order $g^2 N(0)/\omega_D$, where g is the electron–phonon coupling constant. This is a crude estimate of λ and can be of the order of unity in strong coupling superconductors. However, this result was obtained assuming that g is momentum independent. This is certainly not the case in our model, as $Q(q)$, the electron–exciton interaction, is strongly peaked around $q \simeq 0$. Assuming that $Q(q)$ is a constant for $q \leq q_c$, where q_c is an average width of $Q(q)$, and zero outside this range, one finds that the lowest-order vertex correction is

$$\Gamma_1 = \frac{|Q|^2 N(0)}{E}\left(\frac{\Delta E}{E}\right) \qquad (9)$$

where E is a typical exciton energy and $\Delta E \simeq \varepsilon(p_F + q_c) - \varepsilon(p_F)$, $\varepsilon(p_F)$ being the electron energy at the Fermi momentum p_F. Explicit calculations on our model show that $|Q|^2 N(0)/E \simeq 0.3$ and that the factor $\Delta E/E$ reduces this by about an order of magnitude, rendering the vertex corrections small. Thus, the strong momentum dependence of the electron–exciton interaction, which as mentioned previously is due to the separation between the electrons in the spine and the excitons, is responsible for strongly reduced vertex corrections. One should again point out that this is not the case for the three-dimensional s–d model[16] or the two-dimensional metal–semiconductor model,[4] mentioned in Section 2.1.

3.4. Equation for T_c

We shall now prepare the ground for a numerical estimate of the transition temperature in our model. To this end we adopt the method developed by Kirzhnits, Maximov, and Khomskii[36] (to be referred to as KMK). This method applies to a weak coupling superconductor and it results in a simple BCS-like equation for the gap function and for T_c. Its merit is that it brings out explicitly and in a convenient form the relationship between the kernel of this equation and the microscopic properties of the system such as the electron band energies, the exciton band energies, and the electron–exciton coupling matrix elements.

Before we continue our discussion we would like to point out that the equation described in what follows lies within the framework of the mean-field approximation, and the temperature T_c obtained from it is not the actual transition temperature for a system of weakly coupled linear chains. The special problems associated with the dimensionality of the system and the relevance of the mean-field transition temperature to the superconducting behavior of such a system will be discussed in Section 4.

Keeping in mind these words of caution we can now summarize briefly the KMK method in general and then discuss it with regard to the system under consideration. The starting point is the integral equation for the anomalous Green's function F, which at $T = T_c$ reads

$$F(\mathbf{p}, i\omega_n) = -G(\mathbf{p}, i\omega_n)G(-\mathbf{p}, -i\omega_n)$$

$$\times \sum_m \int \frac{d^3k}{(2\pi)^3} V[\mathbf{p}-\mathbf{k}, i(\omega_n - \omega_m)]F(\mathbf{k}, i\omega_m) \qquad (10)$$

where $\omega_n = (2n+1)\pi T_c$. In this equation it is already assumed that the vertex corrections are small, which was shown in the last subsection to be also true in our model. At T_c the "imaginary time" Green's functions G of the superconducting state are replaced by their counterpart in the normal state and approximated by the normal functions G_0

$$G_0(\mathbf{p}, i\omega_n) = \frac{1}{i\omega_n - \xi(\mathbf{p})} \qquad (11)$$

where $\xi(p) = \varepsilon(p) - \varepsilon_F$ is the single-electron energy measured with respect to ε_F. It is this approximation that restricts the present treatment to weak coupling superconductors.

The essential feature of the KMK method is the use of the Lehmann representation for the effective electron–electron interaction. The latter can be written in the form

$$V(\mathbf{q}, \omega) = \frac{V_0(\mathbf{q}, \omega)}{\varepsilon(\mathbf{q}, \omega)} \qquad (12)$$

and the finite-temperature analog of the Kramers–Kronig relation for the reciprocal of the dielectric function leads to the spectral representation

$$V(\mathbf{q}, i\omega_n) = V_0(\mathbf{q})\left[1 - 2\int_0^\infty \frac{\omega'\rho(\mathbf{q}, \omega')\,d\omega'}{\omega_n^2 - \omega'^2}\right] \tag{13}$$

The spectral density $\rho(\mathbf{q}, \omega)$ is related to the dielectric function by

$$\rho(\mathbf{q}, \omega) = -\frac{1}{\pi}\,\mathrm{Im}\left[\frac{1}{\varepsilon(\mathbf{q}, \omega)}\right] \tag{14}$$

It is also convenient to write the anomalous Green's function in the spectral representation

$$F(\mathbf{p}, i\omega_n) = \int_{-\infty}^\infty \frac{f(\mathbf{p}, x)\,dx}{i\omega_n - x} \tag{15}$$

Substituting Equations (13) and (15) into Equation (10) one can perform the frequency sum explicitly, and after some manipulations and plausible approximations, described in Reference 23, one obtains the equation

$$\phi(\mathbf{p}) = -\int \frac{d^3k}{(2\pi)^3}\frac{U(\mathbf{p}, \mathbf{k})\tanh|\xi(\mathbf{k})|/2T_c}{2\xi(\mathbf{k})}\phi(\mathbf{k}) \tag{16}$$

where

$$\phi(\mathbf{p}) = 2|\xi(\mathbf{k})|\int_0^\infty F(\mathbf{p}, x)\,dx$$

and

$$U(\mathbf{p}, \mathbf{k}) = V_0(\mathbf{p} - \mathbf{k})\left[1 - 2\int_0^\infty \frac{\rho(\mathbf{p} - \mathbf{k}, \omega)\,d\omega}{\omega + |\xi(\mathbf{p})| + |\xi(\mathbf{k})|}\right] \tag{17}$$

To obtain an equation similar to the BCS equation, one usually assumes that the material is either isotropic or that the "dirty" limit approximation is valid. This allows one to replace the three-dimensional momentum integral by an integral over the energy variable ξ and over the "angle" variables in momentum space, leading to an integral equation in a single energy variable

$$\phi(\xi) = -\int d\xi'\,\frac{U(\xi, \xi')\tanh(\xi'/2T_c)}{2\xi'}\phi(\xi') \tag{18}$$

where $U(\xi, \xi')$ is essentially $U(p, k)$ integrated over the angle variables of $\mathbf{k}$ in momentum space. Comparison with BCS theory shows that $\phi(\xi) = \mathrm{Re}\,\Delta[\omega = |\xi(p_F)|]$. Without discussing in detail the form of $U(\xi, \xi')$ in the general case, we point out its two most important properties: (a) The kernel $U(\xi, \xi')$ is a smooth function of the variables ξ, ξ', unlike the interaction itself, which has a complicated resonant structure; (b) the magnitude of $U(\xi, \xi')$ decreases when either one of the variables ξ, ξ' departs from the Fermi energy. To make the analogy with the BCS equation complete, one has to separate the contributions of the phonon and the Coulomb interactions. The first has an effective cutoff at ω_D. The latter extends over a much larger energy range; however, it can be replaced by an effective interaction [Equation (5)] with the same cutoff, so that one finally gets a BCS-like equation with the ξ' integration extending from $-\omega_D$ to $+\omega_D$.

One of the significant differences between the phonon and the exciton mechanism is that in the latter case the direct Coulomb interaction and the exciton-induced interaction extend over the same energy so that Equation (18) would have to be integrated over the whole energy band. We note in passing that if the condition $\lambda \leq \mu$ would hold for the exciton mechanism, there would be no hope for a superconducting transition in this case, because when $\lambda \leq \mu$, superconductivity is possible only as a result of the large difference in the cutoffs of the two interactions which replaces μ by the significantly reduced μ^*.

Since we intend to apply the integral equation to a one-dimensional system and since in our case there is no advantage to transforming to energy variables because there is no small-energy cutoff, we prefer to leave Equation (16) as an equation in the momentum variable. Extending the integration over the entire Brillouin zone, we write this equation in one dimension:

$$\phi(p) = -\int_{-\pi/a}^{\pi/a} \frac{dk}{2\pi} \frac{U(p, k)\tanh[|\xi(k)|/2T_c]}{2\xi(k)} \phi(k) \qquad (19)$$

3.5. The Kernel $U(p, k)$

We shall now discuss the kernel $U(p, k)$. The spectral density $\rho(q, \omega)$ in Equation (14) may be separated into the contribution of the exciton-induced interaction ρ_{ex} and the contribution of the Coulomb interaction ρ_C. We can similarly separate $U(p, k)$ into its two components

$$U_{ex}(p, k) = -2\int_0^\infty \frac{V_0(p-k)\rho_{ex}(p-k, \omega)\, d\omega}{\omega + |\xi(p)| + |\xi(k)|} \qquad (20)$$

$$U_C(p, k) = V_0(p-k)\left[1 - 2\int_0^\infty \frac{\rho_C(p-k, \omega)\, d\omega}{\omega + |\xi(p)| + |\xi(k)|}\right] \qquad (21)$$

The numerator in the integrand in Equation (20) is related to the imaginary part of the effective exciton-exchange interaction. We write the latter in the form

$$V_{ex}(q, \omega) = \sum_\alpha |Q_\alpha(q)|^2 D_\alpha(q, \omega) \qquad (22)$$

This corresponds to the exchange of an exciton of momentum q, frequency ω, and band index α (in case there are several exciton bands). $Q_\alpha(q)$ is the coupling strength between the electron and the exciton and $D_\alpha(q, \omega)$ is the exciton propagator. We assume that this propagator has the form of a boson Green's function

$$D_\alpha(q, \omega) = \frac{2E_\alpha(q)}{\omega^2 - E_\alpha^2(q) + i\delta} \tag{23}$$

where $E_\alpha(q)$ is the exciton energy. This assumption deserves a brief discussion. The electron–exciton interaction may be represented by a form similar to the Fröhlich electron–phonon interaction

$$H_{\text{el-ex}} = \sum_{k,q} Q(q) a^\dagger_{k+q} a_k (b_q + b^\dagger_{-q}) \tag{24}$$

where $a^\dagger_k (a_k)$ and $b^\dagger_k (b_k)$ are the creation (destruction) operators of the electrons and excitons, respectively. The exciton propagator has the form given in Equation (23) only when the boson commutation relation, $[b_q, b^\dagger_q] = 1$, is satisfied. For exciton operators this is an approximation and it remains to see how good this approximation is. The propagating exciton is a linear combination of electronic excitations on the single molecules, and we can write

$$b^\dagger_q = \frac{1}{N} \sum_i e^{iqr_i} b^\dagger_i \tag{25}$$

where b_i creates an excitation on the ith molecule. A typical property of the dye molecules discussed by us is that the transition to the lowest electronic excitation exhausts most of the oscillator strength and the higher states play a negligible role. Hence, the excitons in our model propagate in an array of coupled two-state units. For such a system the local destruction and creation operators satisfy mixed commutation relations[37]

$$[b_i, b_j] = [b^\dagger_i, b^\dagger_j] = 0$$

$$[b_i, b^\dagger_j] = 0 \qquad \text{for } i \neq j \tag{26}$$

$$\{b_i, b^\dagger_i\} = 1$$

where { } denotes the anticommutation relation. These are the commutation relations of spin-$\frac{1}{2}$ raising and lowering operators and are therefore called the Pauli commutation relations. It is now easy to calculate the commutation relations for the propagating exciton operators in Equation (25). One gets

$$[b_q, b^\dagger_{q'}] = \delta_{qq'} - \frac{2}{N} \sum_i e^{i(q-q')r_i} b^\dagger_i b_i \tag{27}$$

The second term on the right-hand side is a correction to the boson commutation rule. This correction is small as long as the number of virtually excited excitons is small, which is the case when the electron–exciton coupling is not too strong. In the particular calculation we describe in Section 5 we show that in this case the coupling *is* weak and in fact we find it would be difficult to conceive of a system where the coupling is strong.

It has been argued recently by Chaikin *et al.*[38] the the electron–exciton interaction renormalizes the electron bandwidth resulting in a drastic band narrowing. An examination of the analysis in Reference 25 reveals that a rather strong electron–exciton interaction is required for that to happen. Chaikin *et al.* assumed that the excitons behave like bosons; however, in the region in which their results are valid the deviations from the boson commutation relations are becoming important. It was shown by Bari[39] that when these are taken into account the band narrows at most by a factor of 2.

After this digression we return to the exciton contribution to the kernel in the equation for T_c. On account of Equations (14), (22), and (23), we obtain

$$V_0(q)\rho_{ex}(q, \omega) = \sum_\alpha |Q_\alpha(q)|^2 \delta[\omega - E_\alpha(q)] \tag{28}$$

and the exciton-exchange contribution to the kernel becomes

$$U_{ex}(p, k) = -2 \sum_\alpha \frac{|Q_\alpha(p-k)|^2}{E_\alpha(p-k) + |\xi(p)| + |\xi(k)|} \tag{29}$$

3.6. Effects of the Phonons on the Exciton Mechanism

Our discussion thus far has treated the excitonic system without considering the effects of the phonons. The phonons can contribute in two ways: first, virtual phonons will give rise to an attractive contribution to the net interaction just as in conventional superconductivity, and second, real, thermally excited phonons can also affect the system in a variety of ways. If the phonon contribution to the coupling parameter λ is λ_{ph} and that of the excitons is λ_{ex}, then any limit on λ imposed by the stability of the system would impose a more stringent limit on λ_{ex}. At an early stage, when it was thought that $\lambda \leq \frac{1}{2}$ was a real limit on λ set by the lattice, it was argued[40] that this would limit $\lambda_{ex} \leq (\frac{1}{2} - \lambda_{ph})$ and thus would keep λ_{ex} small and so prevent T_c from attaining a large value. The reason the two contributions can give a limit on the stability of either can be understood as follows. The combined effect of the phonon and exciton interactions can create a charge-density wave in the conduction electron system. The charge-density wave in turn can act back on both the phonon and exciton system and cause each to distort. In this way the exciton and phonon systems are coupled to one another via the electron system. This constraint does not appear to be important for we now know that λ can attain much higher values than $\frac{1}{2}$, so λ_{ex} is not limited significantly by this factor. Moreover, as we shall see in the discussion of the lattice stability (Section 4), the stability is not determined by a single gross parameter like λ but by the details of the terms that contribute to it. How the phonon and exciton terms contribute to the stability of the entire system needs to be determined in each case.

A second objection was raised relating to the damaging effects of phonons on the electron–electron interaction for energy changes just greater than the Debye energy. This followed from the use in the kernel of the integral equation of the Bardeen–Pines form of the phonon-mediated electron–electron interaction,

$$U(p, k) = \frac{g^2 E(p-k)}{[\xi(p)-\xi(k)]^2 - E^2(p-k)} \tag{30}$$

For energy changes $\xi(p) - \xi(k)$ greater than the phonon energy $E(p-k)$ the interaction changes sign and becomes repulsive. However, Ginzburg[9] showed that it is not this interaction that should appear in the kernel of the integral equation but rather an effective interaction of the form of Equation (29). This is a monotonically decreasing function of $|\xi(k)|$ and $|\xi(p)|$, which is attractive at all energies and does not have the resonant form of Equation (30). Because of this all virtual phonons make a positive contribution to the net attraction.

It has also been argued[41] that in the presence of phonons the excitons could not give rise to high values of T_c. This argument was based on the results of solving the BCS integral equation for the gap with a form for the kernel obtained by approximating the phonon and exciton contributions V by two wells, with the phonon contribution cut off at ω_0 and the exciton contribution extending from Ω to Ω'. Solving the gap equation at T_c

$$1 = -\int_0^\infty \frac{N(0)V}{2|\varepsilon_k|} \tanh\left(\frac{\beta_c \varepsilon_k}{2}\right) d^3k \tag{31}$$

then yields a value for $T_c \ll \hbar\omega_D$ and hence in spite of the exciton contribution does not give a large critical temperature. The physical reason for this is that the major contribution to the integral comes from the region where $|\varepsilon(k)|$ is small and in the above approximation for V it was assumed that the exciton system makes no contribution to this region. If the interaction with the exciton system is treated correctly[42] rather than by the above two-well approximation then one finds that the exciton system does contribute in the region of $|\varepsilon_k| \approx 0$ and can lead to high values of T_c.

We see then that the virtual phonons do not affect adversely the superconductivity, but what of the thermal phonons? It was conjectured[43] some years ago that at temperatures greater than or comparable to the Debye temperature the presence of these thermally excited phonons would have a devastating effect on T_c and would probably preclude the possibility of superconductivity occurring by any means at temperatures appreciably above the Debye temperature. This problem has had an interesting history, but we now know from a rigorous treatment by Bergmann and Rainer[44] that this is not the case. To appreciate this point it is instructive to consider first why the customary neglect of the thermal occupation of the phonons in the equation for the gap gives reasonable results. The Eliashberg equation[25] for the gap function, using McMillan's terminology, is as follows:

$$\Delta(\omega) = \frac{1}{Z(\omega)} \int \frac{d\omega'}{\omega'} \operatorname{Re}[\Delta(\omega')] \int_0^{\omega_0} d\omega_q \, \alpha^2(\omega_q) F(\omega_q)$$

$$\times \{[N(\omega_q) + f(-\omega')][(\omega' + \omega_q + \omega)^{-1} + (\omega' + \omega_q - \omega)^{-1}]$$

$$- [N(\omega_q) + f(\omega')][(-\omega' + \omega_q + \omega)^{-1} + (-\omega' + \omega_q - \omega)]\} \qquad (32)$$

where we have simplified the expression by the neglect of the Coulomb interaction. Here $N(\omega_q)$ and $f(\omega')$ are the phonon and electron occupation numbers, respectively. The function $Z(\omega)$ is the renormalization parameter for the electron energies, which is obtained from an equation of somewhat similar form that also contains factors of $N(\omega_q)$ and $f(\omega')$.

We note that the major contribution to the integral for the gap $\Delta(\omega)$ comes from $\omega' \approx 0$, so if we neglect ω' in the curly bracket, the terms in $N(\omega_q)$ cancel pairwise, leaving us with the familiar $\tanh(\beta\omega'/2)$ from the combination of $f(-\omega')$ and $f(\omega')$. The only contribution from the phonon occupation that remains comes from corrections to this approximation and the terms in $Z(\omega)$. These can be shown to be of the same order of magnitude and small. Appel[45] went beyond this approximation to include the effects of phonon scattering on the electron lifetime. He claimed that the dynamical character of the phonon interaction should affect the energy gap and T_c because even though the Fröhlich Hamiltonian commutes with the time reversal operator, this Hamiltonian contains time-dependent phonon operators. Therefore the theorem of Anderson[46] and Maki,[47] which states that T_c should be unaffected by a perturbation that conserves time-reversal symmetry *and* is static, would not apply and T_c should be affected. Appel calculated the effects of phonon scattering on the electron self-energy and showed that it reduced T_c but that this effect would not preclude the possibility of room-temperature superconductivity on these grounds. Allen,[48] working from McMillan's equations, included the residual contribution from the phonon occupation terms but neglected the above lifetime effects, and he also showed an apparent repulsive or destructive effect upon the gap due to thermally excited phonons. In this the temperature dependence of $Z(\omega)$ was ignored. Bergmann and Rainer[44] then showed in a rigorous proof using the Matsubara formulation of the Eliashberg equations, which include all effects of the thermal excitation of the electrons and the phonons, that the thermal phonons are *never* repulsive and that the contribution to T_c of *all* phonon modes is positive. Karakozov *et al.*[49] helped to explain this by noting that for low-frequency phonons the terms in $Z(\omega)$ in (32) exactly cancel the low-frequency contributions to the numerator. An additional factor that has to be considered is renormalization of the electron mass with increasing phonon excitation. The electron self-energy depends explicitly upon the phonon occupation numbers. One can readily show that this leads to a renormalization of the electron mass m^* relative to its band mass m:

$$m^* = m\left\{1 + 2\int_0^{\omega_0} d\omega_q \frac{\alpha^2(\omega_q)}{\omega_q} F(\omega_q)[2N(\omega_q) + 1]\right\} \qquad (33)$$

The electron energies $\bar{\varepsilon}_k$ that result from solving for the normal part of the self-energy $\chi(p)$ in the Nambu formulation[49] of the Eliashberg equation contain this explicit phonon occupation-dependent contribution. This is neglected in the approximation that leads to the gap equation used by McMillan. Including this phonon contribution changes the effective density of states $N(0)$ at the Fermi surface. The effect of this is to offset partially the pair-breaking contribution of the phonons.

Thus we see that in a gross sense treating the phonons and excitons as independent subsystems, the phonons serve only to enhance T_c. At the microscopic level, where one treats the coupled exciton–phonon systems as we do in Section 4 in the "g-ology" picture, we find that the phonons can have a subtle role on T_c and in determining the type of pairing in the system. We discuss that role briefly later.

4. Effects of Limited Dimensionality

We have emphasized in the Introduction our belief that the one-dimensional model proposed for excitonic superconductivity has the best prospects of success. We have already pointed out in Section 3 how the particular one-dimensional structure proposed here helps to reduce vertex corrections and to minimize the harmful exchange-interaction terms. In Section 5 we shall show that the one-dimensional structure is almost essential in order to achieve an effective exciton-induced attraction that is sufficiently strong to overcome the Coulomb repulsion. These arguments follow from the examination of the general properties of excitonic super-conductivity and they clearly favor the one-dimensional models. The price one has to pay for these apparent advantages is that the one-dimensionality poses a series of new questions and difficulties that had previously received scant attention in the context of conventional superconductivity.

Of particular significance is the problem associated with the absence of phase transitions in a one-dimensional system. It was well known[51-53] that phase transitions involving a classical order parameter, such as the liquid–gas transition in a system of particles with finite-range interactions, cannot occur in one-dimensional systems. Stimulated by the suggestion of the possibility of superconductivity in polymeric systems it was shown by Ferrell[54] and more rigorously by Rice[55] and Hohenberg[56] that this could be extended to include the superconducting phase transition. Real systems, however, are not isolated one-dimensional systems but are composed of parallel chains that are in some sense coupled. Because of this such systems can undergo phase transitions at a finite temperature exhibiting three-dimensional long-range order, although this will usually occur at a temperature significantly lower than the mean-field transition temperature. In addition, currents with very long (practically infinite) decay time may occur in one-dimensional systems even without a real phase transition, because the fluctuations that destroy off-diagonal–long-range-order (the signature of the superconducting state) affect differently the conductivity and the order. Thus, the absence of true phase transitions in strictly one-dimensional systems does not exclude the possibility of superconductivity in real filamentary systems, even at high temperatures, as will be discussed in detail. The notion that it did, however, cast a pall on the early development of the ideas of a one-dimensional superconductor, which has only been removed in recent years. The principal effect of this problem was to stimulate interest in other types of systems of two[3] and three[29] dimensions that might be developed to utilize the same exciton mechanism. (These early developments are well reviewed by Keldysh.[7])

The limited dimensionality also plays a key role in the question of stability. The connection between superconductivity and lattice stability was already mentioned in Section 2.3. This connection is demonstrated particularly clearly in the high-T_c compounds of the β-tungsten structure,[57] which usually undergo a transition from a cubic to a tetragonal phase at temperatures somewhat greater than T_c. One possible effect of a structural instability in the neighborhood of T_c is due to the fact that near such an instability one of the phonon modes becomes soft. This, in turn, increases the electron–phonon coupling constant, and T_c may come closer to its maximum value determined by the dependence of Equation (1) on the phonon frequencies, as explained in Section 2.3. Such an effect would account for the relatively high values of T_c in the β-tungsten compounds.

A different point of view, which emphasizes the destructive effect of lattice instabilities, asserts that the structural phase transition occurs because the electron–phonon coupling constant in what would be the high-T_c phase is too large. In the new phase the lattice becomes stable with a lower λ. Therefore, if it were possible in some way to suppress the structural transition, the superconducting transition temperature would be larger than the observed one. This point of view on the restrictions on the values of T_c set by lattice instabilities has been frequently expressed by Matthias.[58] The effect of lattice instabilities on superconductivity becomes even more dramatic in a one-dimensional system. In three dimensions such an instability merely changes the density of states at the Fermi surface, but in a linear chain it produces a gap at the Fermi surface driving the system into the insulating state. Thus, a discussion of superconductivity in a filamentary compound requires a careful examination of other competing instabilities. This problem is discussed in detail in the present section.

Another objection raised against the idea of excitonic superconductivity which was based on dimensionality is that of screening. The screening of the Coulomb interaction in a system of restricted geometry is less effective than in three dimensions, and it was conjectured[59] that in a filamentary system it could never be of sufficient strength to allow the net interaction to be attractive. Although the discussion of this problem involves several general arguments of principle, the ultimate answer depends on detailed calculations for a particular model. The problem of screening will be discussed in Section 4.6.

4.1. Effects of Fluctuations

The absence of phase transitions in one-dimensional systems results from fluctuations of the order in the system. In a one-dimensional system a fluctuation that disrupts the order at one point in the system breaks the long-range order of the system as a whole. At finite temperature it is energetically possible for such a disruption to occur, and hence the system shows no long-range order. In two or three dimensions, however, for the long-range order of the system to be destroyed the order must be disrupted across a line or a plane, respectively. Such a fluctuation is very much more costly in energy and thus less probable than that in the one-dimensional case.

Ferrell[54] extended these classical arguments to the superconducting case. He showed that in a one-dimensional system the zero-point or thermal fluctuations of the superfluid density caused fluctuations of the phase of the order parameter, with the result that the so-called off-diagonal–long-range-order (ODLRO), which is descriptive of the order of the superfluid state, was destroyed. This ODLRO was believed to be an essential characteristic of the superconducting state. This argument was clarified and put on firmer grounds by Rice,[55] and later Hohenberg[56] showed rigorously that ODLRO could not occur in any infinite one- or two-dimensional systems.

These rigorous mathematical results, however, raised a number of physical paradoxes. Using Ferrell's argument, DeWames *et al.*[60] showed that for a *finite* but long one-dimensional system a superconducting phase could exist but its transition temperature would be depressed below that of a short specimen! The paradox that this raised can be seen as follows. Consider a small bead of superconducting metal. According to experience and the above arguments, such a finite sample can superconduct. Consider now a long string of such beads. If this string is made long enough, then according to the arguments of Ferrell, Rice, and Hohenberg such a chain would not superconduct. This raised the problem of how the local conductivity could be changed by the addition of distant beads. This paradox was resolved by the arguments of Little,[61] who considered the effects of fluctuations upon the apparent resistivity of a one-dimensional "superconductor." The arguments relating to the resistivity can be appreciated by first considering Ferrell's arguments. The parameter that describes the order in a superconductor is the function ψ, which is complex, $\psi = \Delta e^{i\phi}$. Fluctuations of the local density of the electrons at some region in the system due to zero-point or thermal motion must be accompanied by a fluctuating in-flow or out-flow of current to or from this region. From the quantum-mechanical current operator,

$$j(x) = \frac{\hbar}{2im}(\psi\nabla\psi^* - \psi^*\nabla\psi) \tag{34}$$

one can show that these fluctuating currents cause fluctuation of the phase ϕ of the order parameter. If one considers two points x and x' in the superconductor, then as one increases the separation between these points, the fluctuation of the relative phase of the order parameter between the two points must increase. Thus, for a sufficiently large separation, the relative phase of the two points will fluctuate over an angle greatly in excess of 2π. In this case the value of the average $\langle\psi(x)\psi^*(x')\rangle$ will clearly vanish. However, the existence of a finite value for this average in the limit that $|x - x'| \to \infty$ is the requirement for ODLRO, so in such a one-dimensional system ODLRO cannot occur.

Rice[55] used the Ginzburg–Landau equations and the above arguments to calculate quantitatively the range of order for both one- and two-dimensional systems. Using this formalism Little[61] then considered the effects of fluctuations on the lifetime of "persistent" currents in a "one-dimensional system" bent into the form of a closed ring. He showed that while such fluctuations destroyed ODLRO, in themselves they did not

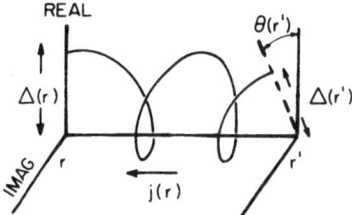

Figure 7. Plot of the real and imaginary parts of the order parameter $\psi(r)$ as a function of position r in a one-dimensional sample carrying a current $j(r)$.

cause the decay of the persistent current. This can be shown on the basis of a topological argument.

If we plot the complex order parameter as a function of position as illustrated in Figure 7, then the value of $\psi(0)$ must equal its value at $\psi(L)$, where L is the perimeter of the ring. This follows because ψ must be single valued. Thermal or zero-point motion causes ψ to fluctuate in amplitude and phase at each point. But, provided it remains finite at all points, then the total phase change round the loop must be an integral multiple of 2π, i.e., $2\pi n$, in order to satisfy the boundary conditions. We can classify the different ψ_n into different subensembles given by the value of this integer n. Within a given subensemble fluctuations can occur that cause the representative path of the order parameter to wind and unwind about the origin line of the Argand diagram. One can readily show, as argued above, that such fluctuations destroy ODLRO, but equally well one can show that the net circulating current is fixed by n (i.e., the total phase change round the ring), and, provided this does not change, no current decay is possible. The essential requirement for the current to decay is for a fluctuation to occur between subensembles such as illustrated in Figure 8. Clearly from the topology of the diagram the only type of fluctuation that can allow this to occur is one in which the order parameter drops to zero at some point, passes through the origin, and emerges on the other side with a different value for n. Because of the finite coherence length ξ_0 in a superconductor (i.e., the effective size of the Cooper pair) the order parameter must be driven to zero over a region whose length is of this order. Consequently, the cost in free energy Δf for such a fluctuation to occur is finite for temperatures below the mean-field T_c, and the probability of such a fluctuation occurring is proportional to $\exp(-\Delta f/kT)$. This becomes

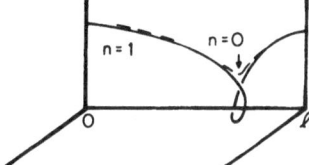

Figure 8. Illustration of the type of fluctuation in the complex-order parameter which can cause the decay of a persistent current in a one-dimensional sample.

exponentially small at low temperatures and consequently the resistivity below T_c should also fall exponentially.

These arguments show that as far as fluctuations are concerned a one-dimensional superconductor cannot exhibit true long-range order (ODLRO) in the strict mathematical sense, nor can it show a sharp phase transition, but nevertheless it can be expected to exhibit a conductivity that approaches an infinite value at temperatures somewhat below T_c.

The rare fluctuations that lead to dissipation also introduce a special problem in considering the statistical mechanics of a one-dimensional superconductor. In most problems in solid-state physics the assumption is made that the physical properties of a material can be described by taking an average over an ensemble of identical systems. Such an ensemble average is equivalent to a time average of one system provided the time average is taken over a sufficiently long time—strictly speaking, over an infinite time period. However, in this particular problem involving the decay of persistent currents the time for fluctuations to occur between the subensembles discussed earlier can be shown to be so long[61,62] that a true time average cannot be obtained even over time periods as long as the age of the universe. It is therefore misleading to use an ensemble average, for this would not describe the properties of the system that would be observable over any reasonable time period. In particular, the proof of the absence of ODLRO in a long but finite one-dimensional system is based on such an ensemble average. If the time evolution of such a system is studied one finds that, in fact, the system does indeed exhibit ODLRO for extremely long periods of time but that over an *infinite* time period it does not. Caution must thus be used in assessing the physical consequences of the limited dimensionality in calculations where conventional methods of statistical mechanics are used.

The above arguments are all based on the assumption that within the superconductor a local order parameter $\psi(x)$ can exist. For the case of a macroscopic bead of superconductor discussed earlier this assumption is reasonable. On the other hand, for a molecular chain of atomic dimensions it is not so obvious that it is valid and requires further justification. One must examine with care the microscopic theory of the condensate. This will be discussed in Section 4.2. However, as stressed earlier any of these results or their extensions that do not take proper account of the possible metastability of any fluctuating condensate can give results that would not correspond to that which would be observed in the laboratory. For example, this metastability imposes upon the microscopic theory the requirement that the expectation values for the gap function or pair condensate used in the gap equation itself be taken within the subensemble and *not* over the whole ensemble. The problems that this raises are not trivial, for it requires the explicit examination of the validity of the ergodic hypothesis

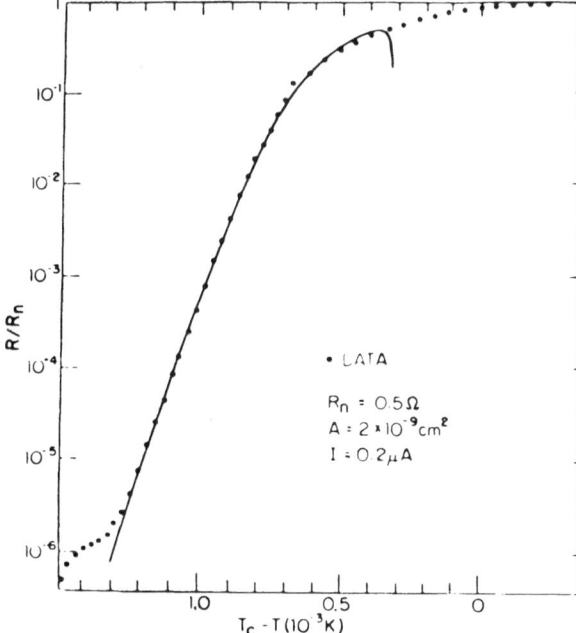

Figure 9. Comparison of McCumber–Halperin theory with experimental data points of Newbower, Beasley, and Tinkham for resistance of tin whisker below T_c due to thermodynamic fluctuations. The small "foot" is believed due to contact effects (Reference 62).

for each system and temperature, and thus a knowledge of the "master equation."[63] For this reason further work is needed to justify the relevance of recent rigorous results to the physical observables.

An offshoot of these sometimes esoteric arguments has been the explanation of the perfect conductivity of a superconductor, a subject that remained obscure long after the development of the microscopic theory. As pointed out by Little[61] and Langer and Ambegaokar,[64] the fluctuations that change the phase of the order parameter by 2π at some point in a wire are the source of dissipation. Langer and Ambegaokar justified the consequences of the electrodynamics of the fluctuations and established a detailed theory of the decay. This work was expanded upon by McCumber and Halperin[65] and has been studied extensively by several experimental groups, notably at Cornell and at Harvard. This experimental work is reviewed by Tinkham.[62] The essential exponential temperature dependence of the resistivity whose origin was outlined above is beautifully illustrated by the results of Newbower et al. (Figure 9).[66]

These arguments and their experimental verification give one confidence that the essence of the problem is understood and that states of very high conductivity could occur in filamentary materials in spite of their limited dimensionality.

4.2. Types of Order in a One-Dimensional Electron Gas

Although a one-dimensional electron system cannot undergo a phase transition at any finite temperature, it may exhibit long-range order at $T = 0$. An understanding of the conditions for the onset of a particular type of order at $T = 0$ and the growth of the corresponding condensate is important for the discussion of the behavior in real filamentary materials. We can base our discussion of this problem on a system described by the electron gas Hamiltonian

$$H = \sum \varepsilon_k a^{\dagger}_{k,\sigma} a_{k,\sigma} + \sum V(q) a^{\dagger}_{k+q,\sigma} a^{\dagger}_{k'-q,\sigma'} a_{k'\sigma'} a_{k\sigma} \qquad (35)$$

where $a^{\dagger}_{k,\sigma}(a_{k,\sigma})$ creates (destroys) an electron with spin projection σ in the state of momentum k and energy ε_k. This Hamiltonian contains only the electron–electron interaction $V(q)$, and it is assumed that the electron–phonon (or electron–exciton) interaction is effectively included in the latter. A discussion of one-dimensional systems based on a Hamiltonian in which the coupling to phonon (or exciton) modes is explicit will be mentioned briefly in Section 4.4. Another line of approach to one-dimensional systems is based on the Hubbard model[67] and it is particularly appropriate in the case of $U/4t \gg 1$ (where U is the interaction between two electrons on the same atomic site and $2t$ is the bandwidth). We shall not discuss this line of approach because the systems we are interested in are more properly represented by a band picture as in Equation (35).

A one-dimensional Fermi system is characterized by a Fermi "surface" consisting of two points at $\pm k_F$. This results in a striking difference between the three- and one-dimensional Fermi system. While in the first case there exist low-energy electron–hole excitations with momenta between 0 and $2k_F$, in the second case, one finds such excitations only in the neighborhood of $q \approx 0$ and $q \approx 2k_F$. This means that the important electron–electron interaction processes are those that involve momentum transfers in the neighborhood of these two values. There are four relevant interaction processes, which are represented in Figure 10. The $q \approx 0$ processes may involve electrons either on one side of the Fermi "surface" (g_4)

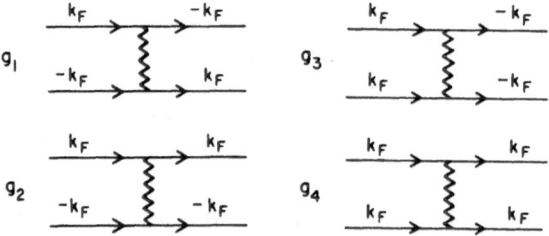

Figure 10. Diagrammatic representation of the four interaction processes in a one-dimensional Fermi system. The forward scattering processes are denoted by g_2 and g_4, the back-scattering by g_1 and g_3. The latter exists only for a half-filled band.

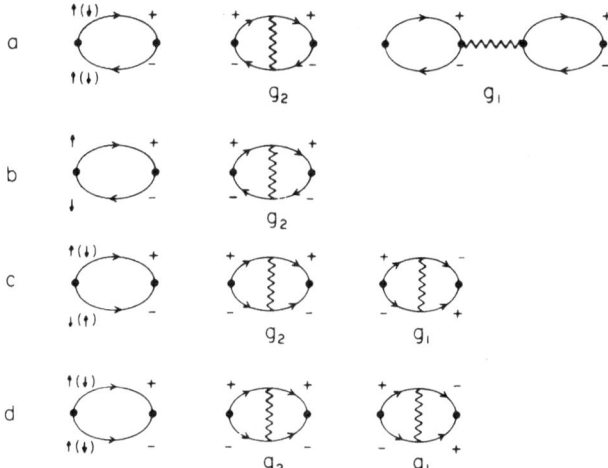

Figure 11. Diagrammatic representation of the zeroth and first-order contributions to the generalized susceptibilities $\chi(2k_F, \omega)$ corresponding to (a) CDW, (b) SDW, (c) SS, and (d) TS. The +, − signs denote the two sides of the Fermi "surface," and the vertical arrows indicate the possible relative orientations of the spins.

or on both sides (g_2). In both cases the electrons do not change their direction of motion as a result of the interaction between them. For this reason, these processes are referred to as forward scattering processes. When electrons on the two sides of the Fermi "surface" exchange momentum $q \approx 2k_F$ they change their direction of motion. This is therefore a backscattering process (g_1). In the case of one electron per atom (half-filled band) there is also an umklapp process in which two electrons on the same side of the Fermi "surface" scatter together to the other side (g_3). The study of instabilities in a model in which the interaction is reduced to these coupling constants is generally referred to as one-dimensional "g-ology." We have mentioned already that the question of order in a single chain can be discussed in terms of its tendency towards long-range order as $T \to 0$. Any type of order would show up in the response of the system to a corresponding generalized external field. There are four possible types of ordering when $T \to 0$: charge-density wave (CDW), antiferromagnetic or spin-density wave (SDW), and singlet and triplet pairing (SS and TS). Other types of order are in principle possible for very large coupling constants,[68] but then the band picture assumed in Equation (35) breaks down. To each of the types of order there corresponds a generalized susceptibility,[69,70] for example,

$$\chi_{CDW}(q, \omega) = -i \int_{-\infty}^{\infty} dt \, e^{i\omega t} \sum_{pp'} \sum_{\sigma\sigma'} \langle T[a_{p\sigma}^{\dagger}(t)a_{p+q\sigma}(t)a_{p'\sigma'}^{\dagger}(0)a_{p'-q\sigma'}(0)]\rangle \quad (36)$$

where T is the time-ordering operator. There is a similar expression for each of the other types of order. A divergence of one of these functions signals the onset of the corresponding type of order. In one-dimensional systems such divergences may occur only at $T = 0$, which is equivalent to the statement that a phase transition cannot occur in such a system at any finite temperature.

As a first step it is instructive to calculate up to first order in perturbation theory the susceptibilities defined in Equation (36). The processes that contribute in each case are shown in Figure 11. The basic effect of one-dimensionality is that the integration over the internal momenta of each electron–electron and electron–hole pair gives a logarithmic contribution of the form $\rho \log(T/D)$ (where ρ is the density of states and D is an energy cutoff) to the susceptibility at $q = 2k_F$ and $\omega = 0$. This is in contrast to the three-dimensional case in which the electron–hole propagator is not singular. The expressions for the four susceptibilities defined in Equation (36) at $\omega = 0$ at $q = 2k_F$ for CDW and SDW and at $q = 0$ for SS and TS are[69,70]

$$\chi_{CDW} = \frac{1}{\pi v} \ln\left(\frac{T}{D}\right)\left[1 + \frac{1}{2\pi v}(2g_1 - g_2)\ln\left(\frac{T}{D}\right)\right] \tag{37}$$

$$\chi_{SDW} = \frac{1}{2\pi v} \ln\left(\frac{T}{D}\right)\left[1 - \frac{g_2}{2\pi v}\ln\left(\frac{T}{D}\right)\right] \tag{38}$$

$$\chi_{SS} = -\frac{1}{\pi v} \ln\left(\frac{T}{D}\right)\left[1 + \frac{1}{2\pi v}(g_1 + g_2)\ln\left(\frac{T}{D}\right)\right] \tag{39}$$

$$\chi_{TS} = -\frac{1}{\pi v} \ln\left(\frac{T}{D}\right)\left[1 + \frac{1}{2\pi v}(g_1 - g_2)\ln\left(\frac{T}{D}\right)\right] \tag{40}$$

where v is the Fermi velocity and $(\pi v)^{-1}$ is the density of states at the Fermi level. In the random-phase approximation one picks from each order in perturbation theory that term which is merely a repetition of the first-order processes, so that the expression for the susceptibilities in this approximation becomes a geometric series which can be summed to give the general form

$$\chi_i = \frac{\chi_i^0}{1 - \lambda_i \ln(T/D)} \tag{41}$$

where the index i denotes one of the possible types of order and the λ_i are the corresponding effective constants, which can be read off from Equations (37)–(40):

$$\lambda_{CDW} = \tfrac{1}{2}(2\tilde{g}_1 - \tilde{g}_2)$$

$$\lambda_{SDW} = \tfrac{1}{2}\tilde{g}_2$$

$$\lambda_{SS} = +\tfrac{1}{2}(\tilde{g}_1 + \tilde{g}_2) \tag{42}$$

$$\lambda_{TS} = +\tfrac{1}{2}(\tilde{g}_2 - \tilde{g}_1)$$

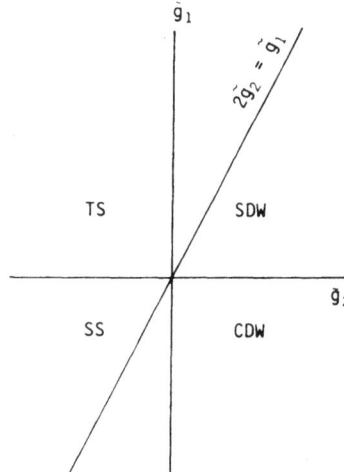

Figure 12. Regions in the (g_1, g_2) plane with the highest mean-field transition temperature to the indicated type of order.

where $\tilde{g}_i$ is g_i in units of πv. In the random-phase approximation one finds a phase transition for negative λ_i given by the BCS-like expression

$$T_c = D e^{1/\lambda_i} \tag{43}$$

which is the mean-field transition temperature. If one now asks in what regions of the (g_1, g_2)-plane is one of these transition temperatures higher than all the others one gets the picture in Figure 12. Although the mean-field treatment is insufficient, the results obtained indicate the regions in the space of the coupling parameters in which the system tends to develop large fluctuations corresponding predominantly to one of the possible types of order, and these regions are remarkably similar to the qualitative picture obtained from the more rigorous treatment.

The basic difficulty in the theoretical treatment of order in one-dimensional Fermi systems is due to the divergence in the bare electron–electron and electron–hole propagators. These cause divergences in the full electron–electron interaction in the particle–particle channel (Cooper channel), which indicate the onset of pairing instabilities and in the particle–hole channel, which indicate the onset of spin- and charge-density wave instabilities. The simplest examples of processes in these two channels are shown in Figure 13. In higher-order processes these two channels become coupled and one has to treat them simultaneously. This was first done by Bychkov *et al.*,[71] who considered a model in which $g_1 = g_2 = g$, neglecting g_3 and g_4. They calculated the effective interaction between two electrons in the parquet approximation. In this approximation one solves the coupled equations for the vertex functions (effective interactions) in the particle–particle and particle–hole channel; however, in each channel one keeps only the terms that appear in the random-phase approximation. Therefore, this is still a mean-field approximation and one finds a phase

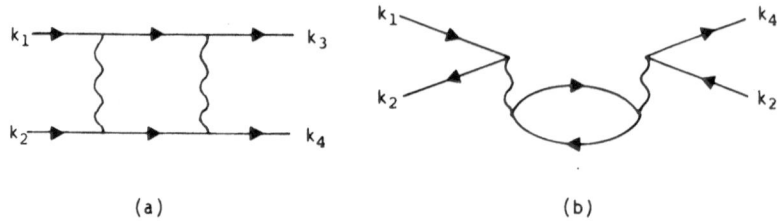

Figure 13. Diagrammatic representation of the simplest interaction processes in the particle–particle channel (a), and in the particle–hole channel (b).

transition at a finite temperature. Bychkov *et al.* found that the particle–particle (Γ_{pp}) and particle–hole (Γ_{ph}) vertex functions are both of the form

$$\Gamma \propto \frac{1}{1 + \tilde{g} \ln (D/T)} \tag{44}$$

which for negative g gives a critical temperature $T_c = D \exp(-1/|\tilde{g}|)$. They concluded that at this temperature the system goes over to an ordered state of condensed quartets of two electrons and two holes. This is a pioneering contribution to the field of one-dimensional systems and attracted a lot of attention; however, it suffers from two significant shortcomings. One is the restriction of the interaction parameters to $g_1 = g_2$, and the other is the use of the mean-field approximations.

As one goes beyond the parquet approximation and includes additional terms which also diverge logarithmically, one finds that all susceptibilities or vertex functions change dramatically from the logarithmic form in Equations (41) and (44) to a power law behavior T^α. A negative value of the exponent α indicates the corresponding type of ordering at $T = 0$. We shall summarize the basic results of the extensive study of the generalized susceptibilities carried out in recent years.

Another treatment of the competition between the superconducting and the Peierls state in the mean-field approximation was done by Levine *et al.*[72,73] They concluded that these two states do not coincide except in very exceptional circumstances.

The first discussion of order in a one-dimensional system that went beyond the mean-field approximation is due to Menyhard and Sólyom.[69,74] They applied the renormalization group method to calculate the susceptibilities corresponding to CDW, SDW, and SS, and found that for $g_1 > 0$, the line $g_1 = 2g_2$ separates the region of superconducting and charge-(spin-)density wave behavior (only g_1, g_2 were taken into account). This approach was extended by Fukuyama *et al.*,[70] who also calculated the TS-response function. The renormalization group method fails for $g_1 < 0$, except in the neighborhood of the origin.

For negative values of the backscattering coupling constant g_1 the results are based on the work of Luther and Emery,[75] who found a remarkable solution for a particular value of negative g_1. They showed that the one-dimensional Hamiltonian can be described in terms of two kinds of degrees of freedom: charge- and spin-density oscillations. The first gives rise to gapless phononlike excitations, while the latter results in an excitation branch with a gap. Luther and Emery were able to diagonalize this Hamiltonian for the case $g_1 = -6/5$ and to calculate explicitly the low-frequency behavior of the susceptibilities. Their results were extended to all values of $g_1 < 0$ by Lee.[76]

Our present understanding of the types of order in a single metallic chain is summarized in Figure 14 taken from Reference 77. In the lower half-plane there is a region in which the CDW and SS susceptibilities both diverge; however the CDW divergence is stronger on the right-hand side of the line $g_1 = 2g_2$ and the SDW divergence is stronger on the left-hand side. Note that the SDW and TS types of order can exist only in the upper half-plane. The picture in the lower half-plane of Figure 14 differs in some details from other treatments of this region (for example, Reference 76) because it contains the effect of the g_4 process (in a real system $g_2 = g_4$).

The general conclusion one can draw from the picture illustrated by Figure 14 is that the regions of the various types of order at $T = 0$ are strikingly similar to the regions determined by the highest mean-field transition temperature (Figure 12). The basic feature of the diagram in the (g_1, g_2) phase is that the sign of $2g_2 - g_1$ distinguishes between superconducting (negative sign) and charge- or spin-density wave (positive sign) behavior. One can understand why this combination of the coupling constants is in one dimension the effective interaction that determines the low-temperature (and low-frequency) behavior of the response functions.

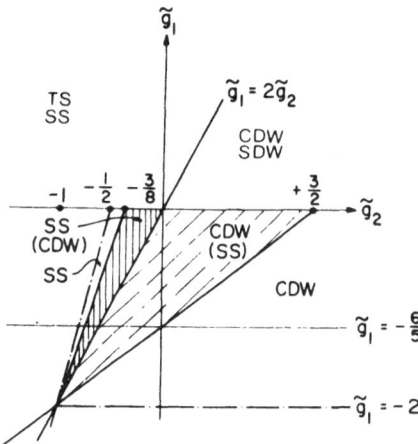

Figure 14. Regions in the (g_1, g_2) plane in which the response function for the indicated types of order diverge. The dotted and dashed lines represent the boundaries of the region in which the model can be solved.

First, one can show that the g_4 process only serves to modify the "sound" velocity of the charge-density oscillations. The g_2 process occurs with a factor of 2 because of the two possible relative spin orientations of the two electrons. The backscattering process is different. If the spins of the two electrons are parallel, then backscattering in one dimension is indistinguishable from a forward scattering process with exchange. Therefore it appears in the effective interaction with a minus sign. Backscattering with antiparallel spins, on the other hand, does not contribute to the low-temperature behavior of the response functions because this process is described by the spin-density degrees of freedom and these have a gap in their excitation spectrum.[75]

4.3. Relevance of "g-ology"

In Section 4.2 we described the possible types of ordering at $T = 0$ in a single metallic chain. This description is based on a very simple model of an electron gas with two coupling parameters g_1 and g_2. These parameters contain all the contributions to the electron–electron interaction without distinction between the direct Coulomb interaction and the interaction induced by phonons or excitons. In the case of phonons the basic difference between these two interactions is the large difference in their energy cutoffs. This problem has been discussed recently,[78,79] in the context of the one-dimensional electron gas, but it should not concern us here, because the exciton-induced interaction has the same cutoff as the Coulomb interaction. Another subtle feature of the phonon- (exciton-) induced electron–electron interaction is the effect of retardation, which is ignored when the interaction is replaced by a static interaction. The simple model described above also neglects any effects that might arise from the presence of impurities. It is not clear how these two effects would modify the details of the picture represented in Figure 14 for single chains, but some work on retardation effects in coupled chains has been done and will be discussed in Section 4.4. Nevertheless, we believe that the gross features of this picture do represent "real physics," in the sense that they indicate the relation between the properties of the interaction and the type of condensate which the system tends to develop.

One reason why it is hard to relate the g-ology picture to real materials is that there is no way known at present to measure the interaction constants g_1. Nevertheless, without being too specific, we believe that all the presently discussed charge-transfer and metallo-organic compounds are represented by points in the (g_1, g_2) plane lying on the right-hand sides of the line $2g_2 = g_1$. Let us now discuss the location of the proposed excitonic system in this plane. The explicit expressions for these two parameters in our model are [see Equation (29)]

$$g_1 = V(2p_F) - \frac{2|Q(2p_F)|^2}{E(2p_F) + 2|\xi(p_F)|} \tag{45}$$

$$g_2 = V(0) - \frac{2|Q(0)|^2}{E(0) + 2|\xi(p_F)|} \tag{46}$$

where the first term is the screened Coulomb interaction. We have already mentioned that the electron–exciton coupling constant decreases rapidly with increasing momentum and is negligible at $q = 2p_F$. This results in a small but positive value of g_1. If the balance between the Coulomb and the exciton-induced interaction at $q = 0$ is favorable, g_2 becomes negative. Calculations on the model discussed in Section 5 show that this is the case and that the system considered is represented by a point on the left-hand side and quite remote from the line $2g_2 = g_1$ (Figure 12). This puts our system in the region in which both the rigorous treatment and simple mean-field theory Equation (42) predict triplet superconductivity.

The occurrence of triplet rather than singlet superconductivity in our model calls for an interesting comparison with He^3. This is one case where superfluidity ("superconductivity" in a neutral fermion system) was predicted successfully by the BCS theory in a regime that is completely different from its usual area of application. If He^3 were a one-dimensional system we would place it in the (g_1, g_2) plane in the neighborhood of the point representing our model of the excitonic superconductor. The interaction in He^3 has a short-range repulsion (hard core) and a long-range attraction. In momentum space this corresponds to negative g_2 and positive g_1, which places it in the region of triplet superconductivity, and indeed, He^3 is a triplet "superconductor."

There are two effects that might change the above prediction of triplet and not singlet pairing. First, if there is also a phonon-induced electron–electron interaction then this will affect mainly the value of g_1. If this interaction is sufficiently strong it will result in a negative value of g_1. The system will then lie in the region of singlet rather than triplet superconductivity. Second, the effect of impurities may result in the suppression of triplet pairing so that singlet pairing will become relatively more favorable. This will be discussed in Section 4.5.

4.4. Interchain Coupling

We have argued that extremely long-lived currents may exist in linear chain systems even without a phase transition. However, if one wants to discuss real phase transitions in such systems one has to include interchain coupling. The nature and strength of this coupling determines the actual transition temperature and may also affect the g-ology diagram in the plane of the single-chain interaction parameters.

One approach to the problem of interchain coupling is to start from exact solutions for the single chain and to introduce interchain coupling as a perturbation. In this approach the longitudinal correlations are treated exactly and the transverse correlations in mean-field theory.[80] A typical expression for any response function in this approximation is

$$\chi(q_{\parallel}, q_{\perp} T) = \frac{\chi_{1d}(q_{\parallel})}{1 - V(q_{\perp})\chi_{1d}(q_{\parallel})} \qquad (47)$$

where χ_{1d} is the exact one-dimensional response function, $q_{\parallel}$ ($q_{\perp}$) are the longitudinal (transverse) momentum components, and $V(q)$ is the interchain interaction.

The problem of weakly coupled linear chains was investigated by Klemm and Gutfreund[81] using the single-chain solutions found by Luther and Emery.[75] Scaling arguments make it possible to generalize these results to the $g_1 = 0$ half-plane. Two types of interactions were considered. One may be referred to in a broad sense as an interchain Coulomb interaction in which the interacting electrons are confined to their chains, and the other is the interchain single-particle tunneling interaction. In the Coulomb interaction, one may distinguish between two processes: interchain forward scattering and interchain backscattering. Inclusion of the first of these two processes still gives an exactly soluble model; however, this type of interchain coupling does not give rise to finite transition temperatures, but merely modifies the exponents of the response functions. It also changes somewhat the boundaries between the regions corresponding to the different types of order. The backscattering Coulomb interchain coupling and single-particle tunneling are represented by the terms

$$\mathcal{H}_{bs} = V \sum_{nn'} \sum_{ss'} \int dx\, \psi_{1s}^{+n}(x)\, \psi_{2s'}^{+n'}(x)\psi_{1s'}^{n'}(x)\psi_{2s}^{n}(x) \qquad (48)$$

$$\mathcal{H}_t = T \sum_{i=1,2} \sum_{nn'} \sum_{s} \int dx\, \psi_{is}^{+n'}(x)\psi_{is}^{n}(x) + \text{H.c.} \qquad (49)$$

where $\psi_{is}^{n}(x)$ is the field operator which destroys an electron with spin component s on one or the other side of the Fermi "surface" ($i = 1$ or 2). The index n denotes the chain and the summation extends over all nearest neighbors. The Coulomb interaction [Equation (48)] may be decoupled into a product of two terms describing the order parameter for a charge-density wave on separate chains, $\langle \psi_{1s}^{\dagger}(x)\psi_{2s}(x)\rangle$. There is no way to get from this term, in first order, coupling between the pairing order parameters, $\langle \psi_{1s}^{\dagger}(x)\psi_{2s}^{\dagger}(x)\rangle$, on two chains. Therefore, this interaction can only give a finite transition temperature to the charge-density wave state. The tunnelling interaction [Equation (49)] may be decoupled, in second order of perturbation theory, to represent an interchain coupling between order parameters of both kinds. Therefore, motion of electrons from one chain to another may result in a phase transition at finite temperature either to a charge density wave or to a superconducting state. In the model of weakly coupled chains one gets a real phase transition to a particular type of ordered state only if the single-chain response function corresponding to this type of order diverges. As a result of this, one finds that for $g_1 - 2g_2 < 0$, the system will undergo only a CDW phase transition. In the region where the one-dimensional SS response function is divergent but the CDW response function is not divergent, the tunneling interaction ensures that the system will be superconducting. In the intermediate region where both one-dimensional responses are divergent but the SS function predominates, there are competing effects, as the Coulomb interactions favour CDW, but the effect of tunneling is to favor the SS transition. The transition temperature itself depends on the strength of the interchain coupling. There is no such detailed treatment of the upper half plane, $g_1 > 0$. Mihaly and Sólyom[82] have discussed this region for the case of interchain Coulomb coupling and they find essentially the same results. A similar treatment which studies the transition between one- and three-dimensional behavior in a system of parallel chains coupled by the Coulomb interaction was presented by Menyhard.[83]

Another approach to the problem of interchain coupling is to consider the coupled chain system as a highly anisotropic three-dimensional system. This anisotropy may be represented by an electron energy dispersion relationship of the form

$$\varepsilon(k) = \varepsilon(k_z) + \eta \varepsilon_F (\cos a k_x + \cos a k_y) \tag{50}$$

where a is the distance between neighboring chains and η is a measure of the interchain coupling, which is given essentially by the ratio of the interchain and intrachain transfer integrals. The advantage of this approach is that for sufficiently strong interchain coupling, that is, when the anisotropy is not too large, the mean-field theory is a good approximation. One can therefore compute the transition temperature in the mean-field approximation and then check *a posteriori*, by calculating the effect of fluctuations, when this approximation breaks down.

Maniv and Weger[84] have estimated along these lines the effect of fluctuations on the superconducting mean-field transition temperature using the Ornstein–Zernike approximation and find that T_c is reduced significantly only when $\eta < 0.1$. This approach was also used to discuss[85] the effects of fluctuations on the Peierls transition in an electron–phonon system described by the Fröhlich Hamiltonian

$$\mathcal{H} = \sum_k \varepsilon_k a_k^\dagger a_k + \sum_q \omega_q b_q^\dagger b_q + \sum_q G_q a_{k+q}^\dagger a_k (b_q + b_{-q}^\dagger) \tag{51}$$

where a_k, b_k are the electron and phonon destruction operators, ε_k and ω_q are their respective energies, and G_q is the electron–phonon coupling constant. The previous approach based on exact solutions for the single chain cannot be applied in this case, because no such solutions are known for the Fröhlich Hamiltonian, which contains retardation effects. We shall summarize the main results of Reference 85. The mean-field transition temperature T_p to the Peierls state is almost independent of η up to a certain critical η_c depending on the nature of the electron dispersion $\varepsilon(k_z)$, at which it drops rapidly to zero. There exists a characteristic tempera.ure $T_0 \simeq \eta T_F/4$, which plays an important role in the model. At this temperature the inverse transverse correlation length crosses the Brillouin-zone boundary. If the mean-field transition temperature T_p happens to be smaller than T_0, the thermodynamic fluctuations will have a small effect, suppressing the actual transition temperature by at most 20%. If, however, the mean-field T_p exceeds T_0, we expect large fluctuations, and mean-field theory breaks down. Whether a given system may be described reasonably well by mean-field theory depends in this model on the values of η and the electron–phonon coupling. This picture was found useful to explain recent pressure experiments on KCP.[86] Pressure is expected to increase η and these experiments show that T_p is in fact enhanced by pressure. The Peierls transition in the Fröhlich Hamiltonian has also been treated in a model in which the interchain coupling is provided by the Coulomb interaction[87] and by the three-dimensional phonons.[88]

The model of the anisotropic electron dispersion relation was also used[89,90] to investigate the competition between the Peierls and the superconducting transition in the region when mean-field theory applies, namely, when η is sufficiently large but there is still nesting of the Fermi surface so that an *uncoupled* channel calculation would yield a Peierls transition. The relevant effective electron–electron interactions are those induced by the exchange of phonons with momentum $q = 0$ and $q = 2k_F$. In the language of g-ology they are described by negative values of g_1, g_2, defined as

$$g_1 = 2G_{2k}^2 N(0)/\omega_{2k_F}$$
$$g_2 = 2G_0^2 N(0)/\omega_0 \tag{52}$$

The calculation was performed in a double Nambu formalism which treats the two transitions on the same footing. It turns out that the boundary line separating the two transitions in the (g_1, g_2) plane depends on the phonon frequency ω_0 (an Einstein phonon was assumed) (Figure 15). The line $2g_2 = g_1$ is obtained as $\omega_0 \to \infty$ corresponding to the static interaction in the electron gas picture. The deviation from this line for finite ω_0 represents the retardation effects of the electron–phonon interaction. It is interesting

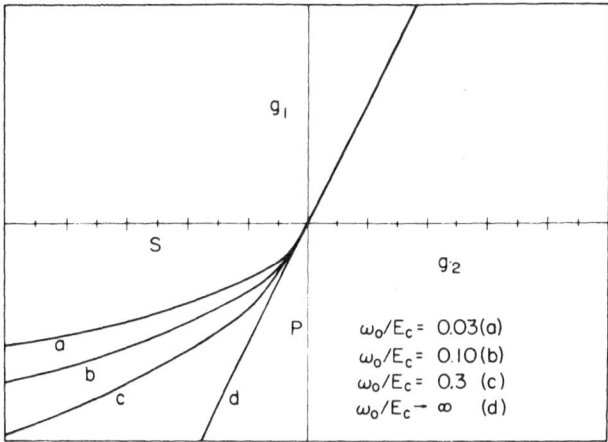

Figure 15. Effects of retardation on the phase diagram of the superconducting (S) and Peierls (P) phases for a coupled linear chain system. E_c is the Coulomb cutoff energy and ω_0 the phonon cutoff.

to note that they tend to increase the region of the Peierls ordering. Another result of this investigation is that high superconducting transition temperatures are possible only for high values of ω_0, of the order of an electronic cutoff energy.

4.5. Localization and Impurities

The discussion in the preceding subsections assumes that the electrons occupy band states with definite momentum and propagate coherently throughout the whole system. However, it is known that any disorder caused by the presence of impurities and defects, which are unavoidable in a real material, results in a spatial localization of all the states in the band.[91] The localization length, namely, the distance over which the amplitude of the wave function drops to e^{-1} of its maximum value, depends on the energy of the state, on the nature of the impurities, and on their distribution.

We shall sketch briefly a proof[92] of localization based on the transfer matrix approach introduced by Borland.[93] Let us assume a simple model of a linear crystal represented by random potentials concentrated around equally spaced lattice sites, so that between two adjacent lattice sites there is a free region in which the wave function is given by

$$\psi(x) = A e^{ikx} + B e^{-ikx} \tag{53}$$

The transfer matrix connects the amplitudes A and B in two consecutive intervals

$$\begin{pmatrix} A_{n+1} \\ B_{n+1} \end{pmatrix} = T_n \begin{pmatrix} A_n \\ B_n \end{pmatrix} \tag{54}$$

Borland has shown that the transfer matrix T_n in the case of equally spaced potentials is

$$T_n = \begin{pmatrix} (1 - i\lambda_n) e^{ikl} & -i\lambda_n \\ i\lambda_n & (1 + i\lambda_n) e^{-ikl} \end{pmatrix} \tag{55}$$

where l is the lattice spacing and λ_n is equal to $\mu_n/2k$, where μ_n is a parameter that characterizes the potential at the nth lattice site and $k = E^{1/2}$. The relation between the wave-function amplitudes in the first and $(n+1)$th interval is

$$\begin{pmatrix} A_{n+1} \\ B_{n+1} \end{pmatrix} = T_n T_{n-1} \cdots T_1 \begin{pmatrix} A_1 \\ B_1 \end{pmatrix} \tag{56}$$

In a band state the wave-function amplitudes in the intervals between different sites are related by phase factors. This is the case when the overall transfer matrix connecting any two such intervals can be diagonalized by a unimodular transformation to give eigenvalues $e^{\pm i\theta}$. For a periodic lattice when all the λ_n are equal the condition for this is $\mathrm{Tr}(T) \leq 2$, where $\mathrm{Tr}(T)$ is the trace of the matrix T. This is the band condition and determines the band energies. When this condition is satisfied by the single-step transfer matrix, it will also hold for the n-step transfer matrix T^n for any value of n. In the case of random values of λ_n the band condition will be violated for a sufficiently long segment of the chain even if every single-step matrix satisfies $\mathrm{Tr}(T_n) \leq 2$. The reason this happens is that the square of deviations from the mean of the parameter λ_n accumulate on the diagonal of the n-step transfer matrix. It can be shown that the trace of the ensemble-averaged n-step transfer matrix grows as one moves along a chain of random potentials. The diagonalization of this matrix then results in the two eigenvalues

$$E_{1,2}^{(n)} = \exp[\pm \mathrm{arccosh}(\overline{T^{(n)}}/2)] \tag{57}$$

so that one gets an exponentially growing and an exponentially decaying state. There is also an extensive literature on localization in one-dimensional systems based on a different approach,[94,95] which relates the localization length to the density of states. We have chosen to describe here this particular proof because it will help us to discuss another point that will come up later.

In discussing the effects of localization on the properties of one-dimensional systems one should first of all worry about their influence on the normal behavior, because localization is also harmful to normal conductivity. It was recently proved rigorously that a one-dimensional system with localized states has a zero static conductivity independent of the localization length.[96,97] However, in real systems there are several effects that allow the flow of electric current. One is interchain coupling, which must have a delocalizing effect because the possibility of hopping from one chain to another provides alternative paths of propagation and enables the electrons to avoid defects or impurities. As far as we know there is no quantitative study of this effect. Another effect that may cause delocalization is the interaction with phonons; this has recently been studied by Gogolin *et al.*[98] In addition, electron–electron interactions may prevent localization by random lattice potentials. In any case there is ample empirical evidence that the filamentary conductors do conduct and can have metallic conductivity in spite of a significant number of impurities and lattice defects.

In addition to the effect of impurities on the conductivity of the normal state one might expect the impurities to favor or inhibit the tendency for the formation of different types of condensates. This problem was first considered by Zavadovski[99] for the charge-density wave and superconductive phases. Zavadovski considered a model of nonmagnetic, weak scattering centers characterized by a potential strength much smaller than the Fermi energy. The key assumption of this model is that impurity scattering does not mix left- and right-going states. Within this approximation he showed that such impurities have no effect on either the Peierls or superconducting transition temperature. The formal reason for this is that the vertex functions whose divergence determines the transition temperature in the Bychkov *et al.* model[71] [Equation (44)] are merely multiplied by phase factors which do not change the location of the poles. This result can be understood by reference to the transfer matrix formulation of the problem of impurities. The basis of Zavadovski's result is that to first order in λ_n one may neglect the off-diagonal terms of the transfer matrix because in diagonalizing this matrix they only occur in the square. The diagonal terms $1 \pm i\lambda_n$ can be absorbed in the exponential and this simply multiplies the state function by a random phase factor at each step. In one dimension this cannot cause interference and hence the correlation functions of the phases are unchanged by such scattering. The error in this approach is the result of the buildup of the diagonal and off-diagonal terms by multiple scattering which eventually results in localization and thus a mixing of the left- and right-going states.

The effect of impurities on the Peierls transition was considered by Patton and Sham[100] using a diagrammatic approach. They showed that impurities tend to suppress the Peierls transition and argued further that this prevented the three-dimensional ordering of the Peierls state between strands of KCP.

Larkin and Mel'nikov[101] studied the effects of impurities on the Peierls and superconducting states in a system of coupled chains. They concluded that impurities tended to suppress the transition to the Peierls state more strongly than the superconducting state. They also suggested that in the region where both condensates tend to develop (for weak, negative coupling constants) deliberate doping could be used to suppress the formation of a static CDW and thus give the superconducting state.

According to the theorem of Anderson[46] nonmagnetic impurities do not affect superconductivity. For scattering within a chain the only effect such impurities should have is a change of coherence length and a change of T_c due to a reduction in the density of states $N(0)$ by such scattering. By the same reasoning nonmagnetic impurities should have a dramatic effect upon the transition temperature of a triplet state superconductor.[102] For this reason triplet state superconductivity may be suppressed by deliberate doping with nonmagnetic impurities to yield singlet superconductivity.

Finally one should comment on the effects of localization by impurities of Cooper pairs. No work on this has been reported to date, but it is plausible to assume that once Cooper pairs are formed in a certain local region they could be delocalized by Josephson tunneling to neighboring regions. One could conjecture that if the localization length is not so small as to prevent the formation of Cooper pairs then complete delocalization would occur at a sufficiently low temperature.

4.6. Effects of Screening

From the earliest work on polymeric systems it was apparent that the degree to which the Coulomb field of the electrons would be screened in such materials would depend upon the direction considered relative to the conductive spine. Rough estimates were made using the Thomas–Fermi approximation, which suggested that the screening length along the spine would be of the order of 3 Å and appreciably greater in the transverse direction. This was criticized by Kuper,[59] who considered a model system consisting of a single isolated spine with appendant side chains filled with a uniform electron gas. On the basis of this model and a detailed application of the Thomas–Fermi method to it he concluded that no significant screening of the Coulomb field would occur and that oscillations of charge in the side chains would not provide a net attractive interaction. However, these conclusions are not generally true and result from the model, which is a poor description of the molecular system originally suggested. In particular, the latter conclusion of the absence of a net attraction results from the treatment of the charge oscillations in the side chains as plasma oscillations in the electron gas and the violation of the condition requiring the separation of the conduction and excitonic system. Treating the electrons

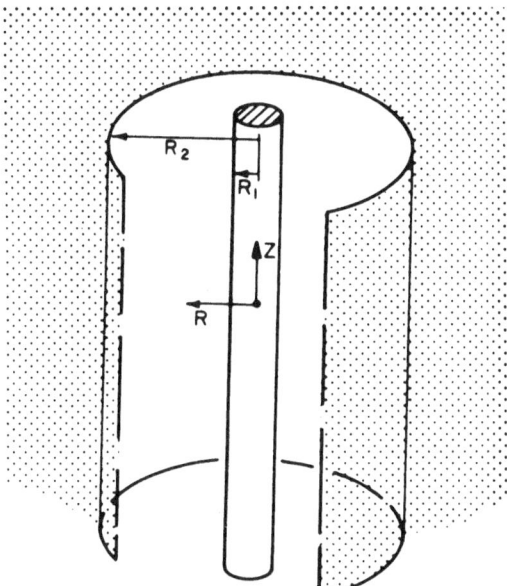

Figure 16. Two-zone geometry used by Davis (Reference 104) to calculate screening in filamentary compounds.

in a polarizable molecule as a plasma grossly overestimates the energy of characteristic absorption of the molecule. Secondly, the absence of substantial screening in the model at long wavelengths results from the filament being treated in isolation. In any real material one would not have an isolated filament but rather a compound with many such molecular filaments parallel to one another. Dzyaloshinskii and Kats[103] showed that in such a configuration screening in the long wavelength limit approaches that of a homogeneous metal. This has the important effect of weakening the Coulomb interaction at small momenta, where, as we shall show, the exciton interaction makes its maximum contribution.

Numerical estimates have been made of the screening in filamentary compounds of the Krogmann type by Davis.[104] He considered a model in which the contribution of neighboring filaments were simulated by replacing them by an outer cylinder of a low-density electron gas separate from the central filamentary spine as shown in Figure 16. Using the Thomas–Fermi approximation he found nearly isotropic screening in the region near the spine of the usual form $V(r) = (e^2/r)\exp(-\kappa r)$ but with a screening length $1/\kappa$ substantially greater than that for a bulk metal. Values for this screening parameter were calculated for a model designed to represent the partially oxidized Pt-chain compounds. His results for KCP are given in Figure 17. In Davis' model only static screening was considered, and clearly this is insufficient for the full calculation of the net effective interaction over an energy range of the order of the exciton energy ≈ 2 eV.

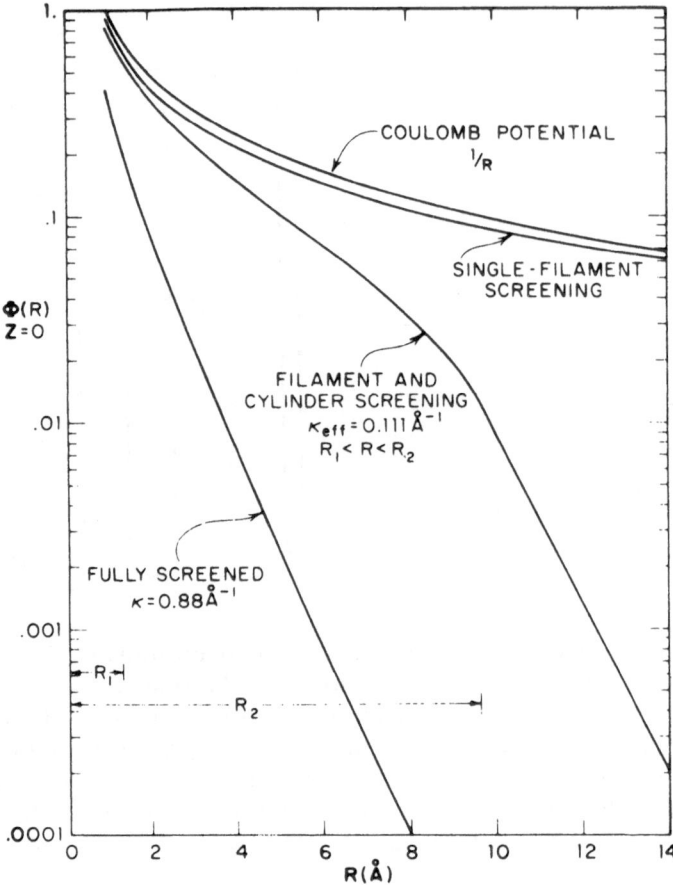

Figure 17. Results obtained by Davis (Reference 104) for the screened potential $\phi(R)$ along a line perpendicular to the filament axis ($Z = 0$) for the model system illustrated in Figure 16.

Bush[105] has extended this model to calculate the dynamic screening of the Coulomb field both in the Thomas–Fermi and the random-phase approximation. He finds that because of the large density of states at low momenta in a one-dimensional metal, extremely strong screening occurs at low Fermi energies and thus at low electron densities. In fact, over a limited range the screening in one dimension is inversely proportional to the electron density. Also, the frequency dependence of the screening has a considerable amount of structure in it due to the various plasmon modes of the interacting filaments. He showed that if the response function of a given spine is known, one may compute the lattice sums exactly to calculate the response of the sample as a whole without resorting to the concentric cylinder approximation of Davis.

Thus, while the essential problem of screening in the filamentary compounds has been solved, many details remain to be clarified. In particular one would expect that structural imperfections, a Peierls distortion, or interrupted strands would severely modify the low-frequency screening in these compounds. It is not clear, though, whether this would have any impact on factors responsible for superconductivity because such materials would still have a large real dielectric constant[106] and the long-wavelength electron–electron interaction would be severely reduced without necessarily being *screened* in the strict sense. In fact, in the model discussed in Section 5 one finds that the principal reduction in the Coulomb field does not come from metallic screening but from the dielectric function of the organic material within which the conductive chain is located. The complex behavior of the dynamic screening in filamentary compounds does not allow one to draw many general conclusions on the strength of the screening but rather requires one to consider each class of materials individually.

In an excitonic system both the polarizable ligands and the conduction electrons would contribute to the screening and each would affect the other, so one really needs to solve the problem of screening in a self-consistent manner and not by treating the screened excitonic interaction and the screened Coulomb field as simply additive interactions. In calculations of the effective interaction in these materials the greatest uncertainty lies in our lack of adequate knowledge of the screening. Measurements that could reveal the strength of the screening by optical, electron energy loss, NMR, or other experiments would be of great value in establishing an experimental reference point against which to compare the theoretical predictions.

The related problem of screening in layer compounds or two-dimensional structures has also been studied in some detail. Interest in this lies in understanding of the inversion or accumulation layers[107] in semiconductors and surface electrons on liquid He or Ne.[108] Again screening is virtually absent in any single isolated layer because the interaction is dominated by contributions from the electromagnetic fields which extend out into the surrounding nonconductive medium. However, in layered structures of many layers one again recovers three-dimensional-like screening with the Yukawa form $(e/r) e^{-\kappa r}$ for the screened interaction, but the screening parameter κ differs from the bulk Thomas–Fermi value. Visscher and Falicov[109] have considered the layered structures. A complete treatment of the full electrodynamic behavior and local screening of both the isolated sheet and the periodic array of such sheets has been given by Fetter.[110,111] The plasma oscillations differ from those of an isotropic bulk material with one branch having a finite energy and all others tending to zero in the long-wavelength limit. Knowledge of this screening would be important in considering any proposed two-dimensional excitonic system.

5. Real Models

The preceding theoretical arguments lead us to the conclusion that there are no known matters of principle that could prevent an excitonic system from superconducting at temperatures at least as high as the Debye temperature. Whether or not this can be realized in practice then depends upon the strength of the coupling that can be realized in a given system. To examine this point one has to be specific about the systems. Thus far only two systems have been discussed in any detail taking into account recent developments in theory. These are the two-dimensional layered system discussed by Allender et al.[4] and a filamentary model discussed by Davis, Gutfreund, and Little (DGL).[33] We have mentioned earlier that the two-dimensional systems look less attractive now because of problems of exchange and the probable failure of the Migdal approximation in the systems considered thus far, and perhaps the most serious problem is the weakness of the interaction. In all systems where reasonably detailed calculations have been carried out it appears that the interaction is insufficient to overcome the Coulomb repulsion.[112,113] The only two-dimensional systems where a reasonably strong excitonic interaction appears to be possible and that would be free of exchange problems are those with a very small Fermi energy.[113] The reason for this is essentially one of a matter of scale. In order for the excitonic system to have modes of low enough energy to give a coupling constant of sufficient size the excitons must have physical dimensions of the order of $10\,\text{Å}$ or thereabouts. For these modes to couple to a significant fraction of the electrons, the Fermi wavelength must be of somewhat similar magnitude. For one dimension, only scattering processes involving very long ($q \approx 0$) and very short ($q \approx 2k_F$) wavelengths are important in scattering between points on the Fermi surface, while in two dimensions all momenta up to $2k_F$ are important. So for the excitons to have a significant effect on superconductivity in the latter case, the coupling constant for all momenta up to $2k_F$ must be large, and this is only possible if k_F is small.

To get a better idea whether high-temperature superconductivity could occur in such two-dimensional systems one needs to define the system more precisely than has been done to date and then to examine in detail the questions discussed above together with questions relating to the screening of the Coulomb field, the strength of the coupling constants, and the stability of the system etc. Based on the calculations done to date[112,113] and the absence of any indication of an excitonic contribution to superconductivity in experiments reported thus far,[5] we feel the prospects for finding such superconductivity in the two-dimensional systems are slim. On the other hand the prospects of satisfying the various theoretical criteria for superconductivity in filamentary systems appear to

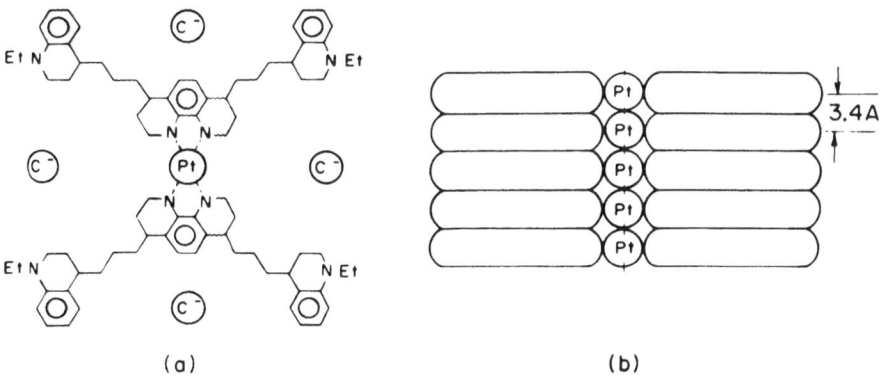

Figure 18. Proposed model of the structure of an excitonic superconductor. (a) Top view of square planar phenanthroline-dye ligands complexed to Pt. Double bonds in the chromophore are omitted for simplicity. "Et" stands for ethyl. (b) Side view of chain.

be much better. The reason for this is simply that one can pack more of the excitonic material more closely round a conducting thread than round a conducting plane and thus in one dimension a stronger net attraction can be expected. In addition, one can satisfy the condition for the validity of the Migdal approximation, and for the stability of the superconducting phase, furthermore, the effects of fluctuation and localization, while important, do not appear to be disastrous to the superconducting state in these systems. In the following section we discuss the essential elements of the model proposed by DGL stressing the essential physical requirements of the model, sketching the method by which T_c was calculated, and discussing the elements of the chemistry that would be required to prepare material of this general class.

5.1. Model of a Filamentary Excitonic Superconductor

The model that was proposed is illustrated in Figure 18. It consists of a chain of Pt atoms surrounded by a sheath of highly polarizable dyelike molecules. Each unit of the conductive chain of Pt atoms is a bis-phenanthroline ligand system complexed to the metal atom. To each of the 1–10 phenanthroline ligands are attached two cyanine-dye chromophoric units at the 4 and 7 positions.

In the proposed structure the polarization of the chromophore results in the movement of a positive hole from one nitrogen atom remote from the Pt site to the nitrogen adjacent to the Pt. Because of the large movement of charge a strong electron–exciton interaction can be expected.

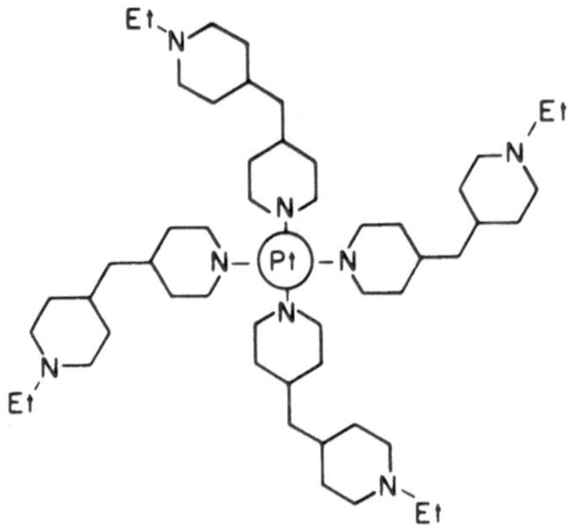

Figure 19. Simplified version of the structure of Figure 18 for which detailed calculations were presented by DGL.

The d_{z^2} orbitals of the Pt atoms of the chain overlap with one another to give a linear conductive pathway. Because of the repulsion between the π electrons of the bulky ligands the Pt atoms along the chain can only come to within about 3.4 Å of one another. It has been assumed that sufficient overlap occurs at this separation to yield a conductive chain. A justification for this is given later.

Four counter ions (Cl) for each square-planar ligand system are required and the Pt chain needs to be partially oxidized to give a partially filled d_{z^2} conduction band as in $K_2Pt(CN)_4Br_{0.3} \cdot 3H_2O$.‡ There is some question whether this is possible in such a system, but this and other variations of the model will be discussed later. Instead of working with the large phenanthroline groups a skeleton structure of the chromophore units alone was used, as illustrated in Figure 19. Calculations were presented on this simplified model. This gave the essence of the results of calculations of the more complex system done earlier. The transition temperature T_c was calculated by the method adopted from Kirzhnits *et al.*[36] This method applies to a weak coupling superconductor and results in the BCS-like equation for the gap function $\phi(\mathbf{k})$ of Equation (19). The interactions were represented by the kernel of Equation (29).

5.1.1. The Interactions

The calculation of the kernel $U(p, k)$ required a knowledge of the electron-band energies $\xi(p)$, the exciton band energies $E_\alpha(q)$, the coupling constants $Q_\alpha(q)$, and the Coulomb interaction between two electrons on the spine. The actual values and variations of the first two quantities where shown by DGL to have only a minor effect on the results within a wide range of reasonable parameters. The interaction energies, however, are crucial and had to be estimated in a reliable way.

‡ The paper by DGL contains an error here in stating that six negatively charged counter ions would be required instead of four.

The electron–exciton coupling parameter has the form

$$Q_\alpha(q) = \langle 1_\alpha q, k - q | V | 0, k \rangle \tag{58}$$

which corresponds to the scattering of an electron from momentum k to $k - q$, accompanied by the creation of an exciton of momentum q and band index α. When the electron states on the spine are described in the tight-binding approximation and the exciton-band states by a linear combination of terms in which only a single molecule is in an excited state, one obtains (DGL)

$$Q_\alpha(q) = \left(\frac{2}{N}\right)^{1/2} \sum_{m,n,\nu} \int |\phi(r_1)|^2 V(r_1, r_2) c_{\alpha n}^\nu(q) e^{iqR_m} \rho_\nu(r_2, R_{nm}) \, d^3r_1 \, d^3r_2 \tag{59}$$

where $|\phi(r_1)|^2$ is the density of the electron on the spine, $\rho_\nu(r_2, R_{nm})$ is the transition density between the ground state and the excited state ν of the dye at site n in the mth unit cell, and $c_{\alpha n}^\nu$ are the coefficients of the exciton band states. The method used for the calculation of ρ was a simple extension of the Pariser–Parr–Pople semiempirical technique,[114,115] which had previously been shown to give accurate values for the molecular energy states and the oscillator strengths.[116] The result for the lowest excitation band of the pyridine cyanine is represented in Figure 20. It shows a clear oscillating dipole pattern. The typical dependence of $|Q_\alpha(q)|^2$ upon q for the mode α which couples most strongly to the spine is illustrated in Figure 21. Its distinctive feature is the sharp falloff in $|Q_\alpha(q)|^2$ with increasing q. This occurs at values of $q \approx 1/d$, where d is of the order of the distance of the nearest terminal group of the dye from the spine as discussed earlier. This single low-lying level provided the excitonic interaction in the model. The other higher excitations of the ligands contributed to ρ_c of Equation (21), as a correction to the Coulomb interaction.

The Coulomb interaction was found to be the hardest quantity to estimate in a reliable way. The starting point was the parameters γ_n, which measure the bare interaction between electrons on two atoms on the spine separated by a distance $r_n = na$, where a is the interatomic spacing. The Nishimoto–Mataga[117] form was used for these parameters:

$$\gamma_n = \frac{e^2}{r_n + b} \tag{60}$$

The parameter b, which corresponds to the interaction of two electrons on the same atom, was derived from experimental values of the ionization energy and the electron affinity, and for platinum this yielded $\gamma_0 = 6.03$ eV giving $b = 2.4$ Å. The bare interaction is modified by the screening of the electrons in the same filament and in neighboring filaments. In addition, it

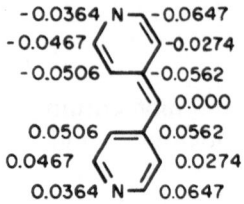

-0.0364 N -0.0647
-0.0467 -0.0274
-0.0506 -0.0562
0.000
0.0506 0.0562
0.0467 0.0274
0.0364 N 0.0647

Figure 20. Linear-combination-of-atomic-orbitals (LCAO) calculated values of the transition density for the principal low-lying absorption band for the pyridine cyanine.

is reduced by the dielectric constant of the surrounding organic medium. The first of these contributions to screening was estimated on the basis of the calculation of Davis[104] on filamentary compounds like KCP. His results were used to obtain the screened parameters

$$\bar{\gamma}_n = e^2 \frac{\exp[-\lambda(r_n + b)]}{(r_n + b)}, \qquad \lambda = 0.14 \text{ Å}^{-1} \tag{61}$$

The Fourier transform of the partially screened interactions [suitably corrected at large q for errors introduced by the discrete atomic representation (see DGL)] are plotted in Figure 22 (upper curve).

The higher-energy excitations were shown to reduce the long-range Coulomb interaction by about a factor of 2. This is shown in the lower curve of Figure 22. For the particular model used this reduction comes from the π electrons alone. The inclusion of the σ-electron contribution would reduce this interaction further. For $q \approx 0$ the total reduction factor

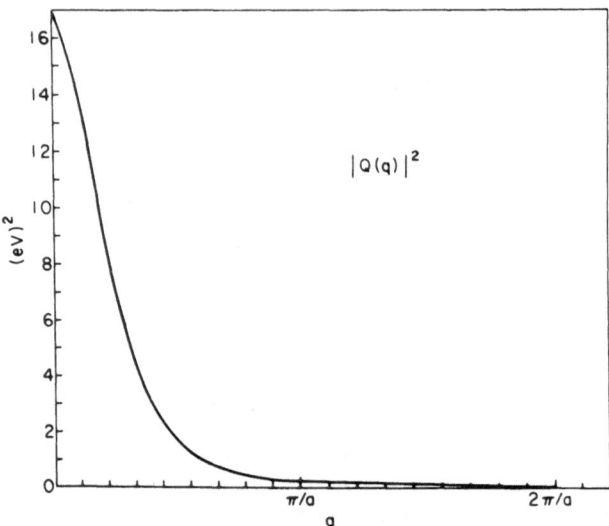

Figure 21. Calculated electron–exciton interaction $|Q(q)|^2$ as a function of momentum transfer q.

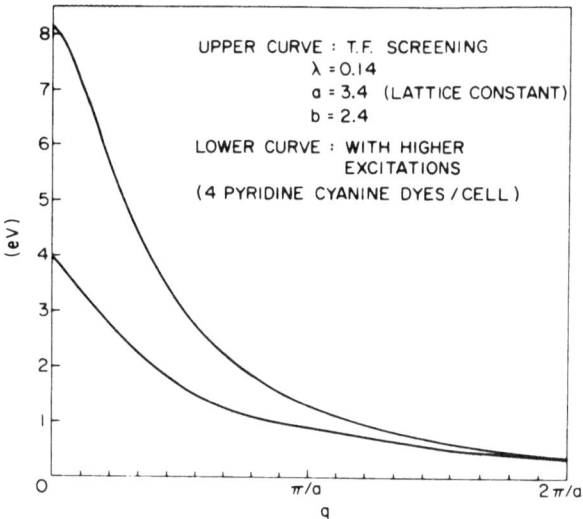

Figure 22. Screened Coulomb interaction calculated using Thomas–Fermi screening due to electrons in the same and neighboring filaments (upper curve), and with the addition of dielectric screening from the neighboring organic environment (lower curve).

should be of the order of the dielectric constant ε of the surrounding medium. This can be estimated from the refractive index n of a similar unsaturated hydrocarbon through the relation $n^2 = \varepsilon$. From this estimate one would expect the Coulomb interaction at small q to be reduced by about 2.6, so perhaps (DGL) have overestimated the Coulomb repulsion by as much as 30%.

5.1.2. Calculation of T_c

Using the geometry of the system illustrated in Figures 18 and 19 DGL calculated numerically the transition temperature T_c. It proved to be convenient to use, instead of Equation (19), the zero-temperature equation for the gap

$$\phi(p) = -\int_{-\pi/a}^{\pi/a} \frac{dk}{4\pi} \frac{U(p-k)\phi(k)}{[\xi^2(k)+\phi^2(k)]^{1/2}} \tag{62}$$

and to obtain T_c from the gap at $T = 0$ through the relation

$$kT_c = 3.5\phi(k_F)\Big|_{T=0} \tag{63}$$

For singlet superconductivity the spins of the electrons of the pair are antiparallel and $\phi(k) = \phi(-k)$. Equation (62) can thus be written as

$$\phi^s(p) = -\frac{1}{4\pi}\int_0^{\pi/a} dk \frac{[U(p,k)+U(p,-k)]\phi^s(k)}{\{\xi^2(k)+[\phi^s(k)]\}^{1/2}} \tag{64}$$

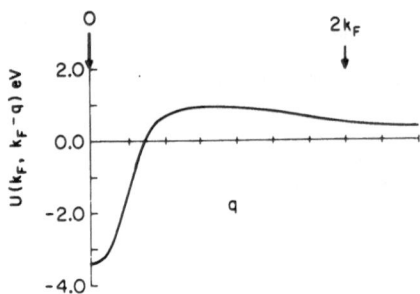

Figure 23. Plot of $U(k_F, k_F - q)$ vs. q for the model system. The major contributions to superconductivity come from the interaction at $q = 2k_F (g_1)$ and at $q = 0$ (g_2).

On the other hand, for triplet superconductivity the orbital symmetry requires $\phi(k) = -\phi(-k)$ and Equation (62) gives

$$\phi'(p) = -\frac{1}{4\pi} \int_0^{\pi/a} dk \frac{[U(p, k) - U(p, -k)]\phi'(k)}{\{\xi^2(k) + [\phi'(k)]^2\}^{1/2}} \qquad (65)$$

It should be noted that because of the denominator, the integral in both cases is dominated by the contribution of the kernel at $k = k_F$ and thus by the terms $U(p_F, k_F)$ and $U(p_F, -k_F)$. In Figure 23 we show a plot of $U(p_F, k_F - q)$ vs. q calculated for the model of Figure 19. We see that $U(p_F, k_F)$ is negative, owing to the strong exciton contribution of Equation (29) resulting from the large value of $|Q_\alpha(q)|^2$ near $q = 0$. On the other hand, $U(p_F, -k_F)$ is positive because of the dominance of the Coulomb repulsion for large momentum $(2k_F)$ transfers. Because of this the combination $[U(p_F, k_F) - U(p_F, -k_F)]$, which occurs in the triplet case, yields a more attractive contribution than does $[U(p_F, k_F) + U(p_F, -k_F)]$, which occurs in the singlet case. Consequently we expect a larger zero-temperature gap and T_c for triplet-state superconductivity compared to that for the singlet case. This is borne out by the numerical result for the two gaps illustrated in Figure 24.

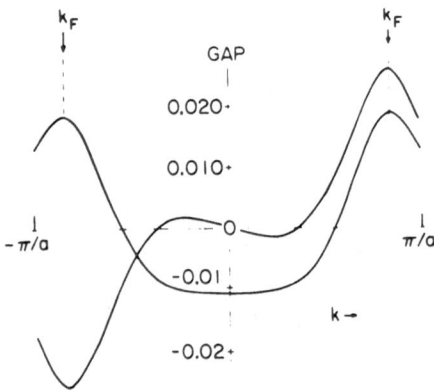

Figure 24. Calculated results for the singlet (symmetric) and triplet (antisymmetric) gap functions for the model system of Figure 1.

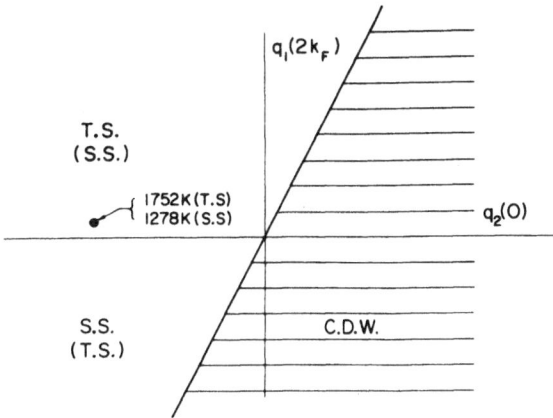

Figure 25. g_1–g_2 plane showing region of charge-density wave instabilities (CDW) and singlet (triplet) superconductivity S.S. (T.S.). The calculated values of T_c for the model are indicated.

This also provides justification for using a single-channel calculation rather than a coupled two-channel approach involving the competition between both superconductivity and charge-density wave condensates, for the representative point on the g-ology plane can be seen from Figure 25 to have $g_1(q = 2k_F)$ small and positive, and $g_2(q = 0)$ large and negative. This puts it well away from the line $g_2 = 2g_1$ which separates the charge-density wave region from the superconducting region and thus in a region where the competition between them should be negligible.

5.1.3. Results

The value of T_c for singlet superconductivity was calculated for bandwidth between 2 and 4 eV and for Fermi energies corresponding to 1/2 and 5/6 filling of the band. These values of T_c ranged between 100 and 1280°K, the larger values occurring with increasing density of states at the Fermi surface. The value of T_c for triplet-state superconductivity was always found to be higher, in agreement with the predictions of g-ology and our discussion of Section 4.2. DGL also noted that no superconducting solution could be found for a similar model with only two polarizable ligands per Pt atom instead of four or when the ligands were moved so that the N atoms were 3 Å rather than 2 Å from the Pt.

Again one should note that these calculations are of the mean-field transition temperature. This is not the temperature of any true phase transition but the temperature at which one can expect the pair condensate to begin to grow and for marked changes in the conductivity to occur. Whether or not bulk superconductivity would occur near this temperature would depend on interchain coupling and the degree of perfection of the individual strands. None of these factors are considered by DGL.

While this calculation was done for a specific model, one can draw some general conclusions from it. First it shows how sensitive T_c is to the distance between the polarizable ligands and the spine. This follows from the form of Equation (59) for the coupling term $Q_\alpha(q)$ and the fact that the superconductivity is determined by the difference between a term proportional to $|Q_\alpha(q)|^2$ and the Coulomb term [Equations (20) and (21)]. Both are large terms. If the exciton term is larger, one gets high-temperature superconductivity, but a small change in $Q_\alpha(q)$ can reverse the inequality and give $T_c = 0$.

Second, it shows that a very dense packing of the polarizable ligands around the spine is essential. Reducing the number of polarizable groups from 4 to 2 weakens the net $Q(q)$ too much to give a finite value for T_c.

Third, the value of $Q(q)$ would be greatly reduced if the conduction-electron charge were not concentrated in a small region close to the polarizable ligands. For this reason it does not appear possible to obtain excitonic superconductivity in a TCNQ type of salt where the π electrons are distributed over a relatively large area.

Fourth, we see that a comparatively narrow band with its large density of states at the Fermi surface is helpful to superconductivity and results in large values for T_c.

The specific model discussed above suffers from a number of shortcomings in the chemistry. First, it appears most unlikely that one could form the bis-phenanthroline structure with a charge of +4 on the ligand system and have it stack. On the other hand, the doubly charged structure would be formed more readily, but *it* would have a significantly smaller transition charge and this would lead to weaker coupling. Second, the model requires the $d_{z^2} - p_z$ band to be partly filled, so the Pt chain must be oxidized. In this process the relatively sensitive double bonds in the chromophore units must not be attacked. This is a general problem and attention must be given to the relative oxidation and reduction potentials of the spine and ligands. Third, the ligand system may not stack in the form postulated and then no metal–metal interaction would occur. This too is a general problem and a better understanding of the factors that are important for stacking is essential for further progress. Fourth, one is relying on a significant metal interaction between the metal atoms even at a metal–metal separation of 3.4 Å determined by the van der Waal's contacts between the ligands. This is substantially larger than the metal–metal separation in KCP. Earlier theoretical estimates,[118,119] however, and a recent band calculation[120] along the lines of an earlier calculation on KCP[121] confirm that a substantial interaction would remain, so this is perhaps less of a problem than was originally thought.

There are a number of alternative structures that retain the essential elements of the above model but avoid some of its limitations. One could replace the charged cyanine chromophores with uncharged merocyanine chromophores. This should increase the probability of stacking and would allow one to compensate for the lack of symmetry between the Pt and ethyl terminal groups. The latter has the effect of reducing the oscillator strength of the transition.[122] Alternately the use of an anion–cation structure as in magnus green salt would favor stacking because of ionic forces.

5.2. Discussion

It is evident from the above that the most important factor in determining T_c is $Q_\alpha(q)$. While it is difficult to estimate accurately the Coulomb term, whatever it is, it can be expected to be more nearly the same for different systems as it is determined, at least in the long wave limit, by the gross features of the system. On the other hand it is the microscopic properties of $Q_\alpha(q)$ that determine its behavior in this same limit. Thus $Q_\alpha(q)$, particularly in the region $q \approx 0$, and $E_\alpha(q)$ in the denominator of $U(p, k)$ [Equation (29)] will determine whether or not a given system will superconduct. For this term to be large the conduction electron density $|\phi(r_1)|^2$ in the spine must be near the positive (negative) part of the transition density ρ_ν and far from the negative (positive) part. [See Equation (59).] In this expression $V(r_1, r_2)$ is the partially screened Coulomb interaction. To maximize $Q_\alpha(q)$, ρ_ν should be concentrated at the extreme end of the polarizable ligand and oriented perpendicular to the chain axis. In the above model the near part of ρ_ν is concentrated in the ring structures adjacent to the Pt. The prospects of getting high-temperature superconductivity would be enormously enhanced if a structure could be devised in which even closer approach of the transition charge and the spine could be achieved. One possibility is the use of some heteroatom immediately adjacent to the metal atom of the spine to concentrate ρ_ν in its neighborhood. An even more promising possibility may exist in the d^8 compounds where the spine atom could act both as the conducting chain and as the terminal group of the polarizable ligand. This may appear to violate much of what we have said in regard to exchange in Section 3.1. However, this is not necessarily so, for if the π electrons of the ligand system are conjugated to the d_{xz} and d_{yz} orbitals of, say Pt, then these electrons will be approximately orthogonal to all the $5d_{z^2} - 6p_z$ states of the conduction band. The orthogonality in this case is maintained by the different symmetry of the $d_{xz}(d_{yz})$ states and both the d_{z^2} and p_z orbitals across the plane of the ligand. So even though the hybridization of the $d_{z^2} - p_z$ states will vary through the band, orthogonality is maintained and the exchange terms will vanish. For the $d_{xz}(d_{yz})$ orbitals to participate with the ligands in this way the Fermi energy would have to intersect both of these levels and the d_{z^2} level. If this situation could be realized in some d^8 system, it would open up a rich field of structures free of many of the obvious chemical and steric difficulties of the model of DGL.

6. Summary

In all superconductors known at present the phonon interaction is believed to be responsible for the superconducting state. In principle, however, other interactions are capable of yielding a similar state, and some might do so at relatively high temperatures. In this review we have examined in detail the possibility of the exciton interaction leading to a relatively high-temperature superconducting phase. We find that it does appear capable of doing so. However, we find that a simple application of the theory of superconductivity to this problem cannot be justified. Instead, one must examine or reexamine each of the basic assumptions customarily used in this theory. These involve questions of stability, vertex corrections, and the form of the gap equation, together with an *ab initio* determination of the interactions. In addition, certain other factors must be considered that are unique to the exciton mechanism, such as the effects of exchange between the exciton and conduction subsystems, and the effects of phonons on the exciton interaction. We have discussed each of these in turn.

We have argued that the most promising systems in which one could hope to find excitonic superconductivity are filamentary or one-dimensional systems. Such systems, because of their limited dimensionality, do not exhibit any true phase transitions, but instead their behavior is dominated by fluctuations that persist down to $T = 0$. In spite of this, high conductivity or other partially ordered phases can occur for certain values of the electron–electron coupling constants. We have discussed these factors and the effect of coupling between the filaments upon these phases. In addition, the effects of impurities and random faults in the filaments have been considered.

Finally, one specific model of a proposed excitonic superconductor is examined, which, if it could be synthesized, can be expected to have a high superconducting transition temperature. We believe that such high transition temperatures should occur and that all known objections to the use of this mechanism to obtain superconductors with substantially higher transition temperatures than those known today have been answered. Several important theoretical questions remain and the field can be expected to continue to generate a rich source of novel problems. The problem of the design of a tractable material and its synthesis which would bring such materials to reality remains a challenge to both the physicist and chemist.

ACKNOWLEDGMENTS

We are indebted to the National Science Foundation (Grant No. DMR-74-00427-A03), the National Aeronautical and Space Agency (Contract No. JPL 953752), and US–Israeli Binational Science Foundation (Grant No. 215) for financial support during the period when much of the above analysis was done. In addition we owe our thanks to the NATO Scientific Affairs division (Grant No. 918) for travel support which made this collaborative work possible. To Professor Heimo Keller, of the Anorganische Chemisches Institute der Universität, Heidelberg, we wish to express our gratitude for the warm hospitality of his Institute, where the review was completed.

References

1. L. B. Coleman, M. J. Cohen, D. J. Sandman, F. G. Yamagishi, A. F. Garito, and A. J. Heeger, Superconducting fluctuations and the Peierls instability in an organic solid, *Solid State Commun.* **12**, 1125–1132 (1973).
2. W. A. Little, Possibility of synthesizing an organic superconductor, *Phys. Rev.* **134**, A1416–A1424 (1964).
3. V. L. Ginzburg, Concerning surface superconductivity, *Zh. Eksp. Teor. Fiz.* **47**, 2318–2320 (1964) [*Sov. Phys.-JETP* **20**, 1549–1550 (1965)].
4. D. Allender, J. Bray, and J. Bardeen, Model for an exciton mechanism of superconductivity, *Phys. Rev. B* **7**, 1020–1029 (1073).
5. M. Strongin, A search for excitonic superconductivity, *Solid State Commun.* **14**, 88 (1974).
6. Proceedings International Conference on Organic Superconductors, W. A. Little (ed.) *J. Polymer. Sci. Pt. C*, No. 29 (1970).
7. L. V. Keldysh, Superconductivity in nonmetallic systems, *Usp. Fiz. Nauk* **86**, 327–333 (1965) [*Sov. Phys. Usp.* **8**, 496–500 (1965)].
8. V. L. Ginzburg, The problem of high temperature superconductivity, *Contemp. Phys.* **9**, 355–374 (1968).
9. V. L. Ginzburg, The problem of high-temperature superconductivity. II, *Usp. Fiz. Nauk* **101**, 185 (1970) [*Sov. Phys. Usp.* **13** (3), 335–352 (1970)].
10. V. L. Ginzburg, The problem of high-temperature superconductivity, *Ann. Rev. Mater. Sci.* **2**, 663–696 (1972).
11. V. L. Ginzburg and D. A. Kirzhnits, On the problem of high temperature superconductivity, *Phys. Rep. Phys. Lett. C* (Netherlands) **4**, 343–356 (1972).
12. L. N. Bulaevsky, V. L. Ginzburg, D. I. Khomskii, D. A. Kirzhnits, Ju. V. Kopaev, E. G. Maximov, G. F. Zarkov, and G. P. Molulevitch, The problem of high temperature superconductivity, Parts I and II, Preprints N45 and N74, Lebedev Physical Institute, Moscow (1974).

13. J. J. Andre, A. Bieber, and F. Gautier, Physical properties of highly anisotropic systems: radical ion salts and charge transfer complexes, *Ann. Phys. (Paris)* **1**, 145–256 (1976).

14. E. B. Yagubskii and M. L. Khidekel, High temperature exciton superconductivity: synthetic aspects, *Russian Chem. Rev.* **41**, 1011–1026 (1972).

15. I. F. Shchegolev, Electric and magnetic properties of linear conducting chains, *Phys. Status Solidi (A)* **12**, 9–45 (1972).

16. H. R. Zeller, Electronic properties of one-dimensional solid state systems, *Adv. Solid State Phys.* **13**, 31–58 (1973).

17. J. S. Miller and A. J. Epstein, One-dimensional inorganic complexes, in *Progress in Inorganic Chemistry*, Vol. 20, pp. 1–151, Stephen J. Lippard (ed.), John Wiley & Sons, New York (1976).

18. H. J. Keller, *Low-Dimensional Cooperative Phenomena*, NATO-ASI Series B7, Plenum Press, New York (1975).

19. H. J. Keller, *Chemistry and Physics of One-Dimensional Metals*, NATO-ASI Series B25, Plenum Press, New York (1977).

20. B. W. Roberts, Superconductivity, in *Handbook of Chemistry and Physics*, 48th Ed., pp. E75–E90, R. C. Weast and S. M. Selby (eds.), The Chemical Rubber Co., Cleveland, Ohio (1967).

21. J. Bardeen, L. N. Cooper, and J. R. Schrieffer, Theory of superconductivity, *Phys. Rev.* **108**, 1175–1204 (1957).

22. S. V. Tiablikov and V. V. Tomachev, The interaction of electrons with lattice vibrations, *Zh. Eksp. Teor. Fiz.* **34**, 1254–1257 [*Sov. Phys. JETP* **34**, 867–869 (1958)].

23. L. R. Testardi, Structural instabilities in A-15 compounds, *Rev. Mod. Phys.* **47**, 637–648 (1975).

24. J. R. Gavaler, Superconductivity in Nb–Ge films above ·22 K, *Appl. Phys. Lett.* **23**, 480–482 (1973).

25. W. L. McMillan, Transition temperature of strong-coupled superconductors, *Phys. Rev.* **167**, 331–344 (1968).

26. P. Morel and P. W. Anderson, Calculation of the superconducting state parameters with retarded electron–phonon interaction, *Phys. Rev.* **125**, 1263–1271 (1962).

27. G. Rickayzen, in *Superconductivity*, Vol. 1, pp. 72–115, R. D. Parks (ed.), Marcel Dekker, New York (1969).

28. W. A. Little, *J. Polymer Sci. Pt. C* **29**, 17–26 (1970).

29. B. T. Geilikman, A possible mechanism for superconductivity in alloys, *Zh. Eksp. Teor. Fiz.* **48**, 1194–1197 (1965) [*Sov. Phys.-JETP* **21**, 796–798 (1965)].

30. M. L. Cohen and P. W. Anderson, in *Superconductivity in d- and f-Band Metals*, D. H. Douglass (ed.), AIP, New York (1972).

31. J. C. Phillips, Superconductivity mechanisms and covalent instabilities, *Phys. Rev. Lett.* **29**, 1551–1554 (1972).

32. D. Pines and P. Nozieres, *The Theory of Quantum Liquids*, W. A. Benjamin, New York (1966).

33. D. Davis, H. Gutfreund, and W. A. Little, Proposed model of a high temperature excitonic superconductor, *Phys. Rev. B* **13**, 4766–4779 (1976).

34. J. R. Schrieffer, *Theory of Superconductivity*, W. A. Benjamin, New York (1964).

35. S. Engelsberg and J. R. Schrieffer, Coupled electron–phonon system, *Phys. Rev.* **131**, 993–1008 (1963).

36. D. A. Kirzhnits, E. G. Maximov, and D. I. Khomskii, The description of superconductivity in terms of dielectric response function, *J. Low Temp. Phys.* **10**, 79–93 (1973).

37. D. B. Chesnut, Exciton renormalization in conducting molecular solids, *Mol. Cryst.* **1**, 351–375 (1966).

38. P. M. Chaikin, A. F. Garito, and A. J. Heeger, Excitonic polarons in molecular crystals, *Phys. Rev. B* **5**, 4966–4969 (1972).

39. R. A. Bari, Excitonic polarons in molecular solids, *Phys. Rev. Lett.* **30**, 790–794 (1973).
40. Yu. M. Balkarei and D. I. Khomskii, Lattice stability in the phononless mechanism of superconductivity, *JETP Lett.* **3**, 181–183 (1966).
41. J. P. Hurault, Superconductivity in small crystallites, *J. Phys. Chem. Solids* **29**, 1765–1772 (1968).
42. V. L. Ginzburg, Manifestation of the exciton mechanism in the case of granulated superconductors, *JETP Lett.* **14**, 396–399 (1971).
43. P. W. Anderson, Editorial comment, *Physics* **2**, 151 (1966).
44. G. Bergmann and D. Rainer, The sensitivity of the transition temperature to changes in $\alpha^2 F(\omega)$, *Z. Phys.* **263**, 59–68 (1973).
45. J. Appel, Role of thermal phonons in high temperature superconductivity, *Phys. Rev. Lett.* **21**, 1164–1167 (1968).
46. P. W. Anderson, Theory of dirty superconductors, *J. Phys. Chem. Solids* **11**, 26–30 (1959).
47. K. Maki, Gapless superconductivity, in *Superconductivity*, Vol. 2, pp. 1035–1105, R. D. Parks (ed.), Marcel Dekker, New York (1969).
48. P. B. Allen, Repulsive effect of low frequency phonons on superconductivity, *Solid State Commun.* **12**, 379–383 (1973).
49. A. E. Karakozov, E. G. Maksimov, and S. A. Mashkov, Effects of the frequency dependence of the electron–phonon interaction spectral function on the thermodynamic properties of superconductors, *Sov. Phys.-JETP* **41**, 971–976 (1976).
50. J. R. Schrieffer, D. J. Scalapino, and J. W. Wilkins, Effective tunneling density of states in superconductors, *Phys. Rev. Lett.* **10**, 336–339 (1963).
51. H. A. Kramers and G. H. Wannier, Statistics of the two-dimensional ferromagnet. Part I, *Phys. Rev.* **60**, 252–276 (1941).
52. L. D. Landau and E. M. Lifshitz, *Statistical Physics*, Pergamon Press, London (1959).
53. L. Van Hove, Sur l'intégrale de configuration pour les systèmes de particles a une dimension, *Physica* **16**, 137–143 (1950).
54. R. A. Ferrell, Possibility of one-dimensional superconductivity, *Phys. Rev. Lett.* **13**, 330–335 (2964).
55. T. M. Rice, Superconductivity in one and two dimensions, *Phys. Rev.* **140**, A889–A1891 (1965).
56. P. C. Hohenberg, Existence of long range order in one and two dimensions, *Phys. Rev.* **158**, 383–386 (1967).
57. M. Weger and I. B. Goldberg, Some lattice and electronic properties of the β-tungstens, in *Solid State Physics*, Vol. 28, pp. 1–178, F. Seitz *et al.* (eds.), Academic Press, New York (1973).
58. B. T. Matthias, Higher temperatures and instabilities, in *Superconductivity in d- and f-Band Metals*, pp. 367–375, D. H. Douglass (ed.), AIP, New York (1972).
59. C. G. Kuper, Little's proposal for a superconducting organic polymer, *Phys. Rev.* **150**, 189–192 (1966).
60. R. E. DeWames, G. W. Lehman, and T. Wolfram, Superconductivity in macroscopic one dimensional systems, *Phys. Rev. Lett.* **13**, 749–750 (1964).
61. W. A. Little, Decay of persistent currents in small superconductors, *Phys. Rev.* **156**, 396–403 (1967).
62. M. Tinkham, The electromagnetic properties of superconductors, *Rev. Mod. Phys.* **46**, 587–596 (1974).
63. K. Huang, *Statistical Mechanics*, p. 203, John Wiley and Sons, New York (1963).
64. J. S. Langer and V. Ambegaokar, Intrinsic resistive transition in narrow superconducting channels, *Phys. Rev.* **164**, 498–510 (1967).
65. D. E. McCumber and B. I. Halperin, Time scale of intrinsic resistive fluctuations in thin superconducting wires, *Phys. Rev. B* **1**, 1054–1070 (1970).

66. R. S. Newbower, M. R. Beasley, and M. Tinkham, Fluctuation effects on the super-conducting transition of tin whisker crystals, *Phys. Rev. B* **5**, 864–868 (1972).

67. V. Emery, Basic aspects in the physics of one dimensional metals, in *Chemistry and Physics of One-Dimensional Metals*, H. J. Keller (ed.), NATO-ASI Series B25, Plenum Press, New York (1977).

68. B. Horovitz, Instabilities of electron systems with nesting fermi surfaces, *Solid State Commun.* **18**, 445–448 (1976).

69. J. Sólyom, Application of the renormalization group technique to the problem of phase transition in one-dimensional metallic systems. II. response functions and the ground-state problem, *J. Low Temp. Phys.* **12**, 547–558 (1973).

70. H. Fukuyama, T. M. Rice, C. M. Varma, and B. I. Halperin, some properties of the one-dimensional Fermi model, *Phys. Rev. B* **10**, 3775–3780 (1974).

71. Yu. A. Bychkov, L. P. Gorkov, and I. E. Dzyaloshinskii, Possibility of superconductivity type phenomena in a one-dimensional system, *Sov. Phys. JETP* **23**, 489–501 (1966).

72. K. Levin, D. L. Mills, and S. L. Cunningham, Incompatibility of BCS pairing and the peierls distortion in one-dimensional systems. I. mean field theory, *Phys. Rev. B* **10**, 3821–3831 (1974).

73. K. Levin, S. L. Cunningham, and D. L. Mills, Incompatibility of BCS pairing and the Peierls distortion in one-dimensional systems. II. fluctuation effects, *Phys. Rev. B* **10**, 3832–3843 (1974).

74. N. Menyhard and J. Sólyom, Application of the renormalization group technique to the problem of phase transition in one-dimensional metallic system. I. invariant couplings, vertex and one-particle Green's function, *J. Low Temp. Phys.* **12**, 529–545 (1973).

75. A. Luther and V. J. Emery, Backward scattering in the one-dimensional electron gas, *Phys. Rev. Lett.* **33**, 589–592 (1974).

76. P. A. Lee, Comments on a solution of a one-dimensional Fermi-gas model, *Phys. Rev. Lett.* **34**, 1247–1250 (1975).

77. H. Gutfreund and R. A. Klemm, Order in metallic chains. I. the single chain, *Phys. Rev. B* **14**, 1073–1085 (1976).

78. S. T. Chui, T. M. Rice, and C. M. Varma, Coulomb effects on the Peierls transition, *Solid State Commun.* **15**, 155–1559 (1974).

79. G. S. Grest, E. Abrahams, S. T. Chui, P. A. Lee, and A. Zawadowski, Two-cutoff scaling for the one-dimensional electron gas, *Phys. Rev. B* **14**, 1225–1232 (1976).

80. D. J. Scalapino, Y. Imry, and P. Pincus, Generalized Ginzburg–Landau theory of pseudo-one-dimensional systems, *Phys. Rev. B* **11**, 2042–2048 (1975).

81. R. A. Klemm and H. Gutfreund, Order in metallic chains. II. coupled chain, *Phys. Rev. B* **14**, 1086–1102 (1976).

82. L. Mihaly and J. Sólyom, Renormalization group treatment of three-dimensional ordering in a system of weakly coupled linear chains, *J. Low Temp. Phys.* **24**, 579–596 (1976).

83. N. Menyhard, The effects of 1D correlations on the phase transition in quasi-1D metallic systems, *Solid State Commun.* **21**, 495–498 (1977).

84. T. Maniv and M. Weger, The superconducting transition in coupled linear chain systems, *J. Phys. Chem. Solids* **36**, 367–376 (1975).

85. B. Horovitz, H. Gutfreund, and M. Weger, Interchain coupling and the Peierls tran-sition in linear-chain systems, *Phys. Rev. B* **12**, 3174–3185 (1975).

86. D. Jerome and M. Weger, Electronic properties of organic conductors: pressure effects, in *Chemistry and Physics of One Dimensional Metals*, H. J. Keller (ed.), NATO-ASI Series B25, Plenum Press, New York (1977).

87. S. Barisic and K. Saub, Selfconsistent calculation of the Peierls instability in quasi-one-dimensional conductors, *J. Phys. C* **6**, L367–370 (1973).

88. W. Dietrich, Fluctuations and three-dimensional ordering in weakly coupled linear conductors, *Z. Phys.* **270**, 239–243 (1974).

89. B. Horovitz and A. Birnboim, Superconductivity and Peierls instability in coupled linear chain systems, *Solid State Commun.* **19**, 91–95 (1976).

90. B. Horovitz, *Solid State Commun.* **18**, 445–448 (1976).

91. N. F. Mott and W. D. Twose, The theory of impurity conduction, *Adv. Phys.* **10**, 107–163 (1961).

92. W. A. Little and H. Gutfreund (to be published).

93. R. E. Borland, Existence of energy gaps in one-dimensional liquids, *Proc. Phys. Soc.* **78**, 926–931 (1961).

94. D. J. Thouless, A relation between the density of states and the range of localization for one-dimensional random systems, *J. Phys. C* **5**, 77–81 (1972).

95. D. J. Thouless, Electrons in disordered systems and theory of localization, *Phys. Rep.* **13**, 93–142 (1974).

96. Yu. A. Bychkov, Frequency dependence of the conductivity of one-dimensional systems, *Sov. Phys.-JETP* **38**, 209–213 (1974) [*Zh. Eksp. Teor. Fiz.* **65**, 427–438 (1973)].

97. V. L. Berezinsky, Kinetics of a quantum particle in a one-dimensional random potential, *Sov. Phys.-JETP* **38**, 620–627 (1974) [*Zh. Eksp. Teor. Fiz.* **65**, 1251–1266 (1973)].

98. A. A. Gogolin, V. I. Melnikov, and E. I. Rashba, Conductivity in disordered chain caused by electron–phonon interaction, *Sov. Phys.-JETP* **42**, 168–178 (1975) [*Zh. Eksp. Teor. Fiz.* **6ᴑ** 327–349 (1975)].

99. A. Zavadovskii, Effect of impurities on superconductivity-like phenomena in one-dimensional systems, *Sov. Phys. JETP* **27**, 767–771 (1968).

100. B. R. Patton and L. J. Sham, Conductivity, superconductivity, and the Peierls instability, *Phys. Rev. Lett.* **31**, 631–634 (1973).

101. A. I. Larkin and V. I. Mel'nikov, Effect of impurities on the phase transitions in the quasi-one-dimensional conductors, *Zh. Eksp. Teor. Fiz.* **71**, 2199–2203 (1976) [*Sov. Phys.-JETP* **44**, 1159–1161 (1976)].

102. I. F. Foulkes and B. L. Gyorffy, *p*-wave pairing in metals, *Phys. Rev. B* **15**, 1395–1398 (1976).

103. I. E. Dzyaloshinskii and E. I. Kats, Superconductivity in quasi-one-dimensional (thread-like) structures, *Sov. Phys.-JETP* **28**, 178–182 (1969).

104. D. Davis, Thomas–Fermi screening in one dimension, *Phys. Rev. B* **7**, 129–135 (1973).

105. B. Bush, Theory of the screened coulomb interaction in quasi-one-dimensional metals, Ph.D. Thesis, Stanford University (1974).

106. A. S. Berenblyum, L. I. Buravov, M. L. Khidekel, I. F. Shchegolev, and E. B. Yakimov, *Zh. Eksp. Teor. Fiz. Pis'ma Red.* **13**, 619–622 (1972) [*Sov. Phys. JETP Lett.* **13**, 440–442 (1973)].

107. F. Stern, Polarizability of a two-dimensional electron gas, *Phys. Rev. Lett.* **18**, 546–548 (1967).

108. T. R. Brown and C. C. Grimes, Observation of cyclotron resonance in surface-bound electrons on liquid helium, *Phys. Rev. Lett.* **29**, 1233–1236 (1972).

109. P. B. Visscher and L. M. Falicov, Dielectric screening in a layered electron gas, *Phys. Rev. B* **3**, 2541–2547 (1971).

110. A. L. Fetter, Electrodynamics of a layered electron gas. I. single layer, *Ann. Phys. (N. Y.)* **81**, 367–393 (1973).

111. A. L. Fetter, Electrodynamics of a layered electron gas. II. periodic array, *Ann. Phys. (N. Y.)* **88**, 1–25 (1974).

112. L. N. Bulaevskii and Yu. A. Kukharenko, Effectiveness of the exciton mechanism of superconductivity in layered compounds with molecules, *Sov. Phys. JETP* **33**, 821–824 (1971).

113. W. A. Little, The effectiveness of the exciton mechanism in superconducting layered compounds, *J. Low Temp. Phys.* **13**, 365–369 (1973).

114. L. Salem, *The Molecular Orbital Theory of Conjugated Systems*, W. A. Benjamin Inc., New York (1966).

115. H. Gutfreund and W. A. Little, Correlation effects of π electrons. II. Low lying excitations of polycyclic hydrocarbons, *J. Chem. Phys.* **50**, 4468–4477 (1969).

116. W. A. Little and H. Gutfreund, Dynamic effective electron–electron interaction in the vicinity of a polarizable molecule, *Phys. Rev. B* **4**, 817–823 (1971).

117. K. Nishimoto and N. Mataga, Electronic structure and spectra of nitrogen heterocyoles, *Z. Phys. Chem.* **13**, 140–157 (1957).

118. A. M. Abarbanel, An energy band calculation of linear chain transition meta. complexes, *Ann. Phys. (N. Y.)* **91**, 356–365 (1975).

119. D. M. Whitmore, A one-dimensional band calculation of a linear, square planar platinum complex, *Phys. Rev. Lett.* **50A**, 55–56 (1974).

120. R. P. Messmer (private communication).

121. R. P. Messmer and D. R. Salahub, Importance of chemical effects in determining the free-electron-like band structure of $K_2Pt(CN)_4Br_{0.3} \cdot 3H_2O$, *Phys. Rev. Lett.* **35**, 533–536 (1975).

122. K. Mees, *The Theory of the Photographic Process*, The MacMillan Company, New York (1966).

JOURNAL DE PHYSIQUE

Colloque C3, supplément au n°6, Tome 44, juin 1983

EXCITONIC SUPERCONDUCTIVITY - THE MISSING LINK

W.A. Little

Physics Department, Stanford University, Stanford, CA 94305, U.S.A.

Résumé- On discutera les facteurs qui déterminent la constante de couplage excitoni-
que et on montrera que, généralement, elle est petite. Néanmoins, on montrera qu'en
choisissant les systèmes et ses niveaux énergétiques soigneusement, un couplage fort
devrait être possible. Dans ce cas, la superconductivité à plus hautes températures
devrait exister.

Abstract- We discuss the factors which determine the strength of the excitonic
coupling constant and show that, in general, it will be small. However, we show
that by proper choice of the system and its energy levels, strong coupling should be
possible. In this case superconductivity at higher temperatures should occur.

It is almost twenty years since the suggestion was first made of the possibility of
synthesizing a polymeric, organic superconductor of high transition temperature.[1]
The high transition temperature was shown to be a consequence of using a different
type of electron-electron coupling mechanism - one involving the exchange of molecular excitations instead of phonons, which has been named the "exciton" mechanism
by Ginzburg.[2] During this period enormous progress has been made in the synthesis
of organic metals,[3] organic superconductors of moderately low transition temperature,[4] polymeric organic metals[5] and polymeric superconductors.[6] At the same
time great progress has been made in the theoretical understanding of these materials
through the study of fluctuations,[7] and the properties of quasi-one and quasi-two
dimensional metals and superconductors.[8],[9] However, virtually no evidence has
been found for the existence of the exciton mechanism. This is puzzling as I believe
it is at least as solidly based on good physics, as the rest of the field! We have
looked into this problem and how have a reasonably good understanding why the effects
of this interaction are so elusive and from this we can make some suggestions as to
what needs to be done to obtain an interaction of sufficient strength to lead to
high transition temperature superconductors.

Let us review briefly how the exciton mechanism comes about. Let A in Figure 1
represent a portion of a conductive polymer along which the electron "a" is moving.
As it moves past the polarizable side chains (B) it induces charges similar to those
at "b" and at "c" on each side-chain. These charges set up a potential at the site
of another electron at "d" which clearly is attractive. These charges result from
the virtual excitation of the molecules of the side-chains. Hence the term "exciton"
for the name of the attractive mechanism. The high transition temperature in the
model comes from the fact that the energy $\hbar\omega$ of these excitations is much greater
than the energy of the phonons; and hence the BCS expression for the transition
temperature $kT_c \simeq \hbar\omega \exp(-1/\lambda)$ yields large values for T_c provided that in this

mechanism the coupling constant λ can be large enough.

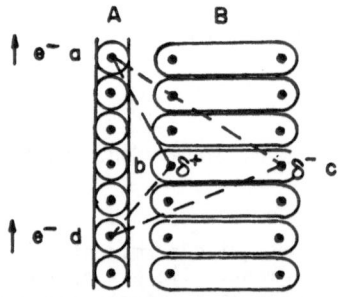

Fig.1   Model of Excitonic Mechanism

This is a somewhat naive explanation of the high T_c but gives physical insight to the more detailed calculations.[10]   One should notice that the strength of the interaction is proportional to $(1/r_b - 1/r_c)$, where r_b is the distance of nearest approach of the transition charge in the side-chain and r_c the distance of furthest movement. Also, because the side-chain lies a few angstroms from the polymer, the induced potential along the polymer will be relatively slowly varying.   Its Fourier components will therefore be peaked at low momenta, with very little weight at high momenta. In terms of the "g-ology" picture, this means that the g_1 coupling term, which represents scattering processes across the Fermi surface will be weak and g_2, that for scattering in the vicinity of the Fermi surface will be strong and attractive (i.e. negative).   This places one then in the superconducting side of the g-ology diagram. See Figure 2.

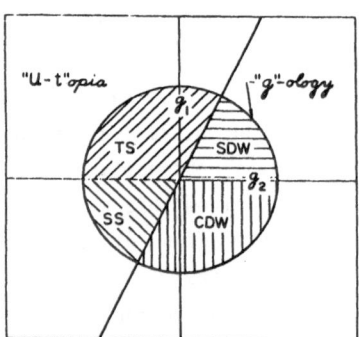

Fig.2   The "g-ology" diagram for a one-dimensional system showing regions of different phases.

Whether or not one has superconductivity at high temperature depends on the strength of the excitonic coupling. In our earliest work[1] we had used an excessively simple molecular orbital treatment of the oscillatory charge in the side-chains. (See Fig. 3). It was assumed that the charge on the nitrogen atoms resonated between them as illustrated in the lower half of the figure.

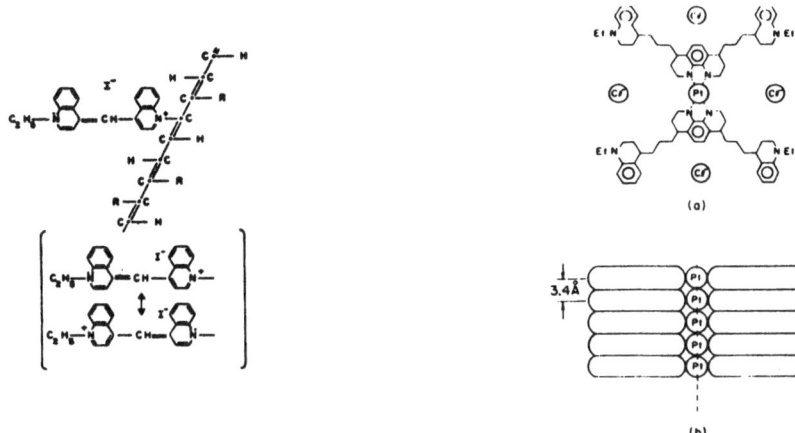

Fig.3 Early Model of Excitonic
Superconductor

(a)

(b)

Fig.4 Later Model

The point of nearest approach to the chain was therefore one bond length - of the order of 1.5A. Our estimate of the resultant coupling constant was large. More detailed molecular orbital calculations[11] however, showed that the charge was much more evenly distributed over all the atoms of the quinoline ring, giving an average distance of nearest approach of the order of 4.5 Å and hence the true coupling constant was much weaker than our original estimate. This led us to the study of more complicated structures containing more, polarisable groups clustered about the chain in a more dense array in order to see if, under these much more favourable conditions the interaction could be large enough to give superconductivity at higher temperatures.[10] (See Fig. 4). We concluded that it could. However, the proposed compound is extremely complicated and would be difficult to prepare. Even if this could be done, it is by no means obvious that the monomers would stack in the manner needed. It was suggested to us by Keller[12] that perhaps a simpler compound could be designed if the transition charge of the excitonic subsystem could be made to move on and off the metal atom itself, so that the distance of nearest approach would be much smaller and, hence, the interaction stronger. We can illustrate this with the following example.

Consider a compound such as KCP in which the conduction electrons move along the transition metal atom chain in the $d_{z^2}-p_z$ orbital band. The d_{xz} and d_{yz} orbitals are, of course, orthogonal to all the states in this band, and in addition, they mix with the orbitals of the ligands as shown in Fig. 5. This admixture accounts for the back-bonding between the metal and the ligand. If now the ligand is really part of a conjugated chromophoric group, then as the charge moves back and forth in this group it will move partially on and off the metal. This very close proximity of the d_{xz} and the d_{z^2} and p_z orbitals will then give strong coupling between the two.[13] This is the point of the approach. There are some problems which arise in the theory from the resultant interaction being more extended in momentum space because of the very close approach of the two systems than in the earlier models, but these problems need only be faced if the interaction is of sufficient strength.[10]

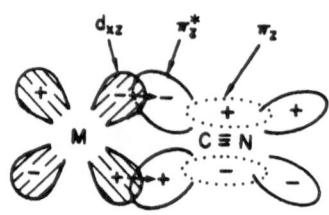

Fig.5 Back-bonding between transition
metal d_{xz}-orbital and the
π-orbitals of the ligand.

We have studied this model in some detail and have found that if one particular criterion is satisfied then indeed very strong coupling can occur even in simple ligand systems, but that if this criterion is not met the coupling will be weak. This can be illustrated as follows: suppose we have a metal such that the energy of the metal orbital is E_m and that this overlaps with an orbital on the ligand of energy E_1. Suppose the overlap energy is V which mixes the metal and ligand orbitals giving a mixed ground and mixed excited state. If E_m is much less than E_1 then in the ground state most of the charge will be on the metal, while in the excited state the charge will be predominantly on the ligand. On the other hand, if E_m is much greater than E_1 then the converse will be true, with most of the charge in the ground state located on the ligand and, in the excited state upon the metal. In both cases, the transition density which describes the oscillatory charge between the metal and ligand will be small. If, however, one makes E_m approximately equal to E_1, then in both the ground and excited states the charges are evenly shared by the metal and the ligand and the transition density is large. In this case, the charge can easily flow between the metal and the ligand, and the resulting excitonic coupling constant will be large. The scale on which the energies of the orbitals are judged to be equal is given by the magnitude of the overlap energy V. This is shown in Fig.6, where the coupling strength is calculated as a function of the energy of the metal orbital while keeping the energy of the ligand orbitals fixed and it is assumed that the ground state is occupied and the excited state unoccupied. From this one can see that unless the energies of the metal and ligand orbitals match each other quite closely the coupling constant will be small. Just how sensitive this criterion is, can be seen in another example involving the solvent sensitivity of the merocyanine dyes.[14] These dyes consist of two dissimilar nuclei (A and B in Fig.7) connected by a conjugated link.

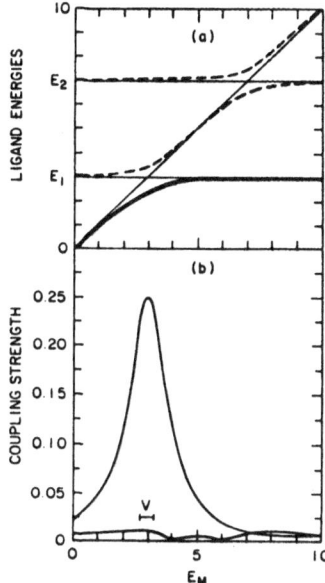

Fig.6(a) Plot of the energies of metal-ligand orbitals as a function of the energy E_m of the d_{xz} metal orbital. E_1 and E_2 are the energies of the ligand orbitals. (b) Resultant excitonic coupling constant from these on electrons in the d_{z^2} -p_z band.

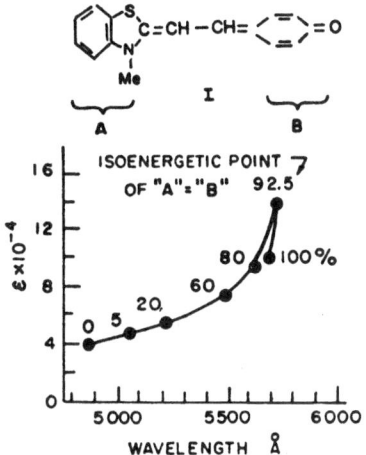

Fig.7 Solvent Sensitivity of the absorption of the merocyanine dye (I) in pyridine-water mixture[14]

In our model one may consider one nucleus, say A, as being analogous to the metal orbital and the other, B to the ligand. If this dye is dissolved in a mixture of water and pyridine, one finds that the relative energies of A and B depend upon the composition of the mixture. The more polar nucleus being favoured in mixtures containing more of the more polar solvent, water; and, the less polar nucleus favoured in the less polar solvent, pyridine. As one increases the concentration of pyridine, the absorption of the dye shifts to lower energy and at the same time the extinction coefficient rises sharply. Once the isoenergetic point is passed, the extinction coefficient falls and the absorption energy rises. This is just the behaviour shown in the previous figure (Fig. 6) where in the upper portion one sees the energy between the ground and excited states is at a minimum when the ligand and metal orbitals are degenerate, and at this isoenergetic point the coupling constant is at a maximum.

This shows that the immediate neighbourhood of the ligand system can have a dramatic effect on the strength of any excitonic coupling which may exist and, that a small change in this neighbourhood can change this coupling. The question also arises as to whether it is ever possible for a ligand orbital energy to be degenerate with the metal orbitals of those metals which can form the square planar structures suitable for the required stacked structure.

137

In order to determine which systems might be most suitable for detailed study we have calculated the overlap energies and collected the metal orbital energies of a number of systems which seem to hold some promise. For this and other more detailed calculations we have used an extended Huckel program.[15] Approximate values for the energies of the metal d-orbitals are given in Table 1 and the overlap energies between some of these metals and ligand heteroatoms in Table 11.

Ni	-14.2 eV	Co	-13.2
Pd	-13.5	Rh	-12.5
Pt	-12.6	Ir	-11.6

Ionization energy of carbon -11.2eV

TABLE I   Approximate energy of Metal d-electrons in eV

	N	O	P	S
Ni	-1.41	-1.31	-1.30	-2.18
Pd	-1.39	-1.26		-2.27
Pt	-1.44	-1.29	-2.06	-1.72

TABLE II   Overlap Energies in eV

For a ligand containing carbon bonded to the heteroatom, the energy of some of the ligand orbitals will lie below the ionisation energy of carbon due to bonding and hence, these could be degenerate with the d-orbital of the metals shown. The nickel-sulphur and palladium-sulphur combination appears to be particularly attractive due to the strong overlap. Underhill[16] has pointed out to us that in the nickel di-thiolates there is evidence that the metal-ligand orbitals are indeed about 50:50 mixtures as required by our model to give a strong excitonic coupling.[17] However, in this case, the antibonding excited state would probably lie too high in energy because of the four overlap interactions between the metal and the sulphur atoms. This energy lies in the denominator in the calculation of the polarisability and consequently if it is large would reduce the coupling. So one needs a balance between sufficient overlap to give a reasonable width to the resonance curve of Fig.6, but not too much to drive the excited state energy too high. A number of systems deserve further study such as the porphyrins, phthalocyanines, dithiolates and other macrocyles. From the arguments we have given above one would expect the metal-ligand admixture to be sensitively dependent upon the substituents on the ligands and consequently, the excitonic coupling or the effective Coulomb U which is related to it, should similarly be dependent on these substituents and their neighbourhood. We believe this is the point which has been missed in earlier considerations of the excitonic coupling strength.

Estimates of the effective coupling strength one could obtain if the above conditions can be met, indicate that the ligands would not need to have the complicated, extended form of our earlier model (Fig.4), but could be smaller and simpler. Detailed calculations on a number of systems are currently in progress and will be reported on at a later time.

The author is indebted to Professor H. Gutfreund of the Hebrew University, Jerusalem, Israel for extensive discussions. The National Science Foundation is gratefully acknowledged for support of this work.

REFERENCES

1) W.A. Little, Phys. Rev. A134, 1416 (1964)

2) V.L. Ginzburg, Contemp. Phys. 9, 355 (1968)

3) Ferraris, J., D.O. Cowan, V. Walatka & J.H. Perlstein, J. Am. Chem. Soc.95,948 (1973)

4) D.Jerome, A.Mazaud, M.Ribault, and K.Bechgaard, J.Phys.(Paris), Lett. 41,L95(1980)

5) Chiang, C.K., Fincher, C.R.Jr., Park.Y.W., Heeger, A.J., Shirakawa H., Louis,E.J., Gau, S.G. and MacDiarmid, A.G., Phys. Rev. Lett. 1977 39, 1098

6) Greene, R.L., G.B.Street & L.J. Suter (1975) Phys. Rev.Lett. 34, 577

7) Proc. Conf. on Fluctuations in Superconductors, Asilomar March 13-15 (1968) Ed. W.S. Goree and F. Chilton, Stanford Research Institute, Menlo Park, CA 94025

8) Chemistry and Physics of One Dimensional Metals Ed. H.J. Keller, Plenum Press, New York (1977)

9) F.R. Gamble, F.J. Di Salvo, R.A. Klemm and T.H. Geballe, Science 168, 568 (1970)

10) D.Davis, H. Gutfreund and W.A. Little, Phys. Rev. B13, 4766 (1976)

11) W.A. Little and H. Gutfreund, Phys. Rev. B4, 817 (1971)

12) H. Gutfreund and W.A. Little, "Prospects of Excitonic Superconductivity in Highly Conducting One Dimensional Solids, J.T. Devreese, R.P. Evrard, and V.E. Doren, Eds(Plenum, New York 1979), p305-372

13) W.A. Little, Int. Jour. Quantum Chem. 15, 545 (1981)

14) L.G.S.Brooker, G.H.Keyes and D.W.Heseltine, J.Am.Chem.Soc., 73, 5332 (1951); and K.Mees, The Theory of the Photographic Process, The MacMillan Co. New York,(1966) p219-221.

15) J. Howell, A. Rossi, D.Wallace,K.Haraki, and R.Hoffman, QCPE XIII,344 (1981)

16) Allen E. Underhill, (private communication).

17) G.N.Schrauzer and H.N.Rabinowitz, J.Am. Chem. Soc. 90, 4279 (1968).

Optical Evidence of an Electronic Contribution to the Pairing Interaction in Superconducting $Tl_2Ba_2Ca_2Cu_3O_{10}$

M. J. Holcomb,[*,†] J. P. Collman,[*] and W. A. Little[†]

Departments of Chemistry and Physics, Stanford University, Stanford, California 94305
(Received 20 May 1994)

We report the results of a study of the temperature-dependent thermal difference reflectance spectra of $Tl_2Ba_2Ca_2Cu_3O_{10}$. At temperatures below T_c, a feature emerges in the spectra with an integrated amplitude that scales as $\Delta_0^2(T)$, where $\Delta_0(T)$ is the temperature-dependent superconducting gap. The temperature dependence and location of this feature can be described by an Eliashberg model with a coupling function that includes both an electron-phonon interaction *and* an interaction located at ~ 1.6 eV. We find remarkably good agreement between theory and experiment based upon this description of the superconducting state.

PACS numbers: 74.25.Gz, 74.20.Mn, 74.72.Fq

The most fundamental problem of high-T_c superconductivity is the determination of the mechanism responsible for the high critical temperatures observed in the cuprate superconductors. If the superconducting state of the cuprates can be described by Eliashberg theory then the problem is one of the precise experimental determination of the energy dependence of the electron-boson coupling function which mediates the pairing of the electrons. Other theoretical models can be expected to contain analogous information on the pairing mechanism. This information is usually obtained by a measurement of the conductance vs voltage of a normal/insulator/superconductor ($N/I/S$) tunnel junction [1]. Such studies have enjoyed much success in conventional low-temperature superconductors, but there has been little success with the cuprates [2]. Though the magnitude of the superconducting gap edge [3], $\Delta(\Delta(\omega = 0), T \approx 0) \equiv \Delta_0$, has been determined with limited accuracy [4], it is the small variations in the tunneling conductance at energies well above the gap edge that contain information on the nature of the pairing interaction [1]. These data are inconclusive and have not revealed the nature of the pairing interaction in the cuprates to date.

The energy dependence of the electron-boson coupling function can also be obtained from the optical properties of a superconductor [5–8]. The transmittance, or reflectance, of a superconductor is a measure of both a superconducting joint density of states function and a joint coherence factor function which results from the coherent nature of the superconducting state [6,9]. In principle, by measuring the changes in the reflectance, or transmittance, of a material when it becomes superconducting it is possible to calculate the electron-boson coupling function responsible for the superconductivity, using the same inversion techniques that have been used so effectively in tunneling experiments [10]. Like $N/I/S$ tunneling, this approach has a long history in the study of conventional low-temperature superconductors [11–19], but because of the inferior signal-to-noise ratios in these measurements as compared to tunneling measurements and the numeri-

cal complexity of the inversion process [17], no rigorous inversion of optical data has been attempted. More recently, significant changes in the optical properties of the cuprate superconductors have been measured over a wide range of energies as the materials enter the superconducting state [20–26]. In most of these studies, an approximate optical gap $2\Delta_0$ has been determined, but again, no rigorous inversion of the optical data has been attempted. Like tunneling measurements, the important information on the detailed nature of the pairing interaction lies in the small variation of the optical properties of the superconductor at energies well above $2\Delta_0$. The difficulty with this measurement is that the changes in the optical properties of the material are of order $[\Delta(\omega)/\omega]^2$, where $\Delta(\omega)$ is the complex energy dependent superconducting gap function [27]. In a standard reflectance measurement these changes may be as small as 0.01% of the total reflectance and thus will be hidden in the experimental linewidth.

We have recently developed an optical technique of sufficient sensitivity to observe such effects. The technique measures the difference in the reflectivity of a sample at two temperatures as a function of photon energy and uses digital averaging and thermal cycling to minimize the effects of $1/f$ noise. The details of the operation and performance of this thermal difference (TD) spectrometer have been described elsewhere [28]. We have been able to observe temperature induced changes in the relative reflectance as small as 0.005% reliably and reproducibly and can do so over a temperature range of 80 to 340 K. We have used it to study the temperature-dependent TD spectra of a thin film $Tl_2Ba_2Ca_2Cu_3O_{10}$ (Tl-2223) sample at photon energies of 0.3 to 5.3 eV. The data are collected by first measuring the difference in the sample's reflectance at $T + 5$ K and $T - 5$ K and then dividing this quantity by the average value of the reflectance at T. Details of the sample preparation and characterization have been described previously [29]. The T_c of the Tl-2223 sample was measured to be ~ 120 K.

In Fig. 1 we plot the TD spectrum of Tl-2223 collected at 300 K from 0.3 to 5.3 eV. The spectrum is dominated

0031-9007/94/73(17)/2360(4)$06.00

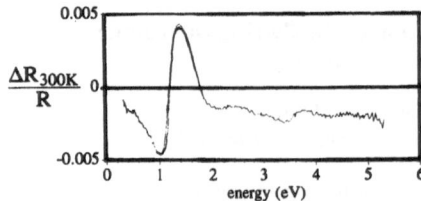

FIG. 1. The thermal difference (TD) reflectance spectrum of Tl$_2$Ba$_2$Ca$_2$Cu$_3$O$_{10}$ (Tl-2223) at 300 K from 0.3 to 5.3 eV. The spectrum was collected by measuring the difference in the reflectance of the sample at $T + \Delta T$ and $T - \Delta T$ and dividing by the average value of the reflectance at both of these temperatures. $\Delta T = 5$ K in all of these experiments.

by a derivativelike structure centered at about 1.25 eV, which results from temperature induced changes in the scattering rate and the plasma frequency of the material. The screened plasma energy [30] is approximately the energy of the *most negative point* of the large derivative structure in the TD spectrum [31]. This lies at about 1.1 eV in agreement with the room temperature reflectance and ellipsometric measurements of Bozovic *et al.* [32]. In Fig. 2(a) we plot the TD spectra of the same sample at 195, 175, 155, and 135 K from 0.3 to 3.0 eV. These spectra has the same form as the 300 K spectrum. We observe that the amplitude of the structure in the TD spectra decreases regularly as the temperature is lowered to 135 K. In Fig. 2(b) we plot the TD spectra at 135, 125, 115, 105, and 95 K. As the temperature is lowered below

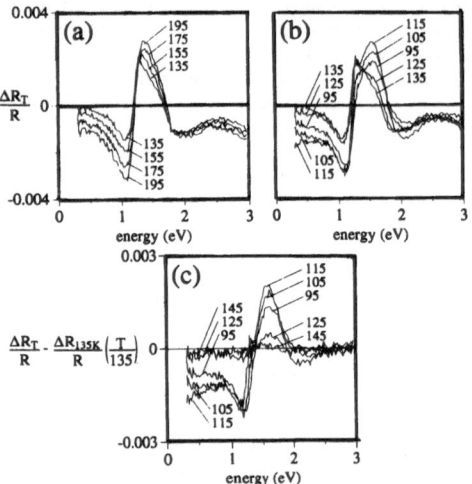

FIG. 2. The temperature-dependent TD reflectance spectra of Tl-2223. (a) The normal state TD spectra at 195, 175, 155, and 135 K. (b) TD spectra of Tl-2223 at 135, 125, 115, 105, and 95 K. The T_c of this sample is approximately 120 K. (c) The TD spectra of Tl-2223 after subtracting the temperature-dependent normal state response from the data.

the T_c of the sample (~ 120 K) a new structure emerges in the TD spectra. The amplitude is a maximum at 115 K and gradually decreases with decreasing temperature. The amplitude of the thermal difference spectra varies approximately linearly with temperature *above* T_c. This allows us to subtract the extrapolated normal state thermal difference response from the spectra collected *below* T_c. In Fig. 2(c) we plot the resulting difference spectra at 145, 125, 115, 105, and 95 K after subtracting from each of the spectra the spectrum collected at 135 K multiplied by a factor $T/135$ to correct for the linear temperature-dependent amplitude of the normal state thermal difference spectrum. For temperatures above 135 K, where no excursion into the superconducting state occurs, this difference signal lies approximately at zero over the entire energy range, while at temperatures below 125 K, a significant structure emerges in the spectra.

Since the normal state thermal difference response of the material has been removed in the subtraction, the data in Fig. 2(c) represent the normalized difference in the reflectance of the material taken between $T + 5$ and $T - 5$ K due to the onset of superconductivity. By then adding these data collected at 10 K intervals we obtain spectra that correspond to the *total* change in the reflectance of the material due to the onset of superconductivity at 120 K. This is equivalent to a numerical integration of the normalized thermal derivative spectra in Fig. 2(c) over temperature from 135 K down to T. We then take the absolute value of the integral over energy, of these *temperature-integrated* spectra as a measure of the total strength of the emerging structure with temperature. This procedure is shown graphically as the inset in Fig. 3, where the area of the shaded region is taken as a measure of the strength of the structure which emerges at temperatures below the sample's critical temperature. These data are plotted as filled circles in Fig. 3 and fitted to the functional forms of $\Delta_0(T)$ and $\Delta_0^2(T)$, where $\Delta_0(T)$ represents the temperature dependence of the superconducting gap edge in the weak

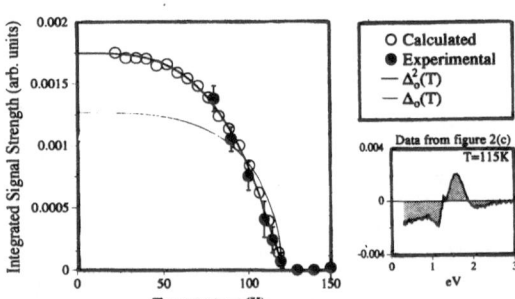

FIG. 3. Amplitude of the normalized, temperature integrated TD spectra from 150 to 80 K, ● experimental, ○ theoretical, with $T_c \sim 120$ K. The shaded area of spectrum in the inset, taken from Fig. 2(c), represents the total change of the sample's reflectance when the material becomes superconducting.

coupling limit [33]. Remarkably good agreement is found with a temperature variation proportional to $\Delta_0^2(T)$, but not with $\Delta_0(T)$.

It is convenient to express these data as the ratio of the reflectivities of the material in the superconducting and normal state R_S/R_N. It can be shown that the TD data collected at $T = 115$ K, with the normal state thermal difference response removed [Fig. 2(c)], are directly related to R_S/R_N of the sample at $T = 110$ K. In Fig. 4(a) we plot the resulting R_S/R_N of the Tl-2223 at 110 K.

The structure in the TD spectra is the result of the temperature variation of the complex dielectric function of the material. In metals, this temperature variation is determined by temperature induced changes of both the electron density and scattering rate [31]. Structural studies of a number of the cuprate superconductors indicate that, though there are subtle structural changes in the material's unit cell at temperatures in the neighborhood of T_c (± 20 K), there are no discontinuous changes in the volume of the unit cell at T_c [34–38]. This is confirmed by our measurements of the energy of the plasma frequency, which remained virtually unchanged with temperature. Thus, we interpret our data as resulting from changes in the scattering rate which arise from the growth of $\Delta(\omega)$ at $T < T_c$. This is equivalent to the strong coupling extension of Mattis-Bardeen theory [39] first developed by Nam [5,12].

We follow Nam [12] and assume a Drude-like [30,40] form for the normal state optical conductivity and express the scattering time in the superconducting state, τ_S, as a function of both the normal state scattering time τ_N and an integral involving $\Delta(\omega)$. This integral [6] consists of both a joint density of states function and a function which results from case II coherence factors appropriate for electromagnetic absorption [9]. The normal state reflectivity at energy ω can be modeled by taking the normal state scattering rate $1/\tau_N$ equal to 0.6ω [40] and using the values obtained by Bozovic et al. [32] on Tl-2223 for the bare plasma frequency (~ 2.6 eV) and the high frequency dielectric constant (~ 4.5). The change in the mass renormalization is small for energies between 0.3 and 5.3 eV and is neglected. These parameters yield

a normal state Drude-like reflectivity which is closely similar to that observed. To calculate the reflectivity in the superconducting state, we *guess* a trial electron-boson coupling function and solve the Eliashberg integral equations to obtain $\Delta(\omega)$. We then solve for τ_S, calculate the reflectivity of the material in the superconducting state, and compare it with the observed data. Changes are then made to the trial function until agreement is obtained. In these calculations, the real and imaginary parts of $\Delta(\omega)$ are related by causality. We find they must be calculated to four decimal places in order to satisfy the sum rules with the appropriate accuracy necessary to calculate the structure in the reflectance ratio.

Our calculations reveal that the measured structure near 1.5 eV in Fig. 4(a) cannot be obtained if the electron pairing is mediated by phonons alone, assuming the phonon energies are less than 100 meV. This result is consistent with the fact that, even in the classic experiments [11,13] and calculations [12] of the optical response of lead, the real part of the optical conductivity in the superconducting state approaches that of the normal state at energies approximately 5 times the energies of the phonons that mediate the pairing. Thus, *irrespective* of the strength of the electron-phonon coupling in Tl-2223, our results cannot be explained in terms of a model that only involves phonons of energy less than 100 meV. The electron-phonon coupling function, and the resulting structure in $\Delta(\omega)$, is too low in energy to alter the reflectance of the material in the superconducting state in the near-infrared and visible region of the spectrum.

Reasonable agreement between the calculations and our data is obtained by adding to the electron-phonon coupling function an additional coupling interaction centered at approximately 1.6 eV, with a half-width of 100 meV. This is shown in Fig. 4(b). Details of the specific form of the electron-boson interaction and the procedures used in the calculations will be given in a later paper.

The temperature dependence of the structure in the reflectance spectra can also be calculated, assuming a critical temperature of 120 K and a weak coupling temperature dependence for $\Delta(\omega)$ [33]. The results are shown in Fig. 3, where the open circles represent the *calculated* temperature-dependent amplitude of the thermal difference optical structure based upon the existence of the additional electron interaction at 1.6 eV. Again, remarkable agreement between experiment and theory is found, for the amplitude of the calculated structure clearly fits the functional form of the weak coupling $\Delta_0^2(T)$.

Fugol et al. [26] have studied the optical absorption of two high T_c superconductors at 1.7 and 1.9 eV and found a similar change upon entering the superconducting state. They interpret their data in terms of a model involving transitions to a narrow hole band near the Fermi energy within which a superconducting gap opens yielding a decrease in absorption proportional to $\Delta^2(T)$. Their model, however, predicts an increase in absorption of similar magnitude proportional to T^2 at $T > T_c$ which is

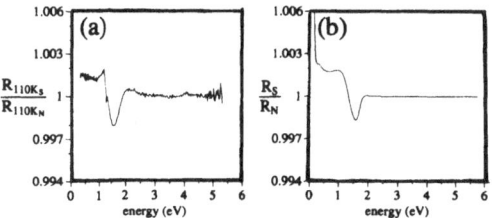

FIG. 4. (a) The R_S/R_N of Tl-2223 at 110 K from 0.3 to 5.3 eV obtained with the TD spectrometer. (b) The calculated R_S/R_N of an Eliashberg superconductor with a coupling function which consists of both an electron-phonon component *and* an interaction centered at 1.6 eV.

not observed, nor can it explain the sign reversal of the differential absorption at certain energies as implied in Fig. 2.

Our results strongly suggest that the superconductivity in the Tl-2223 material can be described within the Eliashberg formalism and that the pertinent electron coupling function consists of an electron-phonon interaction *plus* a previously unobserved interaction at approximately 1.6 eV. We estimate the error in the precise location of the high energy coupling function to be ± 0.15 eV because of the uncertainty in the normal state optical properties and the effects of anisotropy on $\Delta(\omega)$ and τ_N.

Having established the energy scale of the interaction believed to be responsible for the high critical temperature of Tl-2223, we may speculate on its microscopic origin. Optical studies of a variety of the cuprate materials all show an optical transition at approximately 1.7 eV [41–44]. This transition has been tentatively designated as a d^9-$d^{10}\underline{L}$ charge transfer associated with the extended Cu-O network common to all these materials [43]. Because of the close proximity of this to the energy of the feature we observe, we suggest that it provides the interaction responsible for the high temperature superconductivity. We, thus, *predict* that this interaction will be present at this energy for all the high temperature superconductors that possess the extended Cu-O network, and that the distribution of critical temperatures among these materials is not determined by differences in the energy of this interaction but by the *strength* of the coupling to this high energy interaction.

In conclusion, these experiments suggest that $Tl_2Ba_2Ca_2Cu_3O_{10}$ may be described as an Eliashberg superconductor, albeit one with an unusual normal state. Further, the results indicate that electron-phonon coupling is not solely responsible for the high T_c of the material, and that the most likely origin of the excitation responsible for the high T_c is an electronic transition near 1.6 ± 0.15 eV. Future experiments on other high temperature superconductors should confirm or disprove this model.

We are indebted to Dr. Wen Y. Lee of IBM Almaden Research Center, San Jose, for providing thin films of $Tl_2Ba_2Ca_2Cu_3O_{10}$.

*Department of Chemistry.

†Department of Physics.

[1] W. L. McMillan and J. M. Rowell, in *Superconductivity*, edited by R. D. Parks (Dekker, New York, 1969), Vol. 1, p. 561.
[2] J. R. Kirtley, J. Mod. Phys. B **4**, 201 (1990).
[3] D. J. Scalapino, in *Superconductivity*, edited by R. D. Parks (Dekker, New York, 1969), Vol. 1, p. 449.
[4] N. Miyakawa et al., J. Phys. Soc. Jap. **62**, 2445 (1993).
[5] S. B. Nam, Phys. Rev. **156**, 470 (1967).
[6] W. Shaw and J. C. Swihart, Phys. Rev. Lett. **20**, 1000 (1968).
[7] W. Lee, D. Rainer, and W. Zimmermann, Physica (Amsterdam) **159C**, 535 (1989).
[8] N. E. Bickers et al., Phys. Rev. B **42**, 67 (1990).
[9] M. Tinkham, in *Introduction to Superconductivity* (Krieger, New York, 1980), p. 57.
[10] W. A. Little and J. P. Collman, Proc. Natl. Acad. Sci. U.S.A. **85**, 4596 (1988).
[11] R. E. Glover and M. Tinkham, Phys. Rev. **108**, 243 (1957).
[12] S. B. Nam, Phys. Rev. **156**, 487 (1967).
[13] L. H. Palmer and M. Tinkham, Phys. Rev. **165**, 588 (1968).
[14] R. E. Harris and D. M. Ginsberg, Phys. Rev. **188**, 737 (1969).
[15] R. R. Joyce and P. R. Richards, Phys. Rev. Lett. **24**, 1007 (1970).
[16] P. B. Allen, Phys. Rev. B **3**, 305 (1971).
[17] B. Farnworth and T. Timusk, Phys. Rev. B **10**, 2799 (1974).
[18] B. Farnworth and T. Timusk, Phys. Rev. B **14**, 5119 (1976).
[19] D. Karecki, R. E. Peña, and S. Perkowitz, Phys. Rev. B **25**, 1565 (1982).
[20] P. E. Sulewski et al., Phys. Rev. B **35**, 8829 (1987).
[21] Z. Schlesinger et al., Phys. Rev. Lett. **59**, 1958 (1987).
[22] T. Timusk and D. B. Tanner, Physica (Amsterdam) **169C**, 425 (1990).
[23] Z. Schlesinger et al., Nature (London) **343**, 242 (1990).
[24] Z. Schlesinger et al., Phys. Rev. B **41**, 11 237 (1990).
[25] T. Timusk, C. D. Porter, and D. B. Tanner, Phys. Rev. Lett. **66**, 663 (1991).
[26] I. Fugol et al., Solid State Commun. **86**, 385 (1993).
[27] D. J. Scalapino, J. R. Schrieffer, and J. W. Wilkins, Phys. Rev. **148**, 263 (1966).
[28] M. J. Holcomb, J. P. Collman, and W. A. Little, Rev. Sci. Instrum. **64**, 1867 (1993).
[29] W. Y. Lee, J. Appl. Phys. **70**, 3952 (1991).
[30] I. Bozovic, Phys. Rev. B **42**, 1969 (1990).
[31] M. Cardona, in *Modulation Spectroscopy, Solid State Physics Series*, edited by F. Seitz, D. Turnbull, and H. Ehrenreich (Academic Press, New York, 1960).
[32] I. Bozovic et al., Phys. Rev. B **43**, 1169 (1991).
[33] D. J. Scalapino, Y. Wada, and J. C. Swihart, Phys. Rev. Lett. **14**, 102 (1965).
[34] H. You et al., Phys. Rev. B **38**, 9213 (1988).
[35] H. You, U. Welp, and Y. Fang, Phys. Rev. B **43**, 3660 (1991).
[36] M. Okaji et al., Cryogenics **34**, 163 (1994).
[37] S. Huimin et al., Supercond. Sci. Technol. **2**, 52 (1989).
[38] Y. Ono and S. Narita, Jpn. J. Appl. Phys. **31**, L224 (1992).
[39] D. C. Mattis and J. Bardeen, Phys. Rev. **111**, 412 (1958).
[40] Z. Schlesinger et al., Phys. Rev. Lett. **65**, 801 (1990).
[41] M. K. Kelly et al., Phys. Rev. B **40**, 6797 (1989).
[42] D. E. Aspnes and M. K. Kelly, IEEE J. Quantum Electron. **25**, 2378 (1989).
[43] T. Yamamoto et al., Jpn. J. Appl. Phys. **31**, L327 (1992).
[44] S. Uchida, Physica (Amsterdam) **185-189C**, 28 (1991).

MATHEMATICAL BIOSCIENCES **19**, 101–120 (1974)

The Existence of Persistent States in the Brain

W. A. LITTLE

Department of Physics, Stanford University, Stanford, California

Communicated by S. M. Ulam

ABSTRACT

We show that given certain plausible assumptions the existence of persistent states in a neural network can occur only if a certain transfer matrix has degenerate maximum eigenvalues. The existence of such states of persistent order is directly analogous to the existence of long range order in an Ising spin system; while the transition to the state of persistent order is analogous to the transition to the ordered phase of the spin system. It is shown that the persistent state is also characterized by correlations between neurons throughout the brain. It is suggested that these persistent states are associated with short term memory while the eigenvectors of the transfer matrix are a representation of long term memory. A numerical example is given that illustrates certain of these features.

1. INTRODUCTION

In this paper we examine the long term behavior of a neuronal network such as the human brain. We will start from the assumption that the state of the brain at any time may be described by a configuration defined by the set of neurons that have fired within a certain specified recent interval of time and those that have not. We shall examine under what conditions a correlation can exist between states so defined, which are separated by a long period of time. In this context a long period of time is considered to be a time long compared to the refractory period of a neuron. The underlying reason for studying this problem is the belief that the states or configurations defined above are in some way related to the thought processes or experiences sensed by the individual and that in our own experience a long term correlation appears to exist in the latter. If our belief should prove to be valid it would imply a long term correlation between the neuronal configurations. We do not offer any proof that the thought processes and the neuronal configurations are representations of the same thing but suggest that this is a reasonable working hypothesis. Its proof or disproof lie beyond the scope of this paper.

Starting from the state defined above and by using the known behavior of the neuron and the interneuronal connections, we will show how the state of the brain evolves with time. We find that a close analogy exists between this problem and the problem of an interacting spin system which has been studied extensively in statistical mechanics and in solid state physics. The existence of persistent states in the neuronal network corresponds to the occurrence of long range order in the spin problem and both are related to the existence of a degeneracy of the maximum eigenvalue of a certain matrix. This matrix in the neuronal system is determined by the topology of the neuronal network, the size and nature of the synaptic junctions and the various electrochemical potentials in the brain. While the problem of determining the detailed behavior of this matrix is formidable, certain general conclusions can be drawn from this analysis which are interesting. We find that while the number of possible states in the brain as defined above is enormous—of the order of 2^N where N is the number of neurons (of the order of 10^{10} in the human brain)—the number of states which determine the long term behavior is a very much smaller number. This represents a tremendous simplification. If these states could be identified it would provide great insight into the operation of the brain. A second feature that follows from this is that these persistent states are distinguished by the property that a coherence or correlation exists between the neurons *throughout* the entire brain or large portions of it. These states are thus a property of the brain as a whole rather than a localizable entity. Thirdly, one finds that the transformation from the uncorrelated to the correlated state in a portion of, or in the whole brain can occur by the variation of the mean biochemical concentrations in these regions, and that this transformation occurs in a manner closely similar to the phase transition in the analogous spin system. Our results suggest how items of memory stored, in our model, in the topology of the interneuronal connection and the properties of the synaptic junctions, may be recalled to active use as a pattern of firing neurons.

Our results are not exact but have been reached only after making certain simplifying assumptions. While the assumptions themselves appear to be somewhat innocuous we have not been able to dispense with them and, whether our conclusions would remain if we could, has still to be proven. This work should thus be viewed as suggestive rather than definitive.

2. SIGNIFICANCE OF PERSISTENT STATES

Our thesis is based on the argument that the existence of states or behavior in which a correlation exists for long periods of time, are of prime importance to an understanding of the brain. There are many levels at which the significance of these persistent states can be appreciated. Clearly the ability of an animal to entertain one concept such as "flight from a predator" over a period of several minutes has high survival value. Likewise the capability of retaining the details of a recent attack over a longer period so as to analyze and review possible alternative defensive tactics in the future is similarly of survival value. In general, the ability to retain a theme or leit-motif while examining its many consequences over a period of time is of broader value at a more sophisticated level. The ability to meditate over long periods of time without external stimuli is another example of the existence of a long time correlated state of mind. At a much deeper level the conviction each of us has of our own existence is based to some extent on a certain internal continuity and coherence of behavior of ourselves over a long period of time.

On these admittedly imprecise, but suggestive grounds we argue that long term correlations exist within the brain. Assuming that these correlations imply correlations in the neuronal configurations, we ask what the conditions are in order for the latter to occur.

3. PHYSIOLOGICAL CONSIDERATIONS

The basic physiological structure of the brain has been studied extensively [1]. The important building blocks are the neurons which are bulbous nerve fibers illustrated in Fig. 1. Each neuron can be activated by

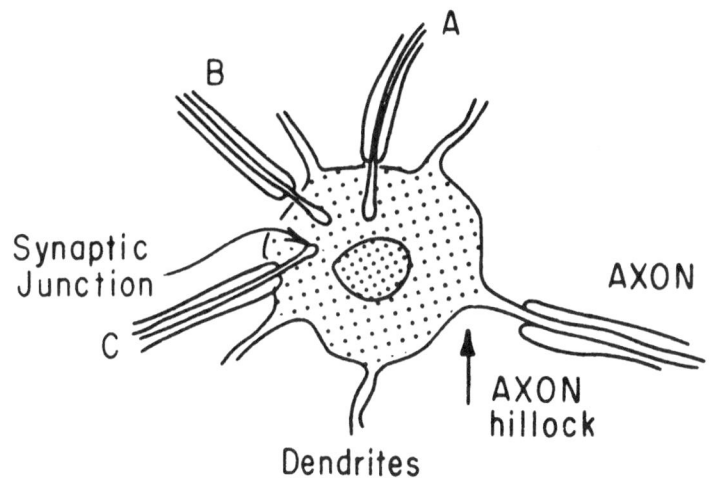

Neuron

FIG. 1. Schematic view of typical neuron.

the flow of an activating chemical across the synaptic junctions from the axon of one neuron either onto the surface or onto the dendrites of another neuron. Alternately, the neuron may be inhibited by inhibitory synapses at which an inhibiting chemical is transmitted across the synaptic gap to the neuron. The transmission of these chemicals causes a change in the ionic concentration within the neuron and this results in a change of its electrochemical potential. These electrical effects are referred to as excitatory postsynaptic potentials, or inhibitory postsynaptic potentials, respectively. If the net potential at the axon hillock resulting from all the excitatory and inhibitory post synaptic potentials exceeds a certain threshold level the neuron "fires" and an action potential propagates down the elongated tail of the neuron termed the axon and normally terminates on the synaptic junction of another neuron. The arrival of the action potential at this synapse triggers the release of the activating or inhibiting chemical, thus activating or inhibiting the next neuron and so the process continues.

The potential of the neuron is determined by the integrated effect of all the excitatory and inhibitory post synaptic potentials delivered to it over an integrating period of several msec. This is the period of latent summation. If the threshold is reached and it fires, a sharp positive pulse appears followed by a negative going excursion [2]. The potential then returns to the resting potential after a few msec. During this latter refractory period the neuron is recovering and cannot fire again.

We note also that the velocity of propagation of the action potential along the axon is about a 10^2 cm/sec in the axons within the brain and the mean length of axon from one neuron to the synaptic junction of another is no more than about 10^{-2} cm. Thus the flight time of a signal from one neuron to the next ($\approx 10^{-1}$ msec) is appreciably less than the refractory period of a neuron.

Our model is based on using in simplified form these various facts in order to determine how the state of the brain evolves with time.

4. MODEL SYSTEM

First, let us consider a neural network in which there are no connections to nerve cells which lie outside the network itself. Thus we consider the network as isolated from external stimuli. Alternately, one may consider the network as part of the brain but situated in a deprived environment receiving no external stimuli. Later we shall consider how this restriction can be removed or relaxed.

Second, we shall suppose that the neurons are not permitted to fire at any random time but rather that they are synchronized such that they can only fire at some integral multiple of a period τ which is of the order of the refractory period of the neuron. We suppose that the net neuron potential at the end of this period is determined by the sum of all the excitatory and inhibitory post synaptic potentials that occur during the period. The value of this potential at the end of the period then determines whether the neuron will fire or will not fire. Further we suppose that the influence of these potentials decays to a negligible value by the time the following opportunity to fire occurs. Thus within each period we may consider each neuron as starting with a clean slate, or in other words that these processes are Markoffian. Later we will consider how this limitation may be relaxed.

Third, we will assume that the connections between the neurons via the axons; and the properties of the synaptic junctions themselves are all fixed and do not change with time. We are not concerned here with learning behavior in which changes might be expected to occur in some one or other of these connections but rather we are interested in the behavior of the network in which these properties are assumed fixed. We believe and later will give arguments to bolster this belief that these given properties represent hereditary information or long term memory while the pattern of firing neurons defined by our "states" are related to short term memory involving the active state of the mind.

These assumptions appear to impose some rather artificial constraints upon the system. They are imposed for reasons of mathematical convenience. By so doing we are able to calculate certain properties of the system. However, we will argue on physical grounds by analogy with other related systems that the properties, which we calculate with these constraints, can be expected to remain even when certain of the constraints are relaxed.

With the above constraints we can define the "state of the brain" by ·the configuration determined by the set of neurons that have fired most recently and those that have not. It is convenient to write this using terminology borrowed from quantum mechanics [3]. We define that state of the brain at time t as the configuration

$$\psi(t) \equiv |s_1, s_2, \ldots, s_N\rangle, \tag{1}$$

where $s_i = +1$ if the ith neuron has just fired and $s_i = -1$ if the ith neuron has not fired, and N is the total number of neurons. There are thus 2^N states defined.

The neurons which have fired will then send a signal down their axons to their terminating synaptic junctions. These signals will trigger the release of the excitatory or inhibitory chemicals which in turn will raise or lower the potential of the neurons to which they are attached. Let us define V_{ij} as the resulting change in potential of the ith neuron due to an activating signal arriving from the jth neuron at the synaptic junction connecting the axon of j to the neuron i. This incremental potential V_{ij} may be positive (excitatory), negative (inhibitory), or may be zero if no synaptic connection exists between j and i. It will also depend upon the size and structure of the synaptic junction and the size or volume of the neuron to which it is attached. We will assume further that the V_{ij} are fixed and do not change with time.

We make use of the fact that the various postsynaptic potentials sum within a period so that the net change of potential of the ith neuron will be given by the sum of the various contributions from the activated synapses. If the existing state of the brain is given by (1) then the sum of the post synaptic potentials of the ith neuron may be written as

$$\sum_j V_{ij}\left(\frac{s_j + 1}{2}\right).$$

We see that if $s_j = +1$ we get a contribution, V_{ij} to the sum but if $s_j = -1$ we get no contribution. If the total potential exceeds some threshold value V_0 the neuron will probably fire. It is convenient to express this mathematically as follows. Let $p(+1)$ be the probability that the ith neuron will fire then $p(+1)$ may be written as

$$p(+1) = \frac{1}{\exp -\beta\left\{\left[\sum_j V_{ij}\left(\frac{s_j + 1}{2}\right)\right] - V_0\right\} + 1}. \tag{2}$$

The behavior of this function is shown in Fig. 3. If the sum is appreciably less than V_0 the exponential is large and $p(+1)$ is small, i.e. the probability of firing is small. On the other hand if this sum is appreciably greater than V_0 then the exponential becomes small and $p(+1)$ approaches unity, i.e. the neuron almost certainly fires. The factor β gives a measure of the uncertainty in the width of the threshold region.

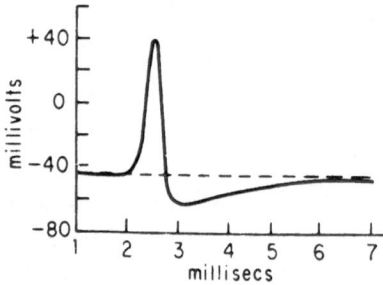

FIG. 2. Axon potential showing positive going signal and refractory period [2].

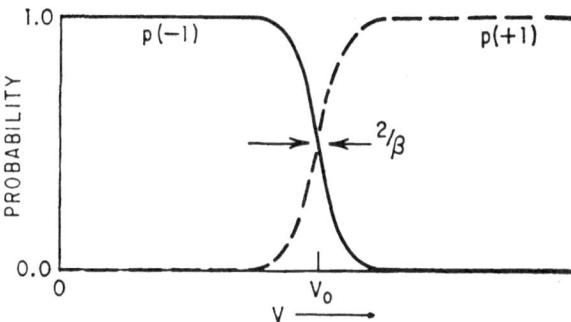

FIG. 3. Probability of neuron firing $p(+1)$ or not firing, $p(-1)$ as a function of the sum of the post-synaptic potentials V relative to the threshold, V_0.

The probability of *not* firing, $p(-1)$ is given by $1 - p(+1)$ and thus can be simplified to give

$$p(-1) = \frac{1}{\exp + \beta\left\{\left[\sum_j V_{ij}\left(\frac{s_j + 1}{2}\right)\right] - V_0\right\} + 1}. \tag{3}$$

Notice that both (2) and (3) may be expressed as

$$p(s'_i) = \frac{1}{\exp - \beta s'_i\left\{\left[\sum_j V_{ij}\left(\frac{s_j + 1}{2}\right)\right] - V_0\right\} + 1}. \tag{4}$$

Strictly speaking β and V_0 should also be considered as dependent on i but to simplify the problem we will treat them as constants throughout the network.

We may use these expressions to compute the probability of obtaining a state $|s'_1, s'_2, \ldots, s'_N\rangle$ given a state $|s_1, s_2, \ldots, s_N\rangle$ immediately preceding it. Using the usual bra and ket notation of quantum mechanics [3] and defining an operator P which yields this probability, Eq. (4) then gives us the result that:

$$\langle s'_1, \ldots, s'_N | P | s_1, \ldots, s_N \rangle$$

$$= \prod_{i=1}^{N} \left(\frac{1}{\exp - \beta s'_i\left\{\left[\sum_j V_{ij}\left(\frac{s_j + 1}{2}\right)\right] - V_0\right\} + 1}\right). \tag{5}$$

This then defines a $2^N \times 2^N$ matrix whose elements give the probability of a particular state $|s_1, s_2, \ldots, s_N\rangle$ yielding after one cycle the new state $|s'_1, s'_2, \ldots, s'_N\rangle$. The primed set refer to the row, $\langle s'_1, s'_2, \ldots, s'_N|$ and the unprimed set to the column $|s_1, s_2, \ldots, s_N\rangle$ of the element of the matrix.

5. ANALOGY WITH A SPIN SYSTEM

Having expressed the problem in this form one can see immediately the close similarity between it and the problem of an Ising system. Kramers and Wannier [4] in their classic paper on the Ising problem showed how the partition function for that spin problem could be expressed in terms of a matrix. The similarity is made more apparent by rewriting (5) in the form:

$$\langle s_1', \ldots, s_N' | P | s_1, \ldots, s_N \rangle$$

$$= \frac{\displaystyle\prod_{i=1}^{N} \exp \beta \frac{s_i'}{2} \left\{ \left[\sum_j V_{ij} \left(\frac{s_j + 1}{2} \right) \right] - V_0 \right\}}{\displaystyle\prod_{i=1}^{N} \sum_{s_i' = \pm 1} \exp \beta \frac{s_i'}{2} \left\{ \left[\sum_j V_{ij} \left(\frac{s_j + 1}{2} \right) \right] - V_0 \right\}} \tag{6}$$

This should be compared with Eq. 3.5 of the review of Ferromagnetism by Newell and Montroll [5].

The methods for handling this type of Ising problem have been discussed at length in many papers and texts. We refer the reader to the review of Newell and Montroll [5], Huang's text on Statistical Mechanics [6] and, for what follows, the paper by Ashkin and Lamb [7]. In the latter the question of the propagation of order in a crystal lattice is discussed. We shall study the analogous problem to this in the neural network. In the crystal problem one considers the configuration of atomic spins on a row of atoms with each atom having a spin of one half. One such configuration can be described by a state analogous to Eq. (1) i.e. $|s_1, s_2, \ldots, s_N\rangle$, where $s_i = +1$ means spin i is "up" and $s_i = -1$ that spin i is "down." The total, N refers to the number of atoms in the row. Due to their interaction with the spins on the next row the probability of a configuration $|s_1', s_2', \ldots, s_N'\rangle$ occurring in the next row can be calculated. An expression somewhat similar to (6) is then obtained. By using the same process again and again one can calculate the probability of obtaining a particular configuration in the mth row. In that problem a question of great importance is whether or not a correlation can exist between a configuration in row q, say, and row r where the distance between q and r becomes very large. When such a correlation does occur we say that long range order exists in the lattice. For a spin system which becomes ferromagnetic it is found that long range order sets in at the Curie point and exists at all temperatures below that. As shown by Lassettre and Howe [8] and discussed further by Ashkin and Lamb the onset of this long range order is intimately associated with the occurrence of a degeneracy of the maximum eigenvalue of the matrix analogous to (6).

In our problem a configuration determines the state of the brain at a particular instant of time. This corresponds to the configuration of spins in *one row* of the lattice in the crystal problem. The configuration which describes the state of the brain after the *next cycle* corresponds to the spin configuration *in the next row*. Thus the existence of a correlation between two states of the brain which are separated by a long period of time is directly analogous to the occurrence of long range order in the corresponding spin problem. We will show that the occurrence of these persistent states is also related to the occurrence of a degeneracy of the maximum eigenvalue of the matrix P given in (5).

We draw the analogy between the neural network and the two dimensional Ising problem. The configuration of N spins in one row corresponding to the configuration of N neurons at one particular instant of time. An analogy could equally well be drawn between a three dimensional Ising problem and the neural network. To do this one would associate the N spins in one *layer* of the crystal with N neurons at one instant of time: the next layer being associated with the next instant of time. Either analogy is equally good, for we note that in the neural problem we have no interaction terms V_{ij} such that i and j are both in the primed set or both in the unprimed set of s_i's. This would correspond in the 2-D spin problem to an interaction between spins on adjacent rows only, with no interaction between spins on the same row. Thus the geometric arrangements of the spins in a row (as in the 2-D analogy) or a layer (as in the 3-D analogy) is irrelevant and we may thus use either.

A nontrivial difference between our problem and the spin problem is that in the spin problem the matrix corresponding to P is symmetric. It is thus diagonalizable. In our case we have no guarantee that the matrix P will be diagonalizable without knowing the neuron connections. Moreover, it is reasonably certain that the matrix P will not be symmetric because the signals very clearly propagate from one neuron down its axon to the synaptic junction of the next neuron and not in the reverse direction. Only by accident would one have an identical path also running in the reverse direction. However, we will make the assumption here that P is diagonalizable. Later we will show how our argument may be extended to the situation in which P is not diagonalizable.

The occurrence of persistent states in the neural network will now be examined using a similar approach to that used for the study of long range order in the spin systems.

6. LONG RANGE ORDER AND PERSISTENT STATES

First, it is useful to note that the probability of obtaining a configuration $|s_1', \ldots, s_N'\rangle$ after two cycles is

$$\sum_{s_1'', \ldots, s_N''} \langle s_1', \ldots, s_N'|P|s_1'', \ldots, s_N''\rangle \langle s_1'', \ldots, s_N''|P|s_1, \ldots, s_N\rangle, \qquad (7)$$

or in matrix notation

$$\langle s'_1, \ldots, s'_N | P^2 | s_1, \ldots, s_N \rangle, \tag{8}$$

and thus after m cycles

$$\langle s'_1, \ldots, s'_N | P^m | s_1, \ldots, s_N \rangle. \tag{9}$$

To contract our notation let us use $\psi(x)$ to represent the state $|s_1, \ldots, s_N\rangle$ and $\psi(x')$ for $|s'_1, \ldots, s'_N\rangle$. Then it is useful to represent these in terms of the eigen vectors φ, of the operator P. There are 2^N such eigen vectors each of which has 2^N components $\varphi_r(x)$, one for each configuration x. Thus we have

$$\psi(x) = \sum_r \phi_r(x), \tag{10}$$

and assuming, we normalize $\varphi_r(x)$ to unity, so

$$\sum_x \phi_r(x)\phi_s(x) = \delta_{rs}, \tag{11}$$

we obtain

$$\langle s'_1, \ldots, s'_N | P | s_1, \ldots, s_N \rangle = \sum_r \lambda_r \phi_r(x')\phi_r(x), \tag{12}$$

where λ_r, is the rth eigenvalue.

We wish to find now the probability of obtaining a particular configuration x'. In general we do not know the initial conditions so we will set up the problem in such a way that the initial conditions play no role. One way to do this is to allow the system to run for a total of M cycles where M is very large and ask for the probability of obtaining a configuration x after m cycles, and then average over all initial configurations and sum over all final configurations. A simpler procedure which gives the same result is to assume that after M cycles the system returns to the initial configuration and we average over all initial configurations. This corresponds to cyclic boundary conditions in the spin problem. Using (9) we obtain the probability $\Gamma(x_1)$ of obtaining the configuration x_1 after m cycles.

$$\Gamma(x_1) = \sum_x \langle x | P^{M-m} | x_1 \rangle \langle x_1 | P^m | x \rangle \Big/ \sum_x \langle x | P^M | x \rangle, \tag{13}$$

which from (12) gives

$$\Gamma(x_1) = \sum_x \sum_{r,u} \phi_u(x)\lambda_u^{M-m}\phi_u(x_1)\phi_r(x_1)\lambda_r^m\phi_r(x) \Big/ \sum_r \lambda_r^m. \tag{14}$$

Using (11)

$$\Gamma(x_1) = \sum_r \lambda_r^M \phi_r^2(x_1) \Big/ \sum_r \lambda_r^M. \tag{15}$$

We notice that this is independent of m and hence gives the probability at any time of obtaining x_1, assuming no constraints or knowledge of the system at an earlier time.

Next we ask what the probability is of obtaining a configuration x_2 after l cycles given that we know we have configuration x_1 after m cycles. This joint probability $\Gamma(x_1, x_2)$ is given by

$$\Gamma(x_1, x_2) = \sum_{\gamma} \langle x|P^{M-l}|x_2\rangle\langle x_2|P^{l-m}|x_1\rangle\langle x_1|P^m|x\rangle \Big/ \sum_{x} \langle x|P^M|x\rangle. \quad (16)$$

$$= \sum_{\gamma} \sum_{r,u,v} \lambda_r^{M-l}\phi_r(x)\phi_r(x_2)\lambda_u^{l-m}\phi_u(x_2)\phi_u(x_1)\lambda_r^m\phi_r(x_1)\phi_r(x) \Big/ \sum_r \lambda_r^M. \quad (17)$$

which again, through the use of (11), gives

$$\Gamma(x_1, x_2) = \sum_{r,u} \lambda_r^{M-l+m}\lambda_u^{l-m}\phi_r(x_2)\phi_u(x_2)\phi_u(x_1)\phi_r(x_1) \Big/ \sum_r \lambda_r^M. \quad (18)$$

If M and $l - m$ are large numbers then the only significant contribution to (18) will come from the maximum eigenvalues. Let us assume first that these eigenvalues are nondegenerate then (18) gives

$$\Gamma(x_1, x_2) = \phi_{\max}^2(x_2)\phi_{\max}^2(x_1). \quad (19)$$

while the probability of obtaining x_1 is obtained from (15) giving

$$\Gamma(x_1) = \phi_{\max}^2(x_1). \quad (20)$$

Thus we have

$$\Gamma(x_1, x_2) = \Gamma(x_2) \cdot \Gamma(x_1). \quad (21)$$

In this situation we see that the joint probability is just the product of the probabilities of obtaining the two configurations independently. In this case the influence of configuration x_1, does not affect the probability of obtaining the configuration x_2. The network then does not have any persistent states.

On the other hand if the maximum eigenvalue of P is degenerate then the degenerate eigenvalues contribute in the sum of (18). Consider the simplest case when $\lambda_{\max}$ is doubly degenerate having eigen functions φ_1 and φ_2. then (18) becomes

$$\Gamma(x_1, x_2) = \{\lambda_1^M\phi_1^2(x_2)\phi_1^2(x_1) + \lambda_2^M\phi_2^2(x_2)\phi_2^2(x_1)$$
$$+ \lambda_1^{M-l+m}\lambda_2^{l-m}\phi_1(x_2)\phi_2(x_2)\phi_2(x_1)\phi_1(x_1)$$
$$+ \lambda_2^{M-l+m}\lambda_1^{l-m}\phi_2(x_2)\phi_1(x_2)\phi_1(x_1)\phi_2(x_1)\} / (\lambda_1^M + \lambda_2^M). \quad (22)$$

and

$$\Gamma(x_1) = (\lambda_1^M\phi_1^2(x_1) + \lambda_2^M\phi_2^2(x_1))/(\lambda_1^M + \lambda_2^M). \quad (23)$$

In this case $\Gamma(x_1, x_2)$ no longer factorizes as in (21). This result also holds even if λ_1 and λ_2 are not strictly degenerate but are sufficiently close in value for $|\lambda_1^M| \approx |\lambda_2^M|$. This is of some importance because of a theorem of Frobenius [7] which shows that the maximum eigenvalue of a matrix whose elements are all positive is nondegenerate. P is such a matrix as can be seen from Eq. (6). However, if the elements are *sufficiently small* then a practical degeneracy such that $|\lambda_1| \approx |\lambda_2^M|$ can still occur and $\Gamma(\alpha_1, \alpha_2)$ will no longer factorize. The probability of obtaining a configuration α_2 is then dependent upon the configuration α_1, and thus the influence of α_1 persists for an arbitrarily long time. This influence results from the two last terms in Eq. (22) and these involve the two eigenvectors associated with the degenerate maximum eigenvalues. We thus have the possibility of states occurring within the brain which are correlated over arbitrarily long periods of time. It is worth noting too that the characteristics of the states which so persist are describable in terms of the eigenvectors associated only with the degenerate maximum eigenvalues. In this sense these persistent states are very much simpler to describe than an arbitrary state of the brain for they involve only that small set of eigenvectors associated with the degenerate maximum eigenvalues, whereas other states of the brain are describably, in general, in terms of the full set of 2^N eigenvectors. This represents in principle, a very beautiful simplification of the behavior of the brain.

We have assumed that the cycle time, τ is of the order of a few msec. A time period t then corresponds to t/τ cycles or powers to which we raise the operator P. A few seconds thus corresponds to about a thousand cycles. For a correlation to exist for even a few seconds then the maximum eigenvalues must be degenerate to within a small fraction of a percent.

Another consequence of a degeneracy of the maximum eigenvalue is that under these circumstances and only under these can a correlation exist between neurons that are widely separated in the brain. To show this we define the topological "distance" between two neurons as the integer n_{ij} equal to the smallest number of synaptic junctions one need cross to get from the one neuron i to the other, j moving always from axon to neuron and not vice-versa. In general we expect this integer to be large for neurons which are widely separated and $n_{ij} \neq n_{ji}$. In order for the firing of i to influence the state of j, at least n_{ij} cycles must occur, so in order for neurons i and j to be correlated we need to have a state which persists for a period of time at least as long as $n_{ij}\tau$. The condition then for large spatial correlations is identical to the condition for persistent states, i.e. a degeneracy of the maximum eigenvalue of P. We see thus that the persistent states are characterized by a coherent or correlated behavior of the neurons throughout the brain or at least within large portions of it. By analogy with the spin system the long range order exists not only from row to row (i.e. in time) but also down the rows themselves (i.e. in space).

Finally we note that whether or not persistent order exists is determined by the properties of the matrix P, which in turn are dependent upon the parameters β, V_{ij}, and V_0 given in (6). In the analogous spin system β would be equal to $1/kT$ where k is Boltzmann's constant and T, the temperature: V_{ij} the interaction energy of the ith and jth spin and $\{\Sigma_j(V_{ij}/2) - V_0\} = H$, the interaction energy of a spin with a static magnetic field. (This can be seen by comparing (6) with Eq. (3.5) of Ref. [5].) The transition to the state of long range order occurs at the Curie point which, in the absence of a magnetic field, is determined by the ratio of V_{ij}/kT. The presence of the field, $H = \{\Sigma_j(V_{ij}/2) - V_0\}$ causes a shift of the Curie point and one finds a phase boundary in the T, H plane separating the ordered phase from the disordered phase. We expect therefore that the transition to the persistent state in the neural network would likewise be determined by the relative magnitudes of V_{ij}, H, and $1/\beta$. In our choice of the simple expression (4) we have lumped all the spread in the uncertainty of firing of the neuron in the parameter β. In the actual system we would expect an uncertainty in the size of V_{ij} and some fluctuation in the magnitude of H, both of which would be dependent on the local physiochemical conditions at the neuron. In our model these are the sources of the fluctuations which give rise to the finite width of the threshold curve. In our approximation we represent it by the single parameter β. By analogy to the spin system we would expect that for fixed values of the set of $\{V_{ij}\}$ a phase boundary could be defined in the H_0, $1/\beta$ plane. Or, more generally, a surface separating the ordered from the disordered state could be defined in the H, $1/\beta$, $\{V_{ij}\}$ space. In the simplest model we may assume that the set of $\{V_{ij}\}$ are scaled by the same factor γ and that the phase boundary is thus described by a surface in the three dimensional space, H, $1/\beta$, γ. We expect therefore that a change in the general physiochemical environment of the neurons which give rise to a shift in H or γ, could thus drive the network or a portion of the network across the phase boundary, transforming that portion from the coherently ordered, persistent state to the disordered state, or vice versa. In the Appendix we illustrate this with a numerical example. Our analogy suggests that such a transition cannot occur continuously but must occur discontinuously just as for other phase transitions in the solid state or for the liquid-gas transition.

One additional point worth stressing is that the expression $\Sigma_\alpha \langle \alpha | P^M | \alpha \rangle$, is directly analogous to the partition function for a lattice of M rows of N atoms in the spin problem so that all the corresponding behavior of the neuronal network can be deduced from it. This describes the time averaged behavior of the network because it involves the sum over configurations during a period of time, $t = M\tau$.

7. DISCUSSIONS OF ASSUMPTIONS OF THE MODEL

Our results have been derived on the basis of five principal assumptions or approximations. These are first, that the network has no external stimuli, second, that the probability of the neurons firing is a Markov process, third, that the neurons are synchronized, fourth, that the transfer matrix, P is diagonalizable, and fifth, that the properties of the synaptic junctions are fixed in time. We will discuss these in turn.

ROLE OF EXTERNAL STIMULI

There are two obvious ways in which one can take into account the presence of external stimuli. If the number of synapses from external sensors is small compared to the number of synapses connected to neurons within the network one could use perturbation theory to calculate the changes in the eigenvectors and eigenvalues of the matrix due to external stimuli. The sum over $\Sigma_j V_{ij}((s_j + 1)/2)$ in (4) would be replaced by

$$\left(\sum_j V_{ij}\left(\frac{s_j + 1}{2}\right) + \sum_k W_{ik}\left(\frac{e_k + 1}{2}\right) \right)$$

where W_{ik} is the postsynaptic potential of the ith neuron arising from a signal at the synapse from the kth external source, and $e_k = \pm 1$ depending whether such a signal is present or not. If we treat the sum over W_{ik} as a perturbation when we can show from standard perturbation theory [3] that to first order these terms cause a shift in the eigenvalues, with the eigenvectors remaining unchanged. For larger W_{ik} changes will occur in the eigenvectors as well. Thus, in principle, one could take these effects into account in this way, however, this procedure does not cast much light on the role the sensory inputs would play and we propose a second way of looking at this aspect of the problem.

We suggest that the input signals play a somewhat different role from the interneuronal signals. We know that a strong external signal results in a rapid series of nerve pulses at the synaptic junction. The effect of such a barrage of signals would be to generate a fairly constant average value for the term $(\overline{\Sigma_k W_{ik}((e_k + 1)/2)})$. Adding this average to V_0 transforms the effective threshold V_0 to a new threshold $(V_0 - \overline{\Sigma_k W_{ik}((e_k + 1)/2)})$ where the bar represents a time average. We suggest that this shift could drive the network or parts of the network across the phase boundary from the ordered to the disordered state or vice versa. Thus the external stimuli could play the role of initiating the onset of this persistent state and similarly other stimuli could terminate this state. If this view is correct our model suggests that the persistent state is a representation of long term memory. Which particular memory trace is uppermost would be determined by the eigenvalues which become the degenerate maximum eigenvalues under a particular form of the external stimuli given by the set of $\{e_k\}$.

We suggest that the different eigenvectors of the matrix P represent certain memories. Under external stimulus or stimulus from some other portion of the brain the eigenvalues are perturbed so two or more become degenerate and larger than any others. A new persistent state will then evolve dominated by the structure of the eigenvectors corresponding to these maximum eigenvalues and with initial conditions determined by the external stimuli, and thus the active state of the brain carry the information contained in the eigenvectors of P. As we pointed out earlier these eigenvectors are determined by the interneuronal connections and the strength of the synaptic junctions as given by the set of V_{ij}'s. Changes in these would change the properties of the eigenvectors. So in this model the process of learning would be any process which resulted in changes in V_{ij} and thus in the eigenvectors of P.

The above model of memory would then require that the sites where information is stored would be highly delocalized. This follows because the eigenfunctions of P are built up of contributions from neurons throughout the entire network. This can be seen from the inverse relationship to (10),

$$\phi_r = \sum_\alpha \psi_r(\alpha), \tag{24}$$

where, in general, contributions to φ_r come from all configurations, α.

MARKOV ASSUMPTION

We have made the assumption that only the most recent signals which reach a particular neuron determine whether it is to fire or not. We may relax this at the cost of greater mathematical complexity. This may be done as follows. Instead of describing the state of the brain by the configuration of neurons which have fired on the last cycle only we can describe it by the combination of the last two or more cycles. The transfer matrix then becomes correspondingly larger but can be handled in exactly the same way so that all our arguments go through as before. This generalization in the spin problem corresponds to including both near neighbors and next nearest neighbors, next next neighbors, etc. This can be handled in the standard way [5] using an expanded transfer matrix.

SYNCHRONIZATION OF THE NEURONS

Our third principal assumption was that the neurons could fire only at certain prescribed times. Clearly the method we have used hangs heavily upon this assumption requiring as it does an evolution of the neuronal configuration in a discontinuous manner. Our method cannot simply be modified to take into account a continuous evolution with time. To do this some other method would need to be devised to determine the nature of the long term correlations between the neuronal configurations. This is a formidable task. However, we suggest that the essential feature derived above, i.e. that a sharp distinction can be drawn between states of persistent order and those without this characteristic, will remain in a model in which the neurons are permitted to fire at arbitrary times. We base this conjecture on the following argument.

We have repeatedly invoked the analogy between the neural network and a two dimensional spin system. The model spin system in turn gives a remarkably good description of the phase transition from the paramagnetic to the ferromagnetic phase and of the liquid-gas phase transition. The two dimensional Ising system has been solved exactly using the matrix method. For its solution by this method one requires a strictly regular array of spins. The problem of a disordered lattice of spins cannot be solved by the matrix method because the interaction of one row of spins with the next cannot be uniquely defined where the concept of the row itself is lost as a result of the disorder. Yet we know from physical measurements that the occurrence of a ferromagnetic phase transition is not strongly dependent upon a high degree of order in the crystalline lattice. Indeed in an amorphous material such transition can still occur. Likewise the actual thermodynamic behavior of the liquid-gas transition is quite well described by the behavior of the analogous transition of the lattice gas [9]. In the lattice gas model the particles are only allowed to occupy mesh points on a regular lattice while in the real gas a particle can, of course, occupy any position. In spite of this difference the phase transition of the lattice gas is remarkably similar to that of a real gas [10]. This shows that the regularity of the lattice is not essential for the occurrence of the phase transition, it merely provides a mathematically convenient way of handling the problem.

In our model our assumption of the strictly regular synchronism of the firing of the neurons corresponds to a strictly regular crystalline array. By analogy with the above we suggest that just as the regularity of the lattice is not essential for the occurrence of the phase transition of the spin system or of the lattice gas, so the regularity in the firing is not an essential requirement for the occurrence of a transition to an ordered persistent state in the neural network. We believe therefore that our conclusions should remain even in a more realistic model.

DIAGONALIZABILITY OF THE CHARACTERISTIC MATRIX

Our fourth principal assumption is that the matrix P is diagonalizable. Without a knowledge of the topology and strength of the various terms in P we cannot tell a priori whether or not P can be diagonalized. We can show, however, that our results can be generalized to an arbitrary matrix, for while a general matrix cannot always be diagonalized it can be reduced to, so called, Jordan Canonical form [11]. In this form eigenvalues occur along the main diagonal with ones or zeros on the diagonal immediately below it. For example

$$P = \begin{vmatrix} \Lambda_1 & 0 & 0 & 0 & 0 & . \\ 0 & \Lambda_2 & 0 & 0 & 0 & . \\ 0 & 0 & \Lambda_3 & 0 & 0 & . \\ 0 & 0 & 0 & \Lambda_4 & 0 & . \\ 0 & 0 & 0 & 0 & \Lambda_5 & . \\ . & . & . & . & . & . \end{vmatrix} \quad \text{where} \quad \Lambda_i = \begin{vmatrix} \lambda_i & 0 & 0 & 0 & 0 & . \\ * & \lambda_i & 0 & 0 & 0 & . \\ 0 & * & \lambda_i & 0 & 0 & . \\ 0 & 0 & * & \lambda_i & 0 & . \\ 0 & 0 & 0 & * & \lambda_i & . \\ . & . & . & . & . & . \end{vmatrix} \quad (25)$$

where $* = 0$ or 1.

A representation of $\psi(\alpha)$ can now no longer be made in terms of eigenvectors alone but must be made in terms of principal vectors [11], $p(\alpha)$ which satisfy the matrix equation:

$$(P - \lambda_r I)^g p_r(\alpha) = 0, \quad (26)$$

where P is the matrix, g, an integer, is the grade of the principal vector, λ an eigenvalue, and I the identity matrix. If $g = 1$, $p_r(\alpha)$ is simply an eigenvector. For $g > 1$, this equation defines the principal vectors. An eigenvector $\psi(\alpha)$ can be derived from (26) as follows:

$$\psi(\alpha) = \frac{1}{(g-1)!}(P - \lambda I)^{g-1} p_r(\alpha). \quad (27)$$

It can be shown [12] that for large m the asymptotic form of

$$P^m p(\alpha) = m^{g-1}\lambda^m \psi(\alpha) + r^{(m)}, \quad (28)$$

where the remainder, $r^{(m)}$ is of order $m^{g-2}|\lambda|^m$. For the particular case of $g = 1$, $r^{(m)}$ is zero. Then we obtain the results used in (12) and (14). For the general case we must use the above asymptotic form to evaluate Eq. (13) and (16). This gives us

$$\langle \alpha'|P^m|\alpha \rangle \simeq \sum_{r,g} m^{g-1}\lambda_r^m \psi_{r,g}(\alpha')\psi_{r,g}(\alpha)$$

where $\psi_{r,g}(\alpha)$ is the eigenvector of eigenvalue, λ_r for the principal vector $p_r(\alpha)$ defined in (27). The conditions for persistent order then are that the maximum eigenvalues must be degenerate ($\lambda_1^M \approx \lambda_2^M$), and that their principal vectors must be of the same grade.

TIME-INDEPENDENCE OF MODEL PARAMETERS

We have assumed that V_{ij}, β, and V_0 are parameters which are fixed in time. It is reasonable to suppose that V_{ij}, in particular, might be influenced by learning. We might suppose that repeated firing of a given synaptic junction might result in a permanent change in its physical and chemical properties and thus in the corresponding value of V_{ij}. Our model has neglected this, however, it appears as if one could extend without great difficulty the model to include such nonlinearities. The basis of this hope is that in humans at least, in order to learn something new so that it becomes part of long term memory one needs to concentrate for at least several seconds. We expect therefore that changes in V_{ij} take a time of this order to occur. On the other hand we have shown that this corresponds to something of the order of a thousand operations of the matrix. So we see that the changes in V_{ij} are likely to be small between each operation of the matrix operator and thus the nonlinear behavior might be approximated by the time-averaged quantities determined in the linear approximation.

8. CONCLUSION

We have argued that in a neural network the occurrence of states in which a correlation persists between neuronal configurations separated by long periods of time can occur if and only if the maximum eigenvalues of a certain transfer matrix are degenerate. By analogy with other systems which show long range order we show that a transition to such a persistent ordered state is analogous to a phase transition. We also show that in the ordered state a correlation occurs between neurons widely separated in the network.

If such persistent states can be identified in the brain, their presence must surely be of considerable significance, for their presence would dominate the average values of any quantity determined by the neuronal configuration just as the crystalline order dominates the average properties of a crystalline solid. It is of some interest to note too that, in general, the degeneracy of the eigenvalues of a matrix reflect some symmetry in the system which it represents, in fact, this is illustrated by a numerical example in the Appendix. This suggests that the capability of having persistent states in a neural network should be shown by some symmetry of the interneuronal network and the properties of the synaptic junctions. It would be interesting to know if this could be seen in the general anatomy of the brain.

APPENDIX

A better appreciation of some of the features of the model can be obtained by a numerical analysis of a simple network. We have considered just four neurons connected in various ways, have computed the 16×16 P-matrix for various values of β, V_{ij}, and V_0, and studied the asymptotic behavior of P^M for large M.

As a simple example we let $V_{ij} = 1.0$ for all i and j and considered powers of the matrix up to P^{32}. For $\beta = 0.2$ and $V_0 = 2.0$ we find that the matrix P^M, for large M, is such as to have identical columns, a characteristic feature of a matrix normalized as P is, with a nondegenerate maximum eigenvalue. On the other hand, for $\beta = 5.0$ and $V_0 = 2.0$ this feature is lost making it possible for P^M to transfer information of the structure of the initial state to the final state. This is the condition for the existence of the persistent state. For these values of V_{ij} and V_0 the transition to the persistent state appears to occur at about $\beta \simeq 1.0$. A sharp transition is not expected, however, because the number of neurons is so small.

For an arbitrary choice of the V_{ij}'s and V_0 we find that we do not always obtain a persistent state even at $\beta = 5.0$. This suggested that in order to obtain a degeneracy in the maximum eigenvalue the matrix P must have some special symmetry. Upon examination of (5) we notice that for the particular choice of conditions such that $\Sigma_j(V_{ij}/2) - V_0 = 0$, the matrix P would be invariant under the operation which changes the sign of all s_i' and all s_j. The choice of parameters of our first example satisfied this symmetry condition. On the other hand we find that for either $V_0 = 4.0$ or $V_0 = 0.0$, both of which violate this condition, no persistent state was found for $\beta = 5.0$. In the vicinity of $V_0 = 2.0$, however, the persistent state is found. The phase boundary in the β, V_0 plane can thus be located.

Having recognized this symmetry principle we examined a more complicated situation with both inhibitory and excitatory synapses described by the values of the V_{ij} given in Table 1. These values of V_{ij} were chosen again to satisfy the above symmetry principle. Again we found a persistent state for $\beta = 5.0$ and $V_0 = 2.0$ but with a structure different from that of our earlier choice of V_{ij}. For $V_0 = 4.0$ and 0.0 at $\beta = 5.0$ the nonpersistent state was found.

TABLE 1

j	i 1	2	3	4
1	-1.0	-1.0	$+1.0$	$+1.0$
2	-1.0	$+1.0$	$+1.0$	$+1.0$
3	$+4.0$	$+2.0$	$+1.0$	$+1.0$
4	$+2.0$	$+2.0$	$+1.0$	$+1.0$

These results illustrate some of the essential features of the model: first, the existence of the persistent state, second, the existence of a phase boundary in the β, V_0 plane, and third, the existence of a symmetry principle for determining regions in which degenerate eigenvalues occur. One might expect the matrix to be invariant under certain other operations for other choices of V_{ij} and V_0. We expect that these would give rise to other types of persistent states.

REFERENCES

1 C. M. Smith, *The Brain*, G. P. Putmanns, New York (1970).

2 A. L. Hodgkin and A. F. Huxley, *Nature* **144,** 710 (1939).

3 L. I. Schiff, *Quantum Mechanics*, 3rd ed., McGraw-Hill, New York (1968).

4 H. A. Kramers and G. H. Wannier, *Phys. Rev.* **60,** 252 (1941).

5 G. F. Newell and E. W. Montroll, *Rev. Mod. Phys.* **25,** 353 (1953).

6 K. Huang, *Statistical Mechanics*, Wiley, New York (1963).

7 J. Ashkin and W. E. Lamb, Jr., *Phys. Rev.* **64,** 159 (1943).

8 E. N. Lassettre and J. P. Howe, *J. Chem. Phys.* **9,** 747, 801 (1941).

9 T. D. Lee and C. N. Yang, *Phys. Rev.* **87,** 410 (1952).

10 M. R. Moldover and W. A. Little, *Phys. Rev. Letters* **15,** 54 (1965).

11 J. N. Franklin, *Matrix Theory*, Prentice-Hall, Engelwood Cliffs, N.J. (1968).

12 Ibid., p. 275.

Physicist's proof of Fermat's theorem of primes

H. Gutfreund [a)] and W. A. Little

Physics Department, Stanford University, Stanford, California 94305

(Received 16 March 1981; accepted for publication 20 May 1981)

The prime numbers play a central role in the theory of numbers. We show that Fermat's theorem on primes may be proved using symmetry properties of Ising-spin configurations; and that similarly this may be extended to certain composite numbers. Our method of proof suggests a "physical" interpretation of the primes.

The prime numbers play a key role in the theory of numbers forming as they do the basis or representation of all composite numbers. In this note we show that there is a connection between the factorization of the integers into primes and the symmetry properties of configurations of Ising-spin systems. This suggests a certain "physical" interpretation of the primes that will be appreciated by physicists, conversant as they are with symmetry arguments.

One of the best-known divisibility theorems of elementary number theory, which played an important role in its development, is Fermat's theorem of primes: For every positive integer a and prime p, such that a is not a multiple of p, $a^{p-1} - 1$ is divisible by p [or, in compact notation $a^{p-1} = 1 \pmod{p}$]. The proof of this theorem, usually given in textbooks on number theory, is based on congruence arithmetic.[1] We present a proof of Fermat's theorem, which is based on the symmetry properties of one-dimensional Ising-spin configurations. For sake of clarity, we demonstrate the proof in three steps.

Let us first prove that for every prime p, $2^p - 2$ is divisible by p. Consider a ring of p sites and assign to each site i, an Ising-spin variable $s_i = \pm 1$. In the space of the 2^p possible configurations, $\alpha = (s_1, s_2, ..., s_p)$ we define a translation operator T that, operating on a given configuration, shifts all the spin variables by one site in, say, the clockwise direction. Let us divide all the configurations into classes such that two configurations α, β are in the same class if for some integer n, $\beta = T^n \alpha$. Clearly, each of the configurations with all spins "up," or all spins "down," is a class in itself. It is obvious that for every α, $\alpha = T^p \alpha$, because T^p corresponds to a complete rotation. We assert that for every α (other than the two trivial ones), $\alpha \neq T^n \alpha$, for $n < p$. Given this, then each configuration belongs to a class of p distinct elements, and therefore the number of all configurations minus the two trivial ones is divisible by p. It is clear that $\alpha = T^n \alpha$ can be valid only when a configuration has an integer number of repeating subperiods of lengths n, and for $n < p$, this would imply that n was a factor of p. For p prime this is impossible, except when $n = 1$, which gives the two trivial configurations.

The significance of p being a prime is that in this case no (nontrivial) configuration has a higher symmetry than the complete rotation. In other words this means that the Abelian group of rotations by the angle $2\pi/p$ has no subgroups.

We note that $2^p - 2$, which is divisible by p, must also be divisible by $2p$, because $2^p - 2$ is even and p odd. Hence, $2^{p-1} - 1$ is divisible by p. Using Ising-spin configurations we can obtain this result directly. Let us define an operator I that, operating on a given configuration, inverts the signs of all the spin variables. Again, we divide the configura-

tions into classes, such that two configurations α, β are in the same class if $\beta = (TI)^n \alpha$ for some integer n. For every α, we have $\alpha = (TI)^{2p} \alpha$, because p, being a prime, must be odd, and therefore a single rotation involves an odd number of inversions. Only after a double rotation does the configuration return to itself. Exactly as before, one now concludes that each configuration (except for the two trivial ones that constitute a class of two elements) belongs to a class of $2p$ distinct elements.

Finally, we generalize the considerations of the above for the case $a = 2$, to a general a, which then yields the normal form of Fermat's theorem. We now consider a spin j with $2j + 1 = a$, and each "Ising" variable s_i assumes one of the $2j + 1$ possible spin projections $(-j, -j+1, ..., j)$. There are a^p configurations from which we have to subtract a translationally invariant configurations. From our discussion, one immediately concludes that for every a, $a^p - a$ is divisible by p. To generalize the second step, we define the operator I so that the operation of I on a given configuration increases the variables s_i by one, with the reservation that when $s_i = j$ is "increased" by 1, it will be changed to $s_i = -j$. Clearly, for every α, $\alpha = (TI)^{ap} \alpha$. To ensure that (except for the trivial configurations) this does not happen for any $(TI)^n$ with $n < ap$, we have to require that a is not a multiple of p. Otherwise we would get $\alpha = (TI)^n \alpha$ for every α. Thus if a is not a multiple of p, each configuration belongs to a class of ap distinct elements so that $(a^p - a)/ap$ or $(a^{p-1} - 1)/p$ is an integer. This completes the proof of Fermat's theorem.

Note that the requirement that a is not a multiple of p, excludes the possibility of a symmetry of configurations higher than that of the product of complete rotations of the site positions and the site variables.

A novel aspect of the method of proof is that it allows one readily to generalize Fermat's proof for primes to a statement about composite numbers as well. As one example, if m is a product of two different primes p_1 and p_2 then one can show that

$$(a^{m-1} - 1) - (a^{p_1 - 1} - 1) - (a^{p_2 - 1} - 1)$$

is divisible by m. The terms which are subtracted here represent the number of configurations that fall into classes smaller than m having cyclic structures. One can generalize this result to more complex composite numbers. Consider a number $m = p_1^{a_1} \cdot p_2^{a_2} \cdots p_k^{a_k}$, where the p_i's are primes and the a_i's integers. Further let a be prime to m, then we will prove that

$$\left[(a^{m-1} - 1) - \sum_i (a^{(m/p_i - 1)} - 1) \right.$$

$$+ \sum_{ij} (a^{(m/p_i p_j - 1)} - 1) - \dots$$

$$\left. \dots + (-1)^k (a^{(m/p_1 p_2 \dots p_k - 1)} - 1) \right]$$

is divisible by m. This is the generalization of Fermat's theorem to composite numbers. The "proof" is again as before. The first term comes from the totality of configurations less the translationally invariant ones. The second term subtracts the cyclic configurations with period (m/p_i), but in so doing, we have subtracted the configurations with period $(m/p_i p_j)$ twice, because this is a subperiod in both the configurations with period (m/p_i) and (m/p_j). Therefore, to correct the sum we must add them once again. Now, however, we have added too much, because the configuration with period $(m/p_i p_j p_k)$ was first subtracted three times [in configurations (m/p_i), (m/p_j) and (m/p_k)] and then added three times [in configurations $(m/p_i p_j)$, $(m/p_i p_k)$, and $(m/p_j p_k)$] therefore we have to subtract it once more. And so on.... Finally, we obtain expression shown, which represents the number of confi rations free of any of the above subperiods. These rema ing configurations may be classified into classes of m me bers each, and hence, this number must be divisible by

We are indebted to Persi Diaconis for bringing to attention a somewhat similar proof of Fermat's theorem Golomb.[2] However, the physical significance of the prin or possible extensions of the theorem to composite nu bers are not discussed in this note.

In conclusion we have found an interesting analogy the primes that is related to a lack of symmetry of cert physical systems.

[a]Permanent address: Institute of Theoretical Physics, The Hebrew U versity, Jerusalem, Israel.

[1]C. S. Ogilvy and J. T. Anderson, *Excursions in Number Theory* (Oxf University, New York, 1966).

[2]S. W. Golomb, Am. Math. Mon. **63**, 718 (1956).

Scaling of Miniature Cryocoolers to Microminiature Size

W. A. Little
RAI
15 Crescent Dr.
Palo Alto, Ca. 94301

The very small size and thermal dissipation of superconducting sensors imposes an extremely small load on the supporting refrigeration system. Present day refrigerators generally are grossly mismatched to such loads. For this reason we have considered the possibility of scaling down a miniature refrigerator to microminiature size. A set of scaling laws is derived. These have been applied to the design of a N_2 - H_2 refrigerator which might use a NbN sensor. Such microminiature refrigerators appear to offer a number of attractive design and manufacturing advantages.

Key words: refrigerator, cryocooler, Joule-Thomson cycle, heat exchanger design.

1. Introduction

During the past decade a host of new superconducting devices have been developed which are based on the Josephson effect. These include supersensitive magnetometer, gradiometers, voltage standards, current comparators, rf attenuators, bolometers and logic elements. In these devices the power dissipated in the cryogenic environment is typically of the order of microwatts. However, the refrigeration systems which are available to maintain the low temperature environment for the devices have a capacity of watts to tens of watts. Such refrigerators are thus poorly matched to the refrigeration requirements of the sensors. For this reason we have considered the problem of scaling down a small refrigerator to micro-miniature size. We have derived a set of scaling laws which allow one to estimate the size and performance of such a refrigerator. The proposed microminiature refrigerators appear to offer a number of attractive features both in manufacture and operation, viz., ease of mass production, compact construction, fast cooldown and long operating time. On this basis one can visualize the superconducting sensor and refrigerator as a single package, the size of a miniature vacuum tube which would be inexpensive, replaceable and disposable.

We have considered three refrigeration systems for miniaturization, the Kirk-Stirling, Gifford-McMahon and Joule-Thomson systems. The first two cycles are more efficient in general than the last but are somewhat more complex in view of their several moving parts. Because of the simplicity of the J-T system and its absence of moving parts we chose it for further study. Our analysis focusses primarily on the properties of the heat exchanges so has some relevance to other cooling cycles.

Miniature J-T refrigerators using N_2, H_2 and He circuits are commercially available [1] and have been described in the literature [2]. These refrigerators typically produce watts of refrigeration at 80K, 20K or 4K. Our purpose was to scale this down by a factor of the order of a hundred.

2. Heat Exchanger Design Considerations

The design of countercurrent heat exchangers is described in two excellent reviews:

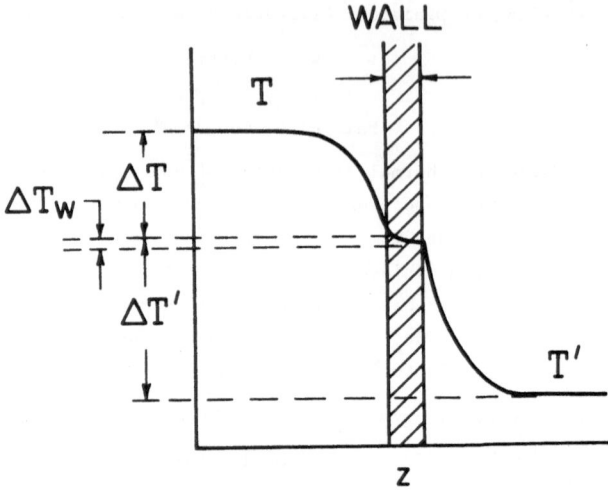

Figure 1. Schematic plot of the temperature profile from one gas to the other
counter flowing gas across the wall of a heat exchanger.

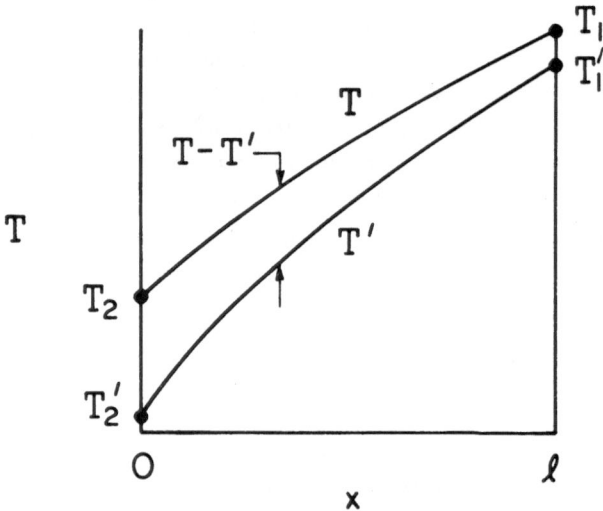

Figure 2. Temperature along the length of a countercurrent heat exchanger. T is t
temperature at each point in the exchanger of the incoming gas and T' th
of the outgoing gas.

"Experimental Cryophysics," Chapter 2 by Hoare, Jackson and Kurti, [3] and J. G. Daunt's article in the Encyclopedia of Physics, Springer Verlag [4]. The temperature profile from the one fluid at temperature T to the other at temperature T' across the wall of the exchanger is illustrated in Figure 1. Here ΔT is the drop in temperature in the gas on one side, ΔT_w the drop across the wall and $\Delta T'$ the drop in the gas on the other side. Daunt states that one can usually neglect ΔT_w, so if the heat transfer/sec/unit area of wall is $\dot q$ then $\dot q = \alpha \Delta T = \alpha' \Delta T'$, where α and α' are certain quantities to be derived later. Then $T - T' = \dot q/\kappa$, where $\frac{1}{\kappa} \approx \frac{1}{\alpha} + \frac{1}{\alpha'}$.

The heat transfer/sec/unit <u>length</u> of the exchanger, $d\dot Q$ will thus be given by

$$d\dot Q = \kappa (T - T') P d\ell = \dot m_1 \, Cp (\frac{\delta T}{\delta \ell}) d\ell \quad, \tag{1}$$

where P is the perimeter of the wall, Cp the heat capacity of the gas/unit mass and $\dot m_1$ is the mass flow of the input gas per unit time.

Likewise for the cold return

$$\dot m' \, C_p' (\frac{\delta T'}{\delta \ell}) d\ell = -d\dot Q \quad . \tag{2}$$

Integrating $d\dot Q$ from (1) we have

$$\int_o^\ell \frac{d\dot Q}{T-T'} = \kappa P \ell \quad, \tag{3}$$

assuming κ is constant, or where it is not, some mean value as discussed by Daunt.

Using the other part of (1) and equation (2) one obtains an expression for T and T' along the exchanger. This is illustrated in Figure (2). It is conventional to define the efficiency of the exchanger as $\varepsilon = 1 - \frac{T_1 - T_1'}{T_1 - T_2'}$ (See Figure 2). From the above integration for a "balanced" exchanger i.e. with no draw off of liquid it can be shown that $\varepsilon = \frac{\kappa P \ell}{\kappa P \ell + \dot m C_p}$.

It is convement to measure $\kappa P \ell$ in units of $\dot m C_p$. Then we have

$$\varepsilon = \frac{(\kappa P \ell / \dot m C_p)}{(\kappa P \ell / \dot m C_p) + 1} = \frac{N}{N+1} \tag{4}$$

and we call the ratio $(\kappa P \ell / \dot m C_p)$ the number of "thermal transfer units," N of the exchanger.

We can calculate α and α' and hence obtain κ as discussed in reference [1]. In general we want $\alpha \approx \alpha'$ so it is sufficient to calculate α. One finds

$$\ell P \alpha = 0 \cdot 10 \, C_p \eta^{0.2} \left(\frac{\dot m}{d}\right)^{0.8} \ell \quad, \tag{5}$$

η = viscoscity. Since $\frac{1}{\kappa} \approx \frac{1}{\alpha} + \frac{1}{\alpha'} \approx \frac{2}{\alpha}$, so

$$\kappa P \ell \approx 0.05 \, C_p \dot m \left(\frac{\eta d}{\dot m}\right)^{0.2} \left(\frac{\ell}{d}\right) \tag{6}$$

where d = diameter of the tube. Reynolds number Re is defined as Re = $(4\dot{m}/\pi\eta d)$. In terms of it

$$N = 0.05 \left(\frac{\pi}{4} Re\right)^{-0.2} \left(\frac{\ell}{d}\right) \tag{7}$$

The pressure drop in the high pressure line can be related to N. Following reference [1] again

$$\Delta p = 0 \cdot 10 \; \rho v^2 \left(\frac{\ell}{d}\right) (Re)^{-0.2} \tag{8}$$

But v = $(4\dot{m}/\rho\pi d^2)$ so

$$\Delta p = 2\rho v^2 \star N \star \left(\frac{\pi}{4}\right)^{0.2} \approx 2N\rho v^2 \tag{9}$$

So $(\Delta p/\rho v^2)$ = M = number of "pressure heads" and therefore M = 2N. Hence

$$\Delta p = 2N \left(\frac{16\dot{m}^2}{\pi^2\rho d^4}\right) \tag{10}$$

3. Scaling Considerations

The amount of refrigeration for a given gas is proportional to $\dot{m}$. For a refrigerator of given exchanger efficiency working between fixed T_1, T_1', T_2 and T_2', N is fixed. For the same efficiency and with the same pressure drop we see from (10) that this requires

$$\boxed{d \approx (\dot{m})^{0.5} \qquad \text{1st Scaling Law}}$$

We see from the definition of Reynolds number taking in to account the above constraint on d that Reynolds number Re is proportional to $(\dot{m})^{0.5}$ and hence from (7) for fixed N, that

$$\boxed{\ell \approx (\dot{m})^{0.6} \qquad \text{2nd Scaling Law}}$$

The intrinsic cooldown time of the exchanger itself is determined by the ratio of mass of exchanger to the refrigeration capacity. If we assume a fixed aspect ratio of the heat exchanger tubing then its mass will be proportional to $d^2\ell$, which from the above is proportional to $(\dot{m})^{1.6}$. Hence the cooldown time is given by

$$\boxed{\text{Cooldown time} \approx (\dot{m})^{0.6} \qquad \text{3rd Scaling Law}}$$

The length of time the exchanger can operate before fouling up with contaminants one would expect to be proportional to ratio of the volume of the exchanger to $\dot{m}$. The volume is proportional to ℓd^2 and hence from the above

$$\text{Operating time} \approx (\dot{m})^{0.6} \qquad \text{4th Scaling Law}$$

This indicates that the microminiature refrigerators would be somewhat more sensitive to contamination (eg. Ne in H_2) and hence would need high purity gas. However, as we shall show the very fast cooldown could allow one to warm up, flush clean and cool down again in less than a minute. This might be an acceptable mode of operation for several instrumentation applications.

As an example of the above scaling we consider a 110 milliwatt liquid nitrogen refrigerator scaled from an Air Products J-T refrigerator [5] which produces 7 watts of refrigeration at 80K using 1360 st.litres/hr of N_2. Typical exchanger efficiency is 95% hence N = 20 and we assume 100 atm inlet pressure and $\Delta p \approx 6$ atm. Using the above scaling and the dimensions of the larger refrigerator we find $d \approx 10^{-2}$ cm, $\ell \approx 24$ cm, Re ≈ 8000 and a cooldown time of 25 seconds. The greatly reduced gas consumption would allow operation of the order of 400 hours per standard cylinder of gas.

Scaling of a similar $N_2 - H_2$ refrigerator would allow operation at 20K and below with a refrigeration capacity of 60 milliwatts.

The very small size of the heat exchanger system suggests the possibility of constructing the heat exchanger, expansion valve and particulate filter using planar photoresist technology similar to that used in the semiconductor industry. The successful development of such a technique would greatly reduce the price of the refrigeration unit and make it feasible to design and construct as an integral package the refrigerator and superconducting circuit elements for use in ambient temperature circuitry. Mercereau has recently demonstrated [6] the operation of a Nb_3Sn quantum interference device at 17K. Similar results can be expected for NbGe at 21K. These temperatures are within range of those attainable with J-T refrigeration alone using the $N_2 - H_2$ cycle.

4. Acknowledgements

I am indebted to Dr. J. Mercereau for several useful discussions on the refrigeration needs of superconducting sensors.

5. References

1] For example; Bendix Instrument and Life Support Division, Hickory Grove Road, Davenport, Iowa 52808; Air Products, Inc., Allentown, Pennyslvania; Santa Barbara Research Center, 75 Coromar Dr., Goleta, California.

2] Geist, J. M. and Lashmet, P. K., Adv. Cryogenic Engineering 5, 73 Plenum Press, N.Y. (1961); Buller, J. S., Advances in Cryogenic Engineering 16, 205 (1970).

3] Hoare, F. E., Experimental Cryophysics, pg. 14, edited by Hoare, F. E., Jackson, L. C., and Kurti, N., Butterworth and Co. (Publishers) Ltd 1961.

4] Daunt, J. G., Encyclopedia of Physics, pg. 1, edited by Flügge, S., Vol. XIV, Low Temperature Physics I, Springer-Verlag, Berlin 1956.

[5] Geist, J. M. and Lashmet, P. K., Adv. Cryogenic Engineering 5, 73 Plenum Press, N.Y. (1961), and data on commercially available units from Air Products Inc.

[6] Palmer, D. W., Notarys, H. A., and Mercereau, J. E., Appl. Phys. Lett. 25, 9 (1974) 527-528.

Organic Superconductivity: The Duke Connection

W. A. Little

I would like to start by reminiscing a little about some of the early days when William Fairbank first came to Stanford in 1959. I had come the previous year as an assistant professor. My responsibility was to install the helium liquefier and a helium pumping system and to set up a laboratory in the old physics building, so that there would be some low temperature facilities available for Bill when he arrived from Duke.

About halfway through that first year we had most of the equipment working and we started on our own rather modest program of research. This was directed to the production of very low temperatures by the adiabatic demagnetization of various salts and metals. It was before the time that ^{3}He was generally available, so it was much more difficult to reach temperatures which are readily attainable now. In thinking back, the work that we were doing was very mundane physics. We were interested in heat transfer, the problems of the Kapitza resistance, various problems of phonon physics, and some work on superconductivity. At that time it was customary to produce large magnetic fields for adiabatic demagnetization by using water-cooled or kerosene-cooled magnets. It was a very unpleasant business! Seeking to avoid these problems I had thought of using a superconducting magnet. In the literature I had come across some work on niobium which indicated that niobium could carry superconducting currents at relatively high fields, that is, above 6 kilogauss; and late in 1959 we built a solenoidal magnet. I believe it was one of the first superconducting magnets that was ever built for a useful purpose. It was wound of niobium wire which the manufacturer had failed to anneal, and as a result it was able to carry a lot more current than we anticipated. It produced a field of about 8 kilogauss. We did not understand why, but it worked, so we used it. However, my opinion of superconductivity as a field of research at that time was rather low, because the superconductive properties were so dependent upon the presence of impurities and dislocations in the metal. It appeared to me to be a dirty field!

It was then that Bill burst on the scene—like a whirlwind, coming in from the South—as an evangelist preaching as he went. He was totally overwhelming. I did not know what he was saying, but he talked, and he talked and he talked! I remember many times thinking I would never get home. As I was leaving the laboratory at 5:30 p.m., Bill would catch me in the corridor and walk with me to my car. We would get to the car and then he would put his foot in the door and talk, and talk, while I got hungrier and hungrier! Eventually I got away—to a cold dinner! But what Bill was talking about was important. He brought with him the concept of "long range order" in superfluids and the idea of the "order parameter." That was not how he said it, as the message was not so clear at that time, but in retrospect that was what he was trying to tell me.

I remember one story which really hit home. This was about a beautiful experiment that he and Michael Buckingham had done on the lambda point of helium. This experiment was inspired by an argument due to Feynman who pointed out that the shape of the lambda transition and its sharpness should be dependent upon the size of the sample no matter how big, "even if it were the size of the sun." (I think this is how Bill put it, with his foot in my car.) I thought about that a great deal. I think I understand it better now, from the work on phase transition by T. D. Lee and C. N. Yang [1], but to me at that time it appeared to be an indication that the helium at one point "knew" what the helium somewhere far distant from it was doing. This was my first introduction to the concept of long range order. But there was much more to it. Bill told us of discussions he had had with Fritz London about condensation in momentum space, order in momentum space, and the existence of long range order in helium and in superconductors.

At the Near Zero conference we have heard of the connection between these ideas and the experiments on flux quantization. The ideas left their mark; but in addition Bill brought to my attention other ideas of Fritz London's, which had evolved from his work on the covalent bond, super-conductivity, and very large organic molecules [2]. As a result I became interested in their possible biological implications, and in the summer of 1962 Tad Day and I started a study of a number of problems in biology, in particular, questions relating to the stability of the genetic code, the origin of life, and the character of enzymatic activity. In the course of that study, Tad brought to my attention a very beautiful piece of work by J. Cairns [3]. This was an autoradiograph of DNA. The striking thing about this was that the DNA was in the form of a closed loop. (See figure 1.) We later learned that if this is labeled with ^{32}P then, on the average, the decay of a single phosphorous atom (the recoil energy of which is sufficient to sever the chain) destroyed the biological activity of the molecule. This immediately brought to my mind what Bill had said concerning London's thinking about the possibility of there being long range order in large organic molecules, specifically the existence of order the whole way round the DNA molecule. This got me interested in the problem of superconductivity in organic molecules.

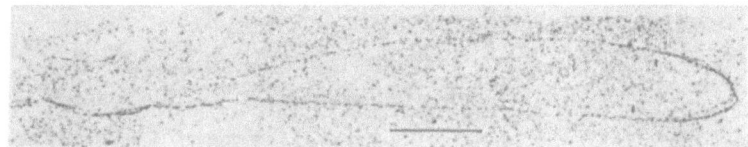

FIGURE 1. Autoradiograph of DNA showing it to be in the form of a closed loop. (See [3].)

In order to investigate this possibility I considered for a model a system which was based on a simplified picture of the DNA structure: a polymer analogous to the sugar-phosphate helix with polarizable side chains, analogous to the base pairs. Using the simplest form of the Bardeen-Cooper-Schrieffer (BCS) theory I then attempted to calculate the transition temperature for a hypothetical system like this. The idea was that the pairing interaction would come about through the interaction of electrons in the polymer or "spine," by the polarization of the side chains. These side chains would give rise to an interaction analogous to the phonon-mediated, electron-electron interaction of superconductors. I found that this hypothetical material should become superconducting, at least in theory, and that the superconducting transition temperature should be enormously high, of the order of 2000 K [4]. (See figure 2.) This led me to think that London was absolutely right and that superconductivity might exist in certain biological materials.

Let me hasten to say that in spite of these arguments, which we still believe to be valid, there is absolutely no evidence of superconducting phenomena in any of these biological materials nor do I believe a need exists

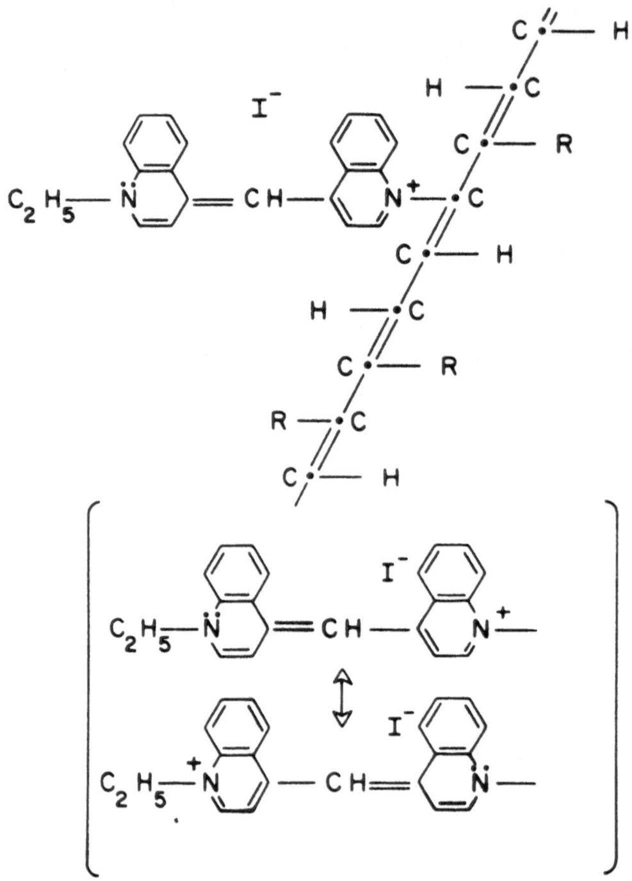

FIGURE 2. Proposed model of an organic superconductor [4].

for such phenomena to explain their biological function. However, it was by this bizarre route that we were led to the idea of a new mechanism in superconductivity, now known as the exciton mechanism, and to the study of complex organic compounds which were totally different from those which had been studied for superconductivity prior to this. The net result has been the development of a new area of physics and materials science, and an associated and coordinated growth of activity in chemistry in the synthesis of new and exciting materials. This new chapter in physics has involved the design, study and theoretical analysis of organic metals and more recently organic superconductors [5]. It has created a new generation of physicists and chemists who can communicate with one another and work together.

I would like to illustrate the developments in this field by sketching first the work in low dimensional physics which has evolved from it. The study of superconductivity in linear chain systems led to discoveries of new phenomena in limited dimension solid state physics. From the study of fluctuations in these systems has come an understanding of the role of noise in superconducting devices.

A linear chain system is at once simple and complex. It is simple because it is one dimensional, and the dynamics are thus simplified, but because of the role of fluctuations it becomes complicated. Let us ignore the fluctuations for the moment and focus on the simple dynamic aspects. Consider the energy of an electron confined to move in a one-dimensional chain, and ignore the interactions. One obtains the typical energy *versus* momentum relationship shown in figure 3a. The electrons interact with the ions of the

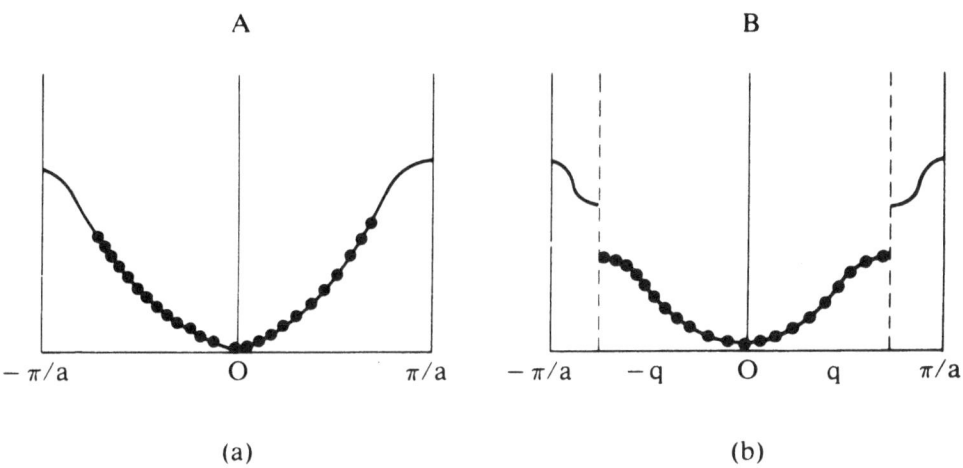

FIGURE 3a and 3b. (a) Energy-momentum relation for free electrons. (b) Energy-momentum relation for electrons in a lattice which has undergone a Peierls' transition.

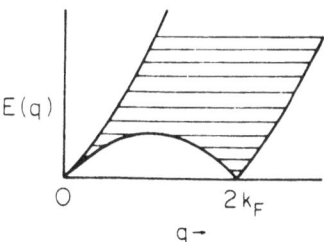

FIGURE 3c. Particle-hole energy as a function of momentum for a one-dimensional system.

lattice. If the lattice is distorted, a periodic potential is created in which the electrons move, and this produces a gap in the energy spectrum. If one now introduces this perturbation so as to put a gap at the Fermi surface, then the states which are filled will be depressed in energy while the states which are not filled will be raised in energy (figure 3b). The net effect is to produce a state of lower energy with a distorted structure. This is the Peierls' transition and it occurs in many linear chain systems.

However, the electron-electron interaction modifies the result. The modification can be understood from perturbation theory. Consider the particle-hole excitations in such a one-dimensional system. There are two regions where one can get low energy excitations (see figure 3c): a region near $q = 0$, involving small momenta, and another near twice the Fermi momentum. This is much simpler than in three dimensions, where one can scatter at any angle from the Fermi surface to another point on the Fermi surface with very small energy changes. It is clear from perturbation theory that these low energy denominators give rise to strong effects in these regions. Often the result is a pairing of the electrons. Different types of pairs can be formed. An electron can be paired with another electron of the opposite spin to give singlet superconductivity; an electron with spin up can be paired with another of spin up to give triplet superconductivity; an electron can be paired with a hole to give either a spin zero charge-density-wave state or a spin one state resulting in a spin-density-wave. All of these types of pairings can be described by a BCS-like expression for the temperature at the transition to the new ordered state,

$$kT_c \simeq \hbar\omega \, \exp(1/\lambda) \quad . \tag{1}$$

The effective coupling constant λ for each of the paired states is given by different combinations of g_1 and g_2, where g_1 corresponds to the interaction for scattering across the Fermi sea, and g_2 that for scattering with small momentum transfer (see figure 4). Which state occurs depends on the relative strengths of g_1 and g_2. One obtains a very simple phase diagram, at least in that region where the electron-electron interaction is weak relative to the hopping probability between elements of the chain. This region we refer to as the region of "g"-ology, where the g-coupling constants play a dominant role. (See figure 5.) Many compounds are well represented by it. In another part of the diagram, which has been referred to as "U-t"opia, the Coulomb interaction U is large compared with the transfer integral t, and there the phase diagram presumably is more complicated. An impressive amount of elegant work has been done in developing this theoretical structure and in correlating the behavior of compounds with it.

Equally impressive have been the achievements of the synthetic chemists, who have developed a number of extraordinarily ingenious ways of synthesizing new materials. When we first proposed the idea of making this

MEAN FIELD T_C IN I.D. SYSTEMS

Ordered State Pairing		State	Coupling Constant
Particle — Particle (Spin 0)		SS	$\lambda_{SS} = \frac{1}{2}(g_2 + g_1)N(0)$
 Particle Particle (Spin 1)		TS	$\lambda_{TS} = \frac{1}{2}(g_2 - g_1)N(0)$
Particle — Hole (Spin 0)		CDW	$\lambda_{CDW} = -\frac{1}{2}(g_2 - g_1)N(0)$
 Particle Hole (Spin 1)		SDW	$\lambda_{SDW} = -\frac{1}{2}(g_2)N(0)$

$$k T_C = h\omega \exp\left\{\frac{1}{\lambda_i}\right\}$$

FIGURE 4. Expressions for mean field critical temperature in I.D. systems. Each paired state [singlet superconductivity (SS), triplet superconductivity (TS), charge-density-wave (CDW) and spin-density-wave (SDW)] has a mean field transition temperature given by a BCS-like expression (bottom). The value of λ for each is a different function of the coupling constants g_1 and g_2, as shown in the right hand column of the table. Here $N(O)$ is the density of states at the Fermi surface.

complex organic superconductor, some of our most gracious colleagues in chemistry said that what we were asking for was at least 50 years ahead of the state-of-art of chemistry and that there was no hope of ever being able

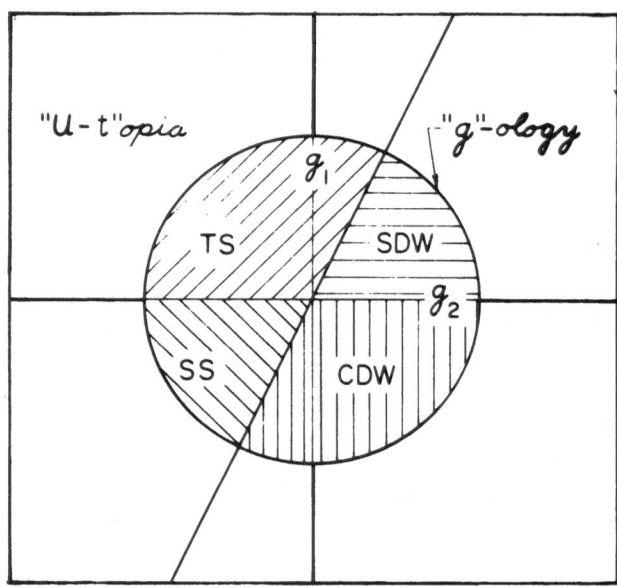

FIGURE 5. Phase diagram of a one-dimensional system as a function of the coupling constants g_1 and g_2.

to realize it. Great progress has nevertheless been made in the course of the last 10 years.

One method of attack has been that used by Gerhard Wegner [7] and his group in Freiberg in West Germany, and by Ray Baughman [8] at Allied Chemicals in this country. What they have done is to produce single crystals of poly-diacetylene. The monomer molecules are designed such that the unit cell dimensions of the crystal structure of the monomer are identical to that which the compound would have if it were polymerized. One starts with a clear single crystal of the basic material, irradiates it with x-rays, ultraviolet, or heat, and thus initiates polymerization. Because the unit cells of the monomer and polymer are identical, the single crystal monomer ends up as a single crystal polymer (figure 6). Thus one evades the fundamental problem of producing a polymer of such great complexity, namely the fact that polymers which are conjugated in the manner proposed in our model (figure 2) are insoluble in nearly all solvents. Having them in solution is essential in order to carry out the polymerization by conventional means.

These materials are very interesting in their own right. They start out being totally colorless, but upon exposure to light polymerize and become metallic looking. I might add that they are not good conductors; we do not understand exactly why. They are extraordinarily strong materials because of the near perfection of the molecular structure. The rate of polymerization can be controlled by the choice of end groups on the diacetylenes.

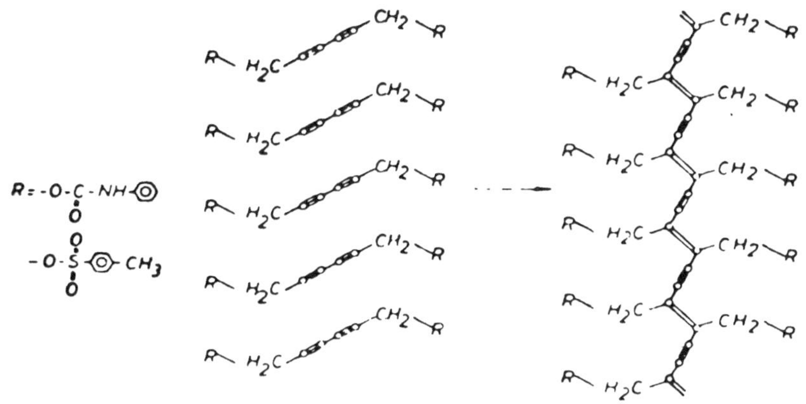

FIGURE 6. Solid state polymerization of diacetylenes.

This has created a novel use for them as "labels" for integrating the time-temperature history of a material [9].

As is well known, many organic polymers such as mylar and polyethylene are among the best insulators that we have. One finds, however, that they become conducting if they are doped, with bromine or with iodine for example, introducing carriers into the bands that were empty, or holes in the bands that were previously filled. Even polyacetylene when it is doped changes in conductivity by eight orders of magnitude. (See figure 7.) Starting out as a colorless material, it becomes transformed, upon doping, into a material which looks very much like aluminized mylar [10]. This is an organic metal and can be formed with traces of bromine, iodine, or various other compounds in it.

On another topic, it was brought to our attention about ten years ago [11] while we were searching for a material able to carry current along a spine-like structure with polarizable organic ligands, that there are a number of platinum compounds which meet this criterion [12]. These form linear chains with the d_{z^2} orbital of one metal atom in the chain overlapping the d_{z^2} orbital of the next. Oxidation gives a linear chain compound with a partially filled band which becomes an extremely good metal. If one looks at it under a polarizing microscope, one finds that for light polarized along the direction of the chain the crystal appears metallic, because conduction can occur along this direction, but a crystal oriented at right angles to the plane of polarization is transparent, because there is no lateral conductivity.

This is just one of a large class of materials which have been discovered in the course of these investigations which have enormously anisotropic electrical, thermal and other properties. Among these are two of the most extensively studied, TCNQ and TTF, which when combined form a

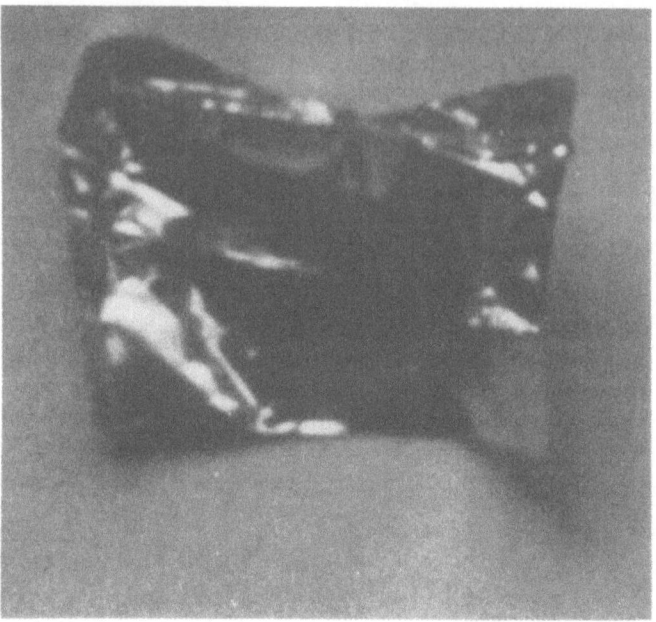

FIGURE 7. Polyacetylene doped with bromine to form an organic metal $(CHBr_{0.05})_x$.

structure in which the individual molecules stack on top of one another, again forming a linear chain compound [13]. (See figure 8.) These linear chain compounds exhibit a number of phase transitions. In studying their conductivity one finds that as one approaches a temperature of about 60 K the conductivity becomes enormously high [14]. At one time it was thought that this might signify the existence of superconductivity, but it does not; it is just that the anisotropy is so high that one has to be extraordinarily careful in making the measurements. The material undergoes a Peierls' transition near 60 K and becomes insulating below it. The study of these materials with various subtle changes of the organic structures and modification of these with selenium have given a number of compounds of extraordinarily interesting properties. Several members of a particular class of compounds $(TMTSF)_2X$ undergo a transition into the superconducting state [15]. So we now know that there are compounds which are organic superconductors, and we know of polymers, the polymer of SN_x for example, which also superconduct [16].

I think this topic is appropriate for the *Near Zero* volume because while our efforts have been directed to a search for a high temperature organic superconductor, and we continue to believe there is a possibility of

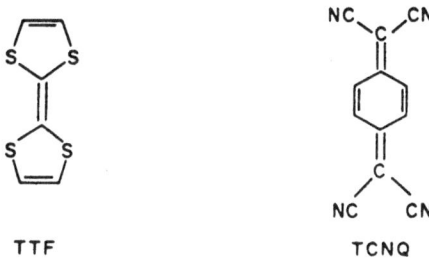

TTF

TCNQ

FIGURE 8a. Construction of highly conductive anisotropic organic metals. The molecules of TCNQ and TTF and various modifications of these are the basis for a series of highly conductive organic metals.

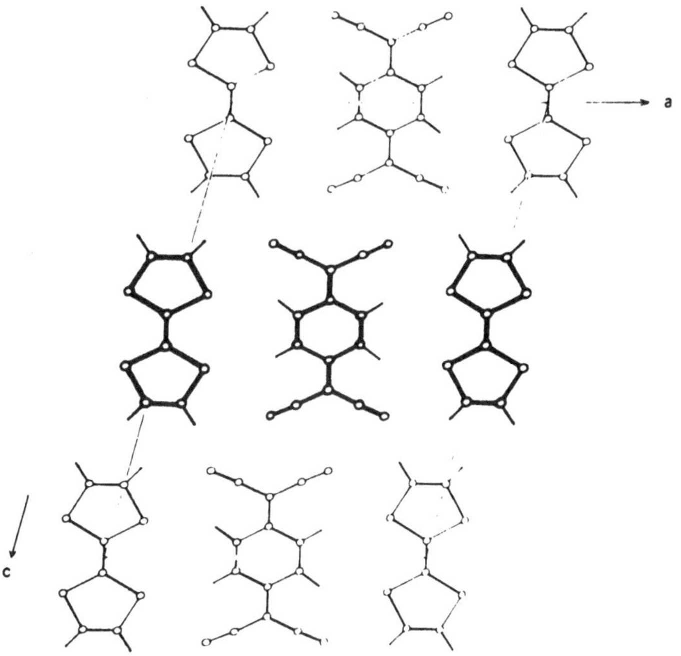

FIGURE 8b. The compound TCNQ-TTF consists of separate one-dimensional stacks of TCNQ and TTF molecules arranged in the crystal as shown.

obtaining high temperature superconductivity, thus far the superconductivity we have observed has been "Near Zero"! In conclusion, I want to repeat that it was largely Bill Fairbank's influence and his enthusiasm and constant interest in the fundamental problems of superconductivity that started us on the road to organic superconductivity. The legacy of all this work has been a new group of physicists and chemists who have done great things, and will continue to do so in the future, I am sure. For this we owe a debt of thanks to Bill Fairbank.

Addendum

The discovery of superconductivity in $YBa_2Cu_3O_{6.9}$ above 95 K has been announced by M. K. Wu *et al.* [17]. The remarkably high transition temperature of this material, together with its structure of one- and two-dimensional components, and the complete absence of an isotope effect on T_c as reported by R. J. Cava *et al.* [18] at the Materials Research Society meeting in Anaheim in April 1987 are strong indications that the electronic or "excitonic" interaction which we proposed for the organic superconductor [4] is operative here. This connection is made in papers presented at the MRS meeting by W. A. Little, J. P. Collman and J. T. McDevitt [19] and submitted to *Physical Review Letters* [20].

References

[1]   C. N. Yang and T. D. Lee, *Phys. Rev.* **87**, 404 (1952); T. D. Lee and C. N. Yang, *Phys. Rev.* **87**, 410 (1952).

[2]   F. London, *Superfluids* (Dover Publications, Inc., New York, 1961), p. 8.

[3]   J. Cairns, *J. Mol. Biol.* **6**, 208 (1963).

[4]   W. A. Little, *Phys. Rev.* **134**, A 1416 (1964).

[5]   For a recent review of the field see *Chemica Scripta* **17**, No. 1–5 (1981).

[6]   A simplified outline of this is given by H. Gutfreund and W. A. Little, in *Highly Conducting One Dimensional Solids*, edited by J. T. Devreese, R. P. Evrard and V. E. van Doren (Plenum, New York, 1979), p. 305.

[7]   G. Wegner and W. Schermann, *Colloid and Polymer Sci.* **252**, 655 (1974).

[8]   R. H. Baughman, in *Contemporary Topics in Polymer Science*, edited by E. M. Pearce and J. R. Schaefgen (Plenum, 1977), Vol. 2, p. 205; R. H. Baughman, *J. Appl. Phys.* **43**, 4362 (1972).

[9]   G. N. Patel, A. F. Preziosi and R. H. Baughman, U.S. Patent 3,999,946, Dec. 28, 1976.

[10]   M. Akhtar, C. K. Chiang, M. J. Cohen, J. Kleppinger, A. J. Heeger, E. J. Louis, A. G. MacDiarmid, J. Milliken, M. J. Moran, D. L. Peebles and H. Shirakawa, *Annals N. Y. Acad. Sci.* **313**, 726 (1978).

[11]   J. Collman, *J. Polymer Science* **C**, Polymer Symposia No. 29 (J. Wiley and Sons, 1970), p. 136.

[12]   K. Krogmann, *Angewandte Chemie* **81**, 10 (1969).

[13]   T. E. Phillips, T. J. Kistenmacher, J. P. Ferraris and D. O. Cowan, *J. Chem. Soc. Chem. Sommun.*, 471 (1973); T. J. Kistenmacher, T. E. Phillips and D. O. Cowan, *Acta Cryst.* **B30**, 763 (1974).

[14]   L. B. Coleman, M. J. Cohen, D. J. Sandman, F. G. Yamagishi, A. F. Garito and A. J. Heeger, *Solid State Commun.* **12**, 1125 (1973).

[15]   M. Ribault, G. Benedek, D. Jerome and K. Bechgaard, *J. Physique Lett.* **41**, L-397 (1980); D. Jerome, *Chemica Scripta* **17**, 13 (1981).

[16]   R. L. Greene, G. B. Street and L. J. Suter, *Phys. Rev. Lett.* **34**, 577 (1975).

[17]   M. K. Wu *et al.*, *Phys. Rev. Lett.* **58**, 908 (1987).

[18]   R. J. Cava *et al.*, *Materials Research Society Proceedings*, Anaheim, 1987.

[19]   W. A. Little, J. P. Collman and J. T. McDevitt, *Materials Research Society Proceedings*, Anaheim, 1987.

[20]   J. P. Collman, J. T. McDevitt and W. A. Little, "Possible Role of the Excitonic Interaction in the New High T_c Superconductors," submitted to *Phys. Rev. Lett.*, 1987.

International Journal of Modern Physics B, Vol. 4, No. 6 (1990) 1181–1199

BLOCH AND THE NEW SUPERCONDUCTORS

WILLIAM A. LITTLE

Department of Physics
Stanford University
Stanford, California 94305

Bob Hofstadter gave a very complete description of the work that Felix had done in many different areas. I would like to touch only on those areas which are connected with superconductivity, and of these, those areas in which I have had personal contact with Felix. This, I hope, will give you a different view of working with Felix, and a view of the kind of interaction that a colleague could have had with him.

Reminiscences

I first met Felix in 1957. I had an invitation to visit Stanford to talk about work which I had done for my PhD at Glasgow University on Nuclear Magnetic Resonance. This was on a study of the Nuclear Overhauser Effect.[1] I came down to Stanford from Canada, where I had a post doctoral position, for this talk. After I left, Felix called me and offered me a job. I declined the offer as I had another year to run at UBC and I wanted to continue with the low temperature work that I was doing there. Fortunately, Felix called me a year later, and offered me a much better job, which I then accepted. My charter was to come to Stanford to set up a Helium liquifier and to establish a low temperature lab in anticipation of the arrival of Bill Fairbank, who was then at Duke University, so that there would be facilities available for him when he would arrive the following year, 1959.

I was extremely fortunate to have an office which was only a few steps from Felix's office, and for the next twenty-five years I had a most wonderful interaction with him on a huge variety of subjects. There was a regular arrangement, especially in the earlier days of the department, to meet at coffee. Very often Felix would come down to coffee and we would talk about some aspect of physics. I would walk back with him to his office as he continued to talk or explain some point, or I would try a new idea on him and get feedback from him. You had to be careful of what you said to Felix. If you carelessly made a statement early on in the discussion he remembered it, and would come back and remind you - "You said this! You said this! You said this!" I'd say "No, no I didn't mean that. I made a mistake...," but it was no use. In this way he would clear away all false statements or correct any incorrect assumptions that you might have made.

I remember the many, many times that I would sit through a colloquium, which could be on any subject at all, and often there would be questions, which I did not understand, on what had gone on in the talk. I would walk back with Felix to his office, and ask him "What did so and so mean about that?" He would then give a most beautiful, crisp, clear explanation, of exactly what that person meant. It was an exhilarating experience. There was not a subject that he could not explain like that. I always felt that I was like a child playing around with toys on the floor, and that every now and then I would look up and I'd ask him "How do you do this," and he would give me help.

There is one thing which is not so well known about Felix, and that is the kind of teacher he was. In these personal interactions, he was a gentle teacher. I could ask him the most stupid question, the answer to which an undergraduate might know. Nevertheless, he would take the time to explain it to me, with such insight that I would understand well the essence of the problem. He explained quantum mechanics to me, he explained many different subjects so that I feel that I learned more physics from him than I learned in any other way. This manner of teaching comes across clearly in the book that Dirk Walecka has written, based on Bloch's notes on statistical mechanics. In this it appears that at some time Felix had been asked a question about the fundamentals of the subject, and in these notes he answers that question, and this comes across just beautifully in the book.

Early Years of Superconductivity

I would like to go back to the time when Felix first was introduced to the subject of superconductivity. As was mentioned earlier, in 1928 he went to work with Wolfgang Pauli, and was asked to try to solve the one remaining problem in condensed matter physics which Pauli considered of importance - superconductivity. It appears to have been a most frustrating year for Felix, as he did not publish anything on the subject in that year, in fact, he did not publish anything on superconductivity until 34 years later. But he did work on superconductivity during that period and he proved a theorem on it to himself, but did not publish it. Several years later however, in 1933 he visited Brillouin in Paris, and during that visit he described to Brillouin his concerns about the problem of superconductivity and his theorem. Brillouin at that time was in the process of writing a paper on superconductivity, and he included a note, added in proof, of a theorem due to Bloch. Brillouin[2] gives Bloch full credit for the argument. I thought it would be useful to present that argument here because it links so well with later discoveries in the field.

To understand Bloch's Theorem of Superconductivity, consider the following: Let ψ be an exact eigenfunction of a state of a super-conductor which is carrying a current described by the momentum of the charge carries p_0. Then let us define another state ϕ, which is related to the first state, ψ by the exponential term in Eq. 1.

$$\phi = \exp(-\frac{\delta p}{\hbar} \sum_i x_i)\psi \qquad (1)$$

The exponential term simply allows each electron to have an additional little bit of momentum δp. Now ϕ is not necessarily an eigenstate, but it is a state whose energy can be calculated. If the energy of state ψ is $E_0 = \overline{T} + \overline{V}$, then that of ϕ is:

$$E = \overline{T} + \overline{V} + \frac{N(p_0 \delta p)}{m} + \frac{N(\delta p)^2}{2m}. \qquad (2)$$

One finds that the energies of the two states are identical except for terms involving δp. Felix pointed out that if you choose δp, small enough so you can neglect the term quadratic in δp, then the energy, E will be less than E_0 if the sign of δp is chosen opposite to p_0, the momentum of the particles in the state you started with, and therefore, the energy of the new state lies lower than that of the initial state. So he could show that for any state that has a current flowing in it, there is another which is lower in energy. This is true for the state ϕ, which is not necessarily an eigenstate, but we know quite generally from variational principles, that a true eigenstate would lie even lower in energy than this. So that he could show that for any state which has a current in it, there is a lower energy state that has no current. So he argued, in equilibrium you could not then get a superconductor, which could carry a persistent current.

That was the famous Bloch Theorem. Eventually that was published but only in his retirement speech, when he retired from the American Physical Society, as the President some 34 years later! In describing superconductivity, in his later years, he remarked that it's study was distinguished by a large number of surprises. The first surprise to him was that a superconductor could exist at all, that is that it could have perfect conductivity. The second, was that it exhibited the Meissner Effect, the ability of a superconductor to exclude a magnetic field from its interior. The third surprise, was the discovery by Bill Fairbank and Bascomb Deaver, and Doll and Näbauer[3] of flux quantization in 1961 and, the fourth, in 1962, the discovery by Brian Josephson of the Josephson effect.[4] Felix thought the experiment of Deaver and Fairbank was so beautiful. They had taken a small cylinder of tin, trapped flux in it below the superconducting transition temperature, and observed that the flux trapped was not uniformly proportional to the field which was applied, but was quantized at a series of individual levels. See Fig. 1.

At that time, C.N. Yang was visiting Stanford as part of a program of the Institute of Theoretical Physics which Leonard Schiff and Felix had set up. Yang and Byers[5] studied the theoretical problem of explaining how one could obtain such flux quantization. In the course of that study, they realized that if there were pairs in the superconducting state, as in the Bardeen, Cooper, Schrieffer (BCS) theory[6] of superconductivity, then the pairs would have quantized angular momentum around the cylinder axis of the superconductor. If initially,

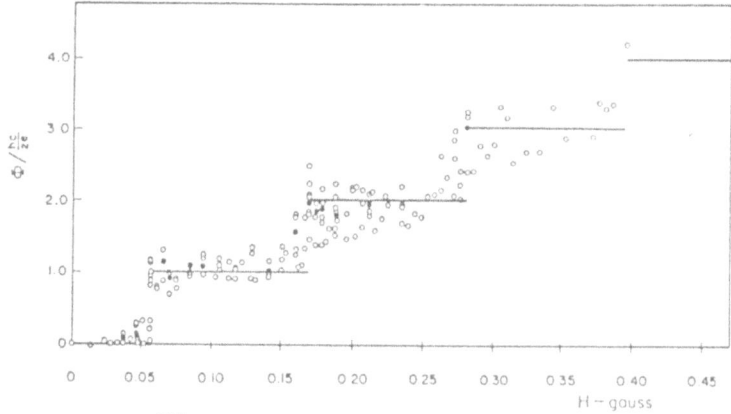

Flux trapped in Sn cylinder.

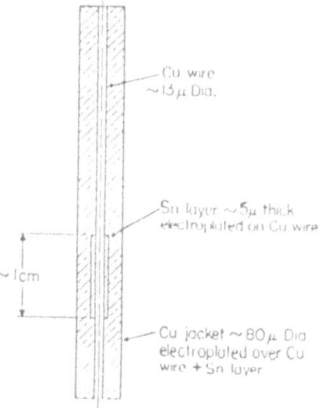

Sample Construction

Fig. 1. Experiment of Deaver and Fairbank on flux quantization.

a superconductor has all the pairs condensed in a state in which the angular momentum is zero, then upon application of a magnetic field, these pairs would be accelerated as in a Betatron and their energy would increase. This is illustrated in Fig. 2.

If instead, you have the pairs in the next angular momentum state, and apply a magnetic field, then depending upon the direction of the field, you either speed up the pairs, or you slow them down, and you find then that eventually there is a magnetic field value, which just brings the pairs to rest. This illustrates Bloch's Theorem, that the

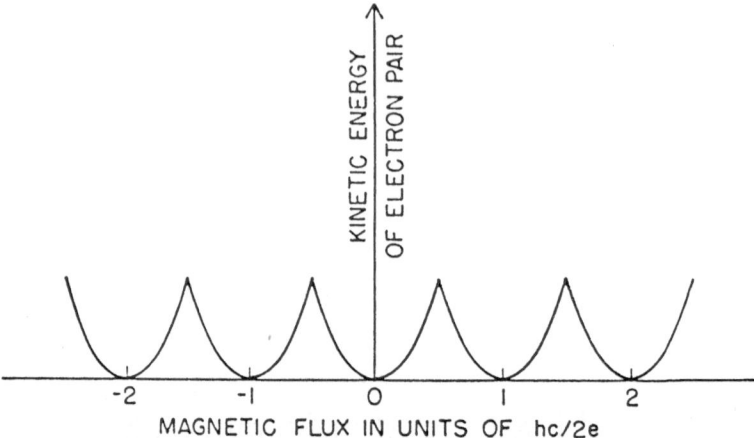

Fig. 2. Kinetic energy vs flux for pairs in a superconducting cylinder.

state of the lowest energy is the state, which, in the absence of a magnetic field, is the one with zero current.

I learned of this from Felix, and Felix liked the explanation which Yang had given for the experiment. He thought this to be a major advance in the understanding of superconductivity. My interest in it came, when I realized that what Yang was saying was that the free energy of the superconducting state would be periodic in the magnetic field for a cylindrical sample. He also had shown that in the normal state the electrons rearrange themselves in such a way that the free energy in this state was independent of the magnetic field, at least at low values of the magnetic field. The consequences of this are shown in Fig. 3.

In a plot of the free energy as a function of the field and temperature, the normal state, is a smooth sheet, but the superconducting state is a corrugated sheet. We know from statistical mechanics and thermodynamics that the condition for equilibruim between two states occurs when the free energy of one is equal to the free energy of the other. In this case, we have a rippled sheet intersecting the smooth sheet of the normal state, which then gives a rippled intersection on the T-H plane. Hence, the transition temperature should be a periodic function of the flux. Fig. 4.

The variations of the transition temperature reflect the variation of the free energy of the superconducting state as a function of the

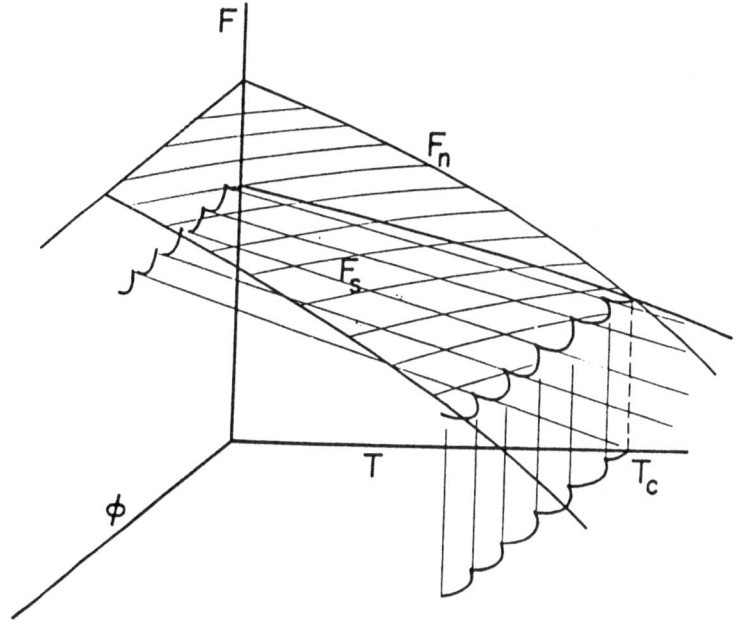

Fig. 3. Plot of free energy vs flux ϕ and temperature for cylindrical superconducting sample.

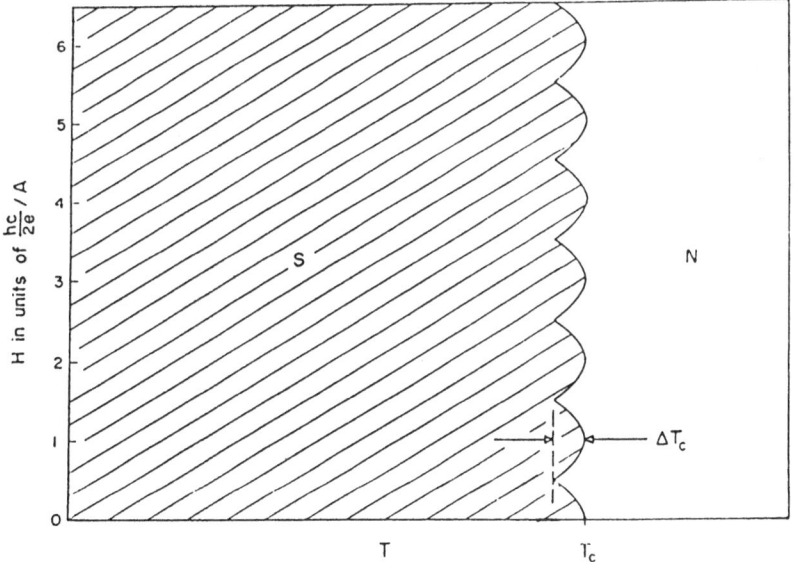

Fig. 4. Periodic variation of T_c for cylindrical superconducting sample in a magnetic field.

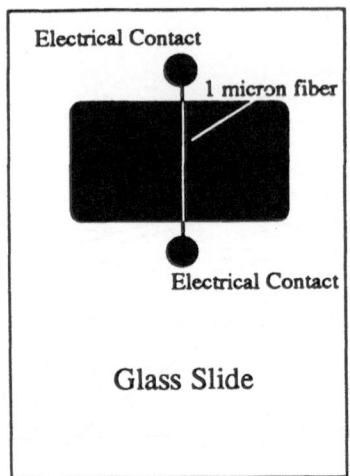

Fig. 5. One micron Sn cylinder for Little - Parks experiment.

magnetic field applied to the cylinder.

We made an estimate of how big the deviation of T_c would be, and we figured that one could get a variation of about 100 μK for a cylinder which was about a micron in diameter. At that time I had my first graduate student at Stanford, Ron Parks, working with me as a post doc. He was an extremely enterprising guy and a delight to work with. Later he went on to the University of Rochester, where as a Professor of Physics he led an active group and edited one of the most successful texts on superconductivity.[7] He died prematurely in 1985. I knew that one could make very small fibers of quartz by blowing them in an oxyacetylene flame.[8] Ron started making these fibers and then trying to cement them down. But every time he cemented them, using a cement, GE7031 a fine spiderweb of glue came away from the joint. Eventually he said "To heck with the quartz, let's put down a little fiber of this cement." In this way he was able to make fibers about a micron in diameter! We then put the fiber in a little barbeque spit-like arrangement in a vacuum system and deposited tin on the outside of the fiber to a thickness of a few hundred angstroms to give a sample shown in Fig. 5.

Then we measured the electrical resistance of this filament in a magnetic field applied along the axis while the sample was immersed in a bath of Helium. This allowed us to determine the variation of the

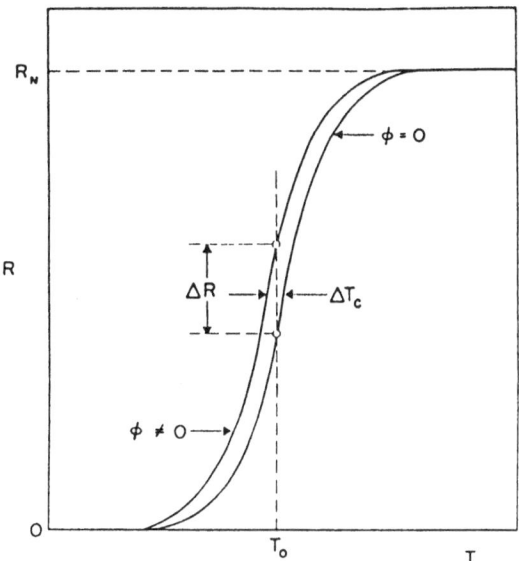

Fig. 6. Variation of resistance of cylinder due to change of T_c with flux ϕ.

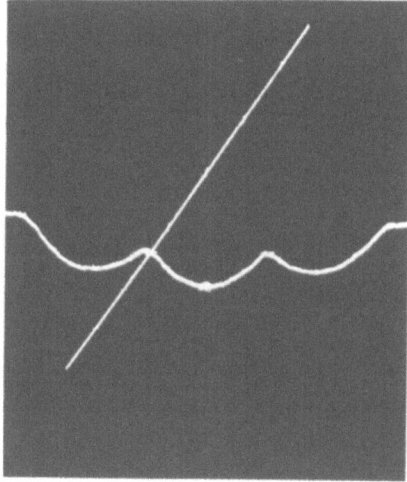

Fig. 7. Variation of resistance (hence T_c) of one micron cylinder of Sn as a function of magnetic field.

transition temperature by looking at the variation of the resistance. Fig. 6.

This worked out rather nicely,[9] and we were able to see just exactly the free energy variations which had been predicted by Yang. Fig. 7.

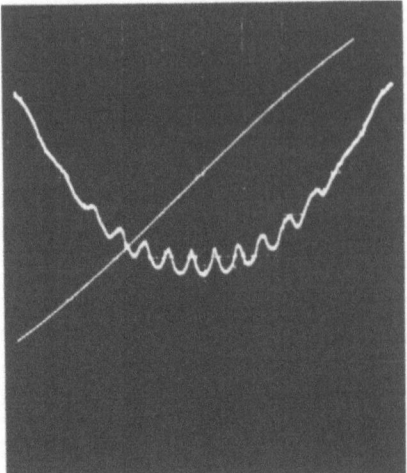

Fig. 8. Periodic variation in T_c of cylindrical Sn sample.

Observing directly on an oscilloscope we could see that the higher angular momentum states were obviously of higher energy than those of the lower states in the absence of a field. One could increase the magnetic field sweep to show more of the scallops. Fig. 8.

Now, that's all very well. It illustrated to us the Bloch Theorem. But the question was, why was it that these materials were able to trap current. If you did cool them still further in the magnetic field, then turn the field off, the current did, indeed remain persistent. We know from the BCS theory that in the superconducting state there is a depression of the free energy of the superconducting state as a result of the accumulation of particles in the condensed state. Now, in our case, we have a series of phases, one with zero angular momentum, another with one unit of angular momentum, another with two units of angular momentum, and so on. If you prepare the system with one particular state occupied, then that state will be depressed in energy because of the coherence effect from the particles in that same state. If you have a larger number of particles in one state than in another state, then this state sags below the energy of the others. As soon as it starts to sag, particles of the other states fall into it and depress its energy still further. The state that starts out by being partially filled draws other particles to it. This is illustrated in Fig. 9.

So what you find, and this has been known in the literature for a long, long time, is that, "Unto every state that hath, shalt be given,

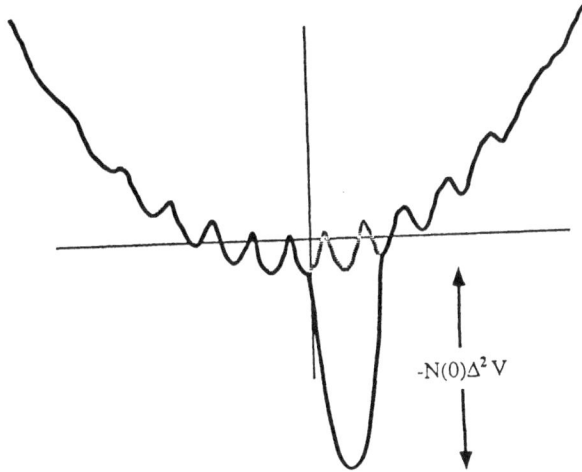

$-N(0)\Delta^2 V$

Fig. 9. Depression of free energy of occupied state. (Artist's impression).

and those that have not, even that which he hath shall be taken from him."[10] Or, the richer get richer and the poor get poorer!

One state is intially occupied, and now as the temperature is lowered more particles fall into it. This lowers the energy by an amount $N(0)\Delta^2 V$, where $N(0)$ is the density of states at the fermi surface, Δ is the amplitude of the gap for this state and V is the volume. The gap function Δ is a measure of the number of pairs in the state. The occupied state now lies lower than the bottom of the zero current state, but that does not violate Bloch's Theorem because, if you take all the particles from the occupied state and put them in the zero current state it will lie still lower. But for that to happen you have to take a finite fraction of all the particles from one state and put them in the other and the probability of that happening is so vanishingly small that it would not happen in the life of the universe, so that the current will continue to persist. This then does not violate the Bloch Theorem, it simply shows that you do not have equilibrium under these circumstances.

We had a good interaction with Felix during this period. As a result I was forced to learn a great deal more about superconductivity and I learned much from Bill Fairbank too, who told me of some

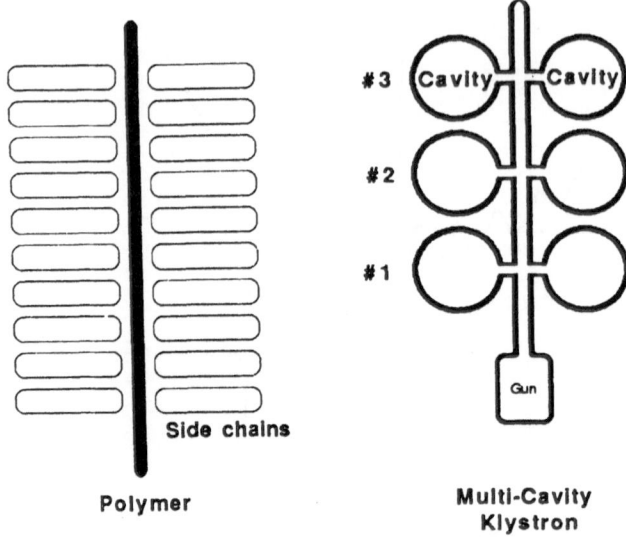

#3

#2

#1

Side chains

Polymer

Multi-Cavity Klystron

Fig. 10.   Fig. 11

interesting ideas of Fritz London contained in London's book "The Theory of Superfluids."[11] In this there is a statement in the introduction suggesting the application of the concept of superfluidlike states to explain the ability of large macromolecules to act as single units. This was analogous to what one sees in a superconductor where all the pairs are in the same state, and the system behaves like a single particle exhibiting macroscopic quantum behavior. I got interested in this and realized that the BCS theory was sufficiently far advanced that one might be able to apply it to a molecule. I thought of taking a long polymer chain with polarizable side-chain molecules attached to it, and studying how an electron moving along this, so called, spine would interact with the polarizable molecules on the side. Fig. 10.

It appeared that one electron would induce a charge on a side-chain and a second electron would get attracted to the polarization cloud produced by the first. I went through the mathematics of this and used the standard canonical transformation method to eliminate the side-chain coordinates to leave one with a net electron-electron interaction, which was an attraction between the particles. At that time I only knew the mathematics of doing this, and I went to Felix and explained what I had done and he had no trouble with understanding it. I did run into trouble when found that the transition temperature of the material appeared to be about 2000 Kelvin. All the known superconductors at that time had transition temperatures less than 20K. Felix suggested that I should think very carefully about that! I thought, of course, of centimeters and meters and things like that! But I came back the next day with an idea that the system that I was dealing with was reminiscent of the multicavity klystrons which were used in high energy physics. In these an electron gun injects electrons down a vacuum path past a series of cavities. Fig. 11. The electrons start out in a continuous stream, but they then excite oscillations in the side-chain cavities, which in turn cause bunching of the beam, and the more cavities you use the more the electrons are forced to bunch. So if you think of it from the point of view of the electron rest frame, the electrons experience a space charge repulsion from one another, but nevertheless they bunch, so they must experience, in addition, a cavity-induced attractive interaction. Felix said that that sounded nice, but why did the electrons bunch in the superconductor then stop at pairs? Why didn't they form clusters of larger numbers of particles? And why was the transition temperature so high? I did not know the answer to these questions. He then sent me off to think about it and suggested we should meet the next day. I went home and I thought and thought and then I realized what I believe is the explanation of the high transition temperature. In a conventional superconductor, according to the BCS theory, the transition temperature is proportional to the phonon frequency times an exponential term containing a coupling constant λ which describes the electron-electron interaction. What I had done unwittingly was to consider a system in which it was not the ions that were producing an ionic polarization to mediate the interaction, but rather it was an electronic polarization. The large electronic frequency then gave the anomalously high transition temperature.[12]

Well, he liked that, he thought that this was a good explanation, and in fact from then on he appeared to believe that one should be able to get superconductivity at much higher temperatures. I have a wonderful videotape of him speaking in 1974 or thereabout, in which he stated that it would not surprise him in the least to see superconductivity at 100K or room temperature, based on such an electron-mediated type of interaction.

Shortly after this, I was asked to write an article for the Scientific American[13] on these ideas. I gave a very cautious title to the paper, but the title was long and was truncated by the editor without my knowledge, and read, "Superconductivity at Room Temperature." Felix didn't like that. He thought that it was misleading. At this time, this was 1965, he knew and I knew that there were serious problems with the model. These were due to the fact that structure which we were considering was one dimensional. Yang had recognized that the superconducting state is distinguished by the existence in it of what he called off-diagonal long range order.[14]

$$Lt_{(r-r')\to\infty} < \psi(r)\psi(r') > \neq 0 \qquad (3)$$

This extends throughout the medium and is responsible for flux quantization in a multi-connected superconductor. But it had been pointed out by several people that the order parameter is a complex function $\psi(r) = \Delta(r)\exp(i\phi(r))$. It has a real part $\Delta(r)$, the energy gap and a phase factor $\phi(r)$. If you have a long thin superconductor, the order parameter at $r = 0$ may be taken along the real axis. But as you move away from $r = 0$, density fluctuations of the electron gas, cause the electrons to move back and forth, causing currents which are related to the gradient of the phase factor. There are thus fluctuations in the relative phase between $\psi(r)$ and $\psi(0)$ as r increases. The amplitude of these fluctuations gets bigger and bigger as you increase r until, a long way away, the representative point in the Argand diagram is oscillating by more than 2π relative to that at $r = 0$. At this point the overlap given by (3) tends to zero and thus one can not get off-diagonal long range order in a one-dimensional material. Felix understood this very well because in 1933 he had discussed spin-waves in ferromagnets and given a beautiful proof that you cannot have a ferromagnet in one dimension because in one dimension the number of such spin-waves grows without limit as the volume increases and this destroys the magnetism. The similarities were not lost on Felix.

I said to him that I did not believe this proof for the superconductor. Felix turned on me and said that I was being stubborn and stupid and I should acknowledge that what I had done was simply wrong and should move on to something else. In fact, I became persona non grata with him, and for the next two or three years I saw

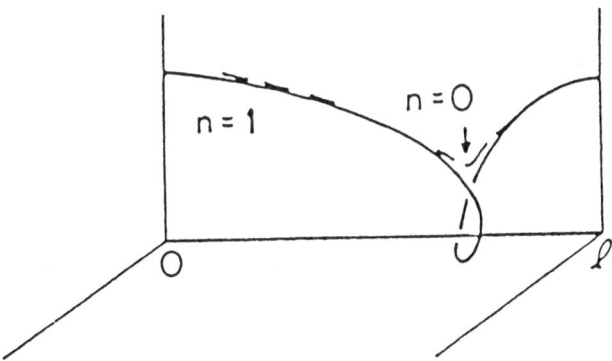

Fig. 12. Type of transition required to cause decay of persistent current. Plot of complex order parameter vs distance along ring.

very little of Felix. He worked on the problem and I worked on the problem and we saw each other at coffee occasionally, but I felt very uncomfortable talking with him. He didn't suffer fools gladly!

However, I caught him at coffee a year or two years later, and told him that I thought I had a rebuttal to the simple argument against superconductivity in one dimension. In a superconducting ring one can describe the current density as the gradient of a scalar term plus the curl of a vector because the ring is connected back on itself. One can show that the density fluctuations are associated with the gradient of the scalar quantity, but the circulating current is associated with the curl of the vector term. The two are independent of one another. The density fluctuations cause oscillations in the phase which appear to destroy the long range order, but these fluctuations do not change the winding number, which is associated with the vector term. This is a topological invariant, and it determines the circulating current. The only way that you can get rid of the persistent current is by a transition which takes one of the terms and pulls it through the origin. Fig. 12.

But if you do that, it means that in one region of space you must depress the order parameter to zero, and that costs you a finite amount of free energy ΔF. That implies that the resistance, which is proportional to the probability of this happening, is proportional to $e^{(\frac{-\Delta F}{KT})}$ and so, at sufficiently low temperatures this process is inhibited. Thus even though such a one-dimensional system does not have long range order, it nevertheless, can have extremely high conductivity at moderately low temperature.[15]

On the strength of this, I got up my courage and spoke to Felix about it. Though we did not interact much during this period he thought I could use some help on the theoretical side and introduced me to Hanoch Gutfreund, who had just completed a post doctoral appointment with him working on the Tomonaga model, which is actually the Bloch model of sound waves in the Fermi gas, and encouraged Hanoch to work with me. This has resulted in a wonderful interaction with Hanoch which has lasted over twenty years and resulted in a collaboration on many different problems.

At this point I felt that Felix was softening a little bit in his attitude to what I was doing, and so I asked him whether he would be willing to speak on the history of superconductivity at a conference that I was organizing on the physical and chemical properties of organic superconductors. He agreed and gave a wonderful talk to which I will refer later. He was enormously supportive thereafter, and encouraged us in our study of many of the subtle problems of this challenging field.

While we believed we understood the essence of the problem of superconductivity in quasi-one-dimensional systems, we lacked a physical example. Fortunately, shortly thereafter in 1975, Richard Greene, who was working with one of my graduate students, Larry Suter, in our department, discovered superconductivity in the polymer $(SN)_x$.[16] This polymer has no metallic atoms in it, but is superconducting at 300 mK, a very low temperature, but nevertheless it is superconducting and indicated to us that superconductivity was indeed possible in polymeric structures. This was followed by the discovery in 1980 by Denis Jerome, Klaus Bechgaard and co-workers of superconductivity in an organic compound.[17] Since then there has been substantial progress, with over 30 organic superconductors now known.

The resolution of the argument over the fluctuations in the superconductor have helped clarify the understanding of noise in superconducting junctions and the decay of a supercurrent. Felix worked on both the problem of flux quantization and the effects of dimensionality in superconductivity in a series of papers towards the end of the sixties. He was trying to establish from first principles, some of the concepts that could derive from the more complicated models, using the BCS or Ginsburg-Landau theories.

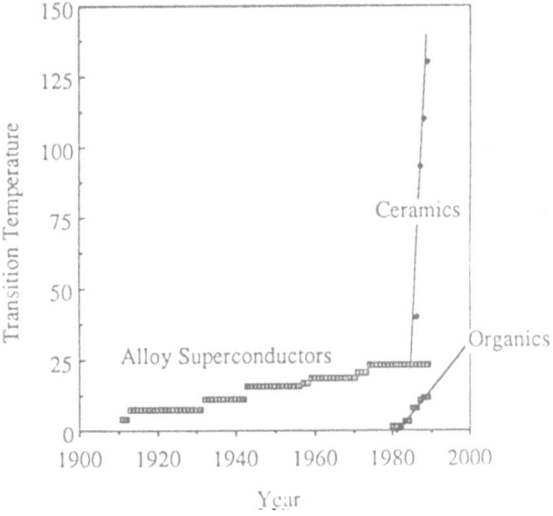

Fig.13. Plot of highest transition temperature superconductor vs year for three classes of superconductors

Let me turn now to high temperature superconductivity. What I think Felix would have classified as the fifth surprise in superconductivity might surely have been the discovery by Bednorz and Müeller of superconductivity above 35K in the cuprate ceramics.[18] Within a few months of this discovery, superconductors with transition temperatures above liquid nitrogen temperatures and eventually above 120K were found.

There are several similarities between these cuprates and the materials we were studying. While the cuprates are not one dimensional, they are, in fact, two dimensional in structure, and many of the arguments which apply to one dimensional systems apply equally well to two dimensional ones. Yet these turn out to be the compounds with the highest transition temperatures today. This goes against the grain of the thinking of many of the superconductor community. There exists at present, a raging controversy among theorists as to what is the proper starting point for understanding the superconductivity of these materials, and what are the appropriate models. Is it based on the magnetism and the heavy correlations which are clearly present, or is it based on conventional, so called, Fermi liquid behavior even in spite of the correlations. We have argued strongly on the basis of the bulk of experimental evidence[19] that these are BCS-like superconductors but with an electronic coupling mechanism similar to what we had postulated for the organics, but involving instead of a polarizable side-chain, a polarizable d-shell of the copper ion.[20] The basis for this lies in the observation of a conventional Fermi surface,[21] states observed within a few millivolts of the Fermi energy in photoemission[22] and energy bands of predominately oxygen character which cross the Fermi surface.[23] In addition, there is now impressive evidence of weak coupling, indicating the energy of the coupling excitations are well above the energy of any phonons. In the normal state the cuprates behave very much like conventional metals. That is not to say that correlations are not important, but rather that the correlations are such as to force the electrons, which actually sit on the oxygen atoms, to behave as a Fermi liquid.

Felix would have derived great pleasure in observing these developments and likewise he would have revelled in the dismay of the bulk of condensed matter theorists upon the observation of high temperature superconductivity. He always felt this community was too smug in its acceptance of the BCS *ansatz* and that they did not appreciate the subtleties which made this possible. If Felix were here today I think he would probably repeat the remarks that he made at the conference on organic superconductors which he attended in Hawaii in 1969. He told a story of a friend who was visiting Washington DC for the first time. He took a cab around the capital to see the many monuments and buildings. He drove past one which housed the archives of the great and on the entrance to this building was an engraved inscription "The Past is but a Prologue." His friend turned to the cab driver and asked the driver what that meant and the driver said, "Sir what it means is, 'You ain't seen nothing yet.'" I think that's what Felix would have said about the state of superconductivity today. "You ain't seen nothing yet!"

References

1) W.A. Little, "The Overhauser Effect in Solids," *Proc. Phys. Soc.* (London) 70B, 785 (1957).

2) L. Brillouin, *J. de Phys. et Rad.* 4, 334 (1933); D. Bohm, *Phys. Rev.* 75, 502 (1949).

3) B.S. Deaver and W.M. Fairbank, *Phys. Rev. Lett.* 7, 43 (1961); R. Doll and M. Näbauer, *Phys. Rev. Lett.* 7, 51 (1961).

4) B.D. Josephson, *Phys. Lett.* 1, 251 (1962).

5) N. Byers and C.N. Yang, *Phys. Rev. Lett.* 7, 46 (1961).

6) J. Bardeen, L.N. Cooper and J.R. Schrieffer, *Phys. Rev.* 108, 1175 (1957).

7) R.D. Parks, "Superconductivity," Vol. 1 and 2, Marcel Dekker, Inc., New York (1969).

8) J. Strong, "Procedures in Experimental Physics," Prentice Hall, Inc., New Jersey (1938) p. 205.

9) W.A. Little and R.D. Parks, *Phys. Rev. Lett.* 9, 9 (1962).

10) The Holy Bible, St. Matthew 25, 29.

11) F. London, "Superfluids," J. Wiley and Sons, Inc., New York (1950) Vol 1.

12) W.A. Little, *Phys. Rev.* 134, A1416 (1964).

13) W.A. Little, *Scientific American* 212, 21 (1965).

14) C.N. Yang, *Rev. Mod. Phys.* 34, 694 (1962).

15) W.A. Little, *Phys. Rev.* 156, 396 (1967).

16) R.L. Greene, G.B. Street and L.J. Suter, *Phys. Rev. Lett.* 34, 577 (1975).

17) D. Jerome, A. Mazoud, M. Ribault, and K. Bechgaard, *J. Phys. (Paris) Lett.* 41, L95 (1980).

18) J.G. Bednorz and K.A. Müller, *Z. Phys.* B64, 189 (1986).

19) W.A. Little, *Science* 242, 1390 (1988).

20) W. Weber, *Z. Phys. B - Cond. Matter* 70, 277 (1988).

21) L.C. Smedskjaer et al., *Physica C* 156, 269 (1988).

22) J.-M. Imer et al., *Phys. Rev. Lett.* 62, 336 (1989).

23) T. Takahashi et al., *Nature* 334, 691 (1988).

PART TWO

William A. Little Symposium and Invited Papers

William A. Little Symposium

Saturday, Sept. 30, 1995

Bloch Auditorium, Room 100

Physics Department

Stanford University

Organizing Committee:
B. Cabrera, J. Collman, S. Fetter, W. Harrison, D. Osheroff and R. Yau

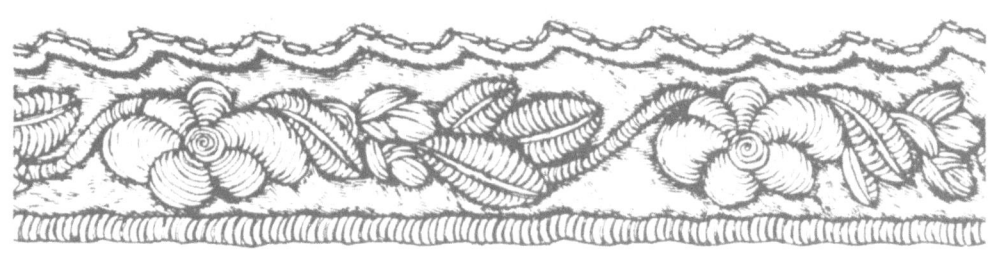

PROGRAM

08:30-09:00	Morning tea & coffee

09:00-10:00   Welcome: Douglas Osheroff,
              Physics Department Chairman
          Vladimir Kresin: 'New Superconducting Systems
                    and the Road to High Tc'

10:00-10:30	Break

10:30-11:30   Pierre-Gilles de Gennes: 'Liquid Crystals'

12:00-02:00   Lunch and tour of Rodin Sculpture Garden

02:00-03:00   Hanoch Gutfreund: 'Organic Superconductors'

03:00-03:15   Break

03:15-04:15   Bernardo Huberman: 'Neural Networks'

04:15-04:45   William Little: 'Recent Advances in Cryogenics'

*   *   *   *

06:00-07:00   Reception

07:00   Dinner
          After dinner speaker: Prof. Walter Meyerhof

ROAD TO HIGH T$_C$ AND SEARCH FOR
NEW SUPERCONDUCTING SYSTEMS

Vladimir Z. Kresin

Lawrence Berkeley Laboratory
University of California
Berkeley, CA 94720

INTRODUCTION

This symposium is a big event not only for Bill Little, but for all of us. It is truly a big event for me because all of my scientific life has been greatly affected by his creative work. I first met Bill Little at the International Low Temperature Conference in Moscow in 1966. He was a major star at that convention, because everybody knew of the famous paper [1] he published in 1964.

This paper has been a real milestone in the field of superconductivity, and a historical perspective can enhance one's appreciation of this fact. Indeed, the dream of high-temperature superconductivity, culminating in the ideal case of room temperature superconductivity, was born a long time ago, even before the appearance of the famous Bardeen-Cooper-Schrieffer theory [2]. BCS theory was created in 1957, and one of the first questions this theory confronted was that of the upper limit of the critical temperature. In other words, it is essential to understand whether superconductivity is a low-temperature phenomenon or is not subject to such a restriction. An answer was provided relatively quickly and it was, unfortunately, not optimistic. By the early 1960's everybody knew that superconductivity was a low-temperature phenomenon and was in principle unobservable at room temperature.

Let me remind you of the logic underlying this conclusion. The famous Fröhlich Hamiltonian contains terms describing a free electron gas, a free acoustic phonon field, and the electron-phonon interaction. An analysis of the electron subsystem performed, for

From High Temperature Superconductivity to Microminiature Refrigeration
Edited by Cabrera *et al.*, Plenum Press, New York, 1996

example, in [3-6], allows one to look at pairing (key ingredient of the theory), evaluate the energy gap, etc. But in addition, one can also study the effect of superconducting pairing on the phonon system. It can be shown [7],see e.g.[8], that electron-phonon coupling leads to a change in the phonon frequency ($\Omega_0(q) \rightarrow \Omega(q)$) which can be described by the following simple relation:

$$\Omega(q) = \Omega_0(q) \, [\, 1-2\lambda]^{1/2} \tag{1}$$

(λ is the coupling constant)

We see that according to this relation the coupling constant may not exceed the value $\lambda=0.5$. Indeed, if $\lambda > 0.5$, the frequency becomes imaginary, which implies that the lattice becomes unstable. In other words, the lattice is unable to tolerate the superconducting transition if the interaction strength exceeds the limit $\lambda = 0.5$.

If we now consider the famous BCS equation for the critical temperature,

$$T_c = \Theta \exp(-1/\lambda) \tag{2}$$

and use $\lambda=0.5$, we come to the conclusion that the maximum value of the critical temperature is about an order of magnitude less than the Debye temperature. Typical values of the Debye temperature are on the order of a few hundred Kelvin; therefore one concludes that the highest possible critical temperatures can be at most 40-50 K. It follows that we should forget the dream of high-temperature superconductivity.

This spirit changed dramatically after the appearance of the Little paper [1]. This seminal paper is important from several points-of-view:

  (1) It revived the dream of high-temperature superconductivity and initiated the search for high-temperature superconductivity;

  (2) It generated enormous interest in the physics of low-dimensional systems; and

  (3) It predicted the possibility of superconductivity in organic materials.

Let me comment briefly on each of these aspects.Everybody knows the famous Little model of an organic polymer. This organic polymer contains two groups of electrons. While in the BCS model the pairing is provided by the phonon sub-system, here the situation is different. Conduction electron pairing is provided by virtual transitions in the second electron subsystem (side chains). While the pairing interaction still may be weak ($\lambda < 0.5$), we are now dealing with a different scale of virtual transitions (ΔE_{el}). One sees directly from a BCS-like equation that this can lead to a great increase in the value of T_c.

Model calculations based on quantum chemistry resulted in an impressive result: $T_{c;max}=2200K$ (!). Of course, this is a bit too high: the above result is based on a number of approximations. Nevertheless, Little's remarkable and realistic conclusion was that"... superconductivity could and should occur in structures such as this even at room temperatures".

As I mentioned above, this paper generated an intensive study of low-dimensional systems, raising a number of interesting problems: phase transitions in one- and two-dimensional systems, correlation between the Peierls and Cooper instabilities, charge density waves, etc.

And, finally, the Little model involved an organic polymer. Not a single organic superconductor was known then. The paper [1] provided a stimulus for the search for organic superconductivity. It is remarkable that only 15 years ago there were no organic superconductors at all. The first one, carrying the "simple" name "bistetraelenafulvalene-hexafluorophosphate" was discovered in 1981 [9]; at present these superconducting materials form a large new family (see, e.g.[10-12]). Thus Little's prediction has been confirmed. The new family is growing and already includes a material with T_C around 15 K; this time derivative is very impressive. The organic superconductor family is very promising. I shall not go into detail about these materials, since several contributions to this book focus on organic superconductivity.

SEARCH FOR NEW SYSTEM

Many publications, particularly in the mid-1960 s, were inspired by the paper [1].Reduced dimensionality and a rigorous analysis of the interplay of various instabilities were at the focus of a paper by the Russian scientists Y.Bychkov, L.Gor'kov and I. Dzyaloshinskii [13]. The study of these issues has continued to the present day. A detailed analysis of the Little model by Monte-Carlo simulations was carried out by Hirsch and Scalapino [14].

Various non-phonon models started being proposed. For example, V.Ginzburg (Moscow) proposed the so- called "sandwich" model, in which pairing in a two-dimensional film is provided by virtual excitations in an insulator placed on the film surface ("excitonic" mechanism) [15]. A metal-semiconductor system with excitonic exchange was described by Bardeen [16] and his colleagues. B.Geilikman (Moscow) studied the electronic mechanism in 3D; this case also can lead to high-temperature superconductivity [17]. Geilikman considered several scenarios, each involving a system containing different groups of electrons. For example, if in addition to the usual electron gas there is a subsystem with localized electronic states, then virtual transitions within this subsystem can provide a pairing channel.

Bill Little himself was actively involved in these developments. A fruitful collaboration between the IBM Laboratory in San Jose and Stanford University led to the synthesis of the superconducting polymer S-N [18]. It is interesting that each of the components (S or N) is separately not metallic, and yet the polymer is not only metallic, but even superconducting. Its T_C is very low ($\approx$1K), but, nevertheless, this material was the very first superconducting polymer, and this work has remained a very important contribution.

Bill Little had the good fortune to meet Hanoch Gutfreund and collaborate with him; jointly they published a number of very interesting papers, including the well known review article [19]. As usual, Bill stayed involved in theoretical as well as experimental research.

This raises an interesting question about Little: is he a theorist or an experimentalist? I have friends, particularly in Europe, who would not believe me when I told them that Bill was involved in experiments. They were convinced (correctly) that he is a top-class theorist. This can be inferred from his papers, his presentations, or from fruitful discussions. But then I would remind my friends, for example, about the famous Little-Parks experiments. The usual response was "Is that the same Little?". The answer to the question raised in beginning of this paragraph is actually straightforward: he is both a theorist and an experimentalist. It is hard to imagine that in our time of narrow specialization there exists a scientist who can invent and produce a tiny refrigerator and also write papers evaluating collective excitations in polyatomic molecules. And yet we indeed have among us such a unique scientist who is at the top of both professions.

VERY STRONG COUPLING.

As was mentioned above, according to the Fröhlich model, phonon-mediated superconductivity appears restricted to a coupling strength of $\lambda < 0.5$. As a result, the critical temperature would not exceed 40-50K. However, life is full of surprises. By the end of the 1960's (this happened after the appearance of the Little paper), it was demonstrated [20,21] that the pessimistic conclusion about the upper limit on T_C for phonon-mediated superconductivity was incorrect. This does not mean that the evaluation of Eq. (1) contains a mistake. The important point is that this conclusion was based on the Fröhlich Hamiltonian, which is not a correct Hamiltonian, at least as far as the problem of lattice instability is concerned. It assumes that the Hamiltonian can be written as a sum of a purely electronic term, a term describing the free acoustic phonon field, and an interaction term. It turns out that such a Hamiltonian cannot be rigorously formulated, and therefore should not be used. The rigorous approach based on adiabatic theory leads to a different conclusion. Namely, it shows that higher values of λ are realistic and do not lead to lattice instability. In fact, many superconductors with coupling constants exceeding 0.5 are known today. They form a big group of so-called strong-coupled superconductors. Examples include lead ($\lambda = 1.4$), mercury ($\lambda = 1.3$), and others. Eventually, the lattice does become unstable, but this happens for much greater coupling strengths.

An interesting question concerns the expression for the critical temperature. As is known, the famous BCS expression for T_C, Eq.(2) is written in the so-called weak-coupling approximation. In other words, it assumes that $\lambda \ll 1$. If λ is greater than 1 but less than 1.5, one can use the McMillan-Dynes expression [22]. In the case of very strong coupling ($\lambda \gg 1$), the critical temperature can exceed the characteristic phonon temperature ($T_c > \Theta = \Theta_D$). This case was studied in [23,24], and an expression entirely different from the BCS equation (2) was obtained. Later, I had a happy occasion to work with Bill Little and Hanoch Gutfreund; we studied this case of very strong coupling and derived analytically an expression for T_C. A very elegant matrix method developed in [25] was used, and the Coulomb repulsion was also included . The expression has the following form:

$$T_c = 0.18 \, \lambda_{eff}{}^{1/2} \, \Theta \; ; \; \lambda_{eff} = \lambda \, (1 + 2.6 \, \mu*)^{-1} \qquad (3)$$

It is interesting to note the change in the interplay between the attraction (described by the coupling constant) and the repulsion. The BCS expression contains difference λ-$\mu*$. The limit of very strong coupling, on the other hand, involves the ratio λ $(1+2.6\mu*)^{-1}$. This equation is valid for $\pi T_c >> \Theta$ $(\lambda>5)$. Of course, it is valuable to look for an expression valid for any value of λ , including the region between $\lambda >1.5$ and $\lambda <5$, which cannot be described either by the McMillian-Dynes equation or by equation (3) which is valid for very large λ .

A general equation can be obtained by means of the same matrix method [25] which was used by Little, Gutfreund and me; this equation has the following form [26]:

$$T_c = 0.25 \ \Theta \ (e^{\ 2/\lambda \ eff.} - 1)^{-1/2}, \tag{4}$$

where

$$\lambda_{eff.} = (\lambda - \mu*) \ [\ 1+2\mu*+\lambda\mu* \ t \ (\lambda)] \ ^{-1};$$

and t (λ) is defined in [26].

Note that this equation is not an interpolation between the weak and very-strong coupling expressions, although it does reduce to all the right limits. It represents a direct solution of the Eliashberg equation and is valid for all values of λ.

HIGH T_C OXIDES. UPPER LIMIT OF T_C.

In 1986 we entered a new era of superconductivity which began with the paper by George Bednorz and Alex Müller [27]. They discovered a new class of high-temperature superconducting materials, copper oxides. This discovery gave rise to a period of uniquely intense research. The number of papers published in the last ten years is counted in thousands. The family of the oxides has grown, and the critical temperature has reached values on the order of 150K.

It is not my intention to retell this exciting development in the field of superconductivity; there exist many nice review articles and books about this subject. Instead, I would like to present some new results. I have been told of Bill's request that this Symposium be a scientific meeting rather than a general celebration. That's why, in accordance with his request, I would like to present a new result, obtained jointly with Dr. Stuart Wolf (Naval Research Laboratory, Washington, DC) and Dr. Yurii Ovchinnikov (L.D. Landau Institute for Theoretical Physics, Moscow).

Our recent work has led to the conclusion that there exists an upper limit to the critical temperature in copper oxides. This upper limit lies in the region of 160-170K. Let me briefly describe our arguments.

In describing the properties of the high T_c oxides, one encounters a parameter which has not played any significant role in conventional materials. This parameter is the carrier concentration n. The cuprates are doped materials, with T_c strongly depending on the carrier concentration. This dependence $T_c(n)$ has a rather sharp maximum at $n=n_{opt}$. In the so-called overdoped region ($n>n_{opt.}$) one observes a sharp decrease in T_c. For example,: in the overdoped $Tl_2Ba_2CuO_{6+x}$ compound T_c drops from 90K to 14K [28]. A similar decrease has been seen in the Bi-based cuprate [29]. We have arrived at the conclusion that this decrease is mainly caused by the pair-breaking effect of magnetic impurities . The influence of magnetic impurities (pair-breaking, the appearance of gaplessness at certain values of magnetic impurity concentration, the depression of T_c, etc.) was studied by Abrikosov and Gor'kov [30] and then by De Gennes [5] , see also the reviews [31,32]. The interaction between localized magnetic moments and the singlet-state Cooper pairs destroys pair correlation and is accompanied by spin-flip scattering (this scattering provides conservation of the total spin). The pair-breaking effect of magnetic impurities in the cuprates has indeed been observed experimentally (see, e.g, [33,34]).

The appearance of a peak in the dependence $T_c(n)$ is due to a peculiar interplay of two factors. First of all, doping increases n, and consequently T_c, since the coupling constant $\lambda \sim n$ [35]. However, if T_c is increased by the usual means of additional doping, then the addition of oxygen is accompanied by the generation of new local magnetic moments whose presence depresses T_c. The observed maximum thus corresponds to a crossover of these two trends. If we want to increase the value of T_c further, we need to employ a different type of doping, e.g., pressure-induced doping, the photoinduced channel, etc.[36-38]. For example, external pressure would decrease the c lattice parameter, leading to additional charge transfer and thereby raising T_c, without introducing additional magnetic moments .

We began with a study of the unusual temperature dependence of the critical field. A dependence of H_{c2} on T drastically different from the conventional picture was recently observed recently in overdoped high T_c oxides [28,29]. Conventional bulk superconductors display a linear temperature dependence of H_{c2} near T_c, a quadratic behavior for T-->0, and a negative curvature over the entire temperature region . Contrary to this picture, the layered cuprates exhibit a linear dependence near T=0K and a positive curvature for all temperatures. Furthermore, $H_{c2}(0)$ greatly exceeds the value which follows from conventional theory [39].

We have shown [40] that such an unusual dependence of $H_{c2}(T)$ is due to a relatively weak temperature dependence of the spin-flip relaxation time τ_s. The latter is due to the presence of magnetic impurities and their correlations at low temperatures.

We are dealing here with a new phenomenon which can be called "strengthening of superconductivity". Let me describe here the qualitative picture. Spin-flip scattering leads to a depression of the superconducting state, as reflected by the relatively small values of T_c and H_{c2}. The magnetic impurities can be treated as independent [30-32], and the impurity spin-flip scattering provides conservation of the total spin. However, at low temperatures (in the region near T=1K), because of the correlation of the magnetic moments, the ordering trend of the moments becomes important, and this frustrates the spin-flip scattering. Pairing is now less depressed, which leads to a large increase in H_{c2} and, consequently, to a positive curvature in H_{c2} vs. T (for a more detailed discussion, see [40]). Note that a similar effect has been observed in Sm-Ce-Cu-O [41a], La-Sr-Cu-O [41b], and recently in Y-Ba-Zn-Cu-O

[41c] systems. Impurity ordering has been observed in the Sm-Ce-Cu-O compound by direct magnetic susceptibility measurements [41a]. A susceptibility increase, reflecting correlation of the moments, has also been observed in [41b]. Note also that the effect of ordering on the curvature of Hc2(T) has also been observed in the Ni/V system [42]. The positive curvature appears when $T_c > T_{Cu}$, where T_{Cu} is the Curie temperature.

An interesting question concerns the location of the magnetic impurities. We believe that the magnetic moments are localized on the Ba-O layer which is adjacent to the Cu-O layer. Localized magnetic moments can be formed on the apical oxygen site. Probably, we are dealing with the formation of the paramagnetic radical O_2^-, composed of the apical oxygen and an additional oxygen in the Tl-O (Bi-O) layers.

The presence of magnetic moments on the apical oxygen site is the key factor responsible for the large difference in the values of T_c of 2201 and 1223 (or 2223) structures. Indeed, the properties of the $Tl_2Ba_2CuO_{6+x}$ compound, and in particular its T_c, are greatly affected by the two O^- apical ions. In the case of $Tl_2Ba_2Ca_2Cu_3O_{10-x}$, the pair-breaking effect is much weaker since the apical oxygens are located outside of the total set of three planes. As a result, the middle plane is hardly affected by the magnetic ions and the effect on the two other planes is not as strong as in the 2201 compound, where the Cu-O plane is affected from both sides. As a result, the Tl-based 2223 compound has T_c=125K, which is much closer to the "intrinsic" limit $T_{c;int}$=160K, see below). For compounds with more than three Cu-O planes, it is difficult to adequately dope the inner planes and T_c saturates or decreases.

An analysis of the overdoped state, and of the dependence $H_{c2}(T)$ in particular, has allowed us to introduce an important new parameter, the so-called "intrinsic" Tc, that is, the value of the critical temperature in the absence of magnetic impurities. Our calculations have yielded the value $T_{c;int} \approx 160$-170K for the Tl-based compound. This value greatly exceeds the experimental value $T_{c;m}$=90K, observed at the optimum ambient pressure doping (e.g., for the 2201 Tl-based cuprate $T_{c;m}^{dop}$= 90K). This reveals that the materials contain magnetic impurities even at optimum doping, which depresses the critical temperature $T_{c;m}^{dop}$ relative to the 'intrinsic" value $T_{c;int}$.

We have calculated the value of $T_{c;int}$ for various cuprates. It is remarkable that it turns out to be approximately the same for all high T_c oxides and lies in the region $T_{c;int}$ =160-170K. We believe that this is not an accidental coincidence. The fact of the matter is that even though there are many high temperature oxides, all of them share one common feature. Namely, all of them contain a key structural unit, the CuO plane. In other words, there is really only one high-temperature superconductor: the copper oxide plane Depending on particular conditions (doping level, lattice dynamics, etc.), one can observe critical temperatures ranging from 0K to 160-170K, the latter being the upper limit of T_c for this class of materials. Note that the critical temperature of Hg-based cuprates with T_c on the order of 150K is not far from this upper limit.

This conclusion about the upper limit of T_c may be upsetting, but is really not so surprising. Indeed, the amount of effort spent in the field of superconductivity in the last ten years has greatly exceeded that of the previous 70 years. As a consequence, it is not unexpected that we should have arrived almost to the saturation point for the critical temperature.

CONCLUDING REMARKS.

One may wonder about future developments. What about our dream to produce a room-temperature superconductor? I do not think that there is any reason to be pessimistic. The road to high T_C continues but it should follow the main lesson given to us by Bednorz and Müler. Indeed, for a long time Nb played a special role in the field. Eventually, people came to the conclusion that any new superconductor with a higher T_C must be based on Nb. Bednorz and Müller's breakthrough came in an entirely different direction: the cuprates do not contain Nb. But the lesson has not been very well taken, because now almost everybody is working with the oxides. We think that it is time to start looking for other classes of materials. This will continue the remarkable road to high T_C. Of course, nobody knows the exact trajectory, and each opinion is subjective. My personal view is that the future room-temperature superconductor will be an organic compound. The discovery of this class of superconductors has confirmed Bill Little's prediction. It is significant that organic superconductors display great similarities with the copper oxides. Stuart Wolf and I have pointed out this amazing similarity [43]. In both cases we are dealing with low-dimensional systems, low carrier concentrations, small Fermi energies, etc. But there are also essential differences, for example, in the number of nesting states. We think that the great variety of organic materials makes this direction extremely promising.

To nobody's surprise, Bill Little was also attracted by various aspects of high temperature superconductivity in the cuprates. He has published a number of very interesting papers. They include a study of a possible excitonic contribution to the origin of high $4T_C$ in the cuprates [44a], a very elegant experiment demonstrating S-wave symmetry [44b], a description of a long-range proximity effect [44c]. Recently he has focused on the optical properties of the cuprates [45]. According to many experts, the experiment [45] is a high example of experimental art.

The paper [45] demonstrates two facts. First, it contains several very interesting results (e.g., the observation of an electronic peak at ≈ 1.5 eV). And second, this paper clearly illustrates that Bill is in great shape and we can expect many more remarkable achievements from him in the future.

Let me conclude by quoting a romantic piece written by Bill Little for the *Scientific American* magazine [46]. This piece reflects his dream about high temperature superconductivity:

"The magnet floats freely above the sheet supported entirely by its magnetic field. The field is unable to penetrate the superconductor and so provides a cushion on which the magnet rests. It is easy to imagine hovecraft of the future utilizing this principle to carry passangers and cargo above roadways of superconducting sheet, moving like flying carpet without friction and without material wear or tear. We can even imagine riding on magnetic skis down superconducting slopes and ski jumps - many fantastic things would become possible."

I would like to wish Bill and all of us someday to take a ride on such a carpet.

REFERENCES

1. W.A. Little, *Phys. Rev.* 134, A1416 (1964).
2. J. Bardeen, L. Cooper, and J.R. Schrieffer, *Phys, Rev.* 108, 1175 (1957)
3. N. Bogoluybov, N. Tomachev, and D. Shirkov, *A new method in the theory of superconductivity*. Cons. Bureau, New York (1959).
4. L. Gor'kov, *Sov. Phys. - JETP*, 7, 505 (1958)
5. P. de Gennes, *Superconductivity in Metals and Alloys*, Benjamin, NY (1966).
6. G.M. Eliashberg, *Sov. Phys. - JETP* 11, 696 (1960).
7. A. Migdal, Sov. Phys. - JETP 7, 996 (1958).
8. A. Abrikosov, L, Gor'kov, and I. Dzyaloshinskii, *Methods of quantum field theory in statistical physics*. Dover, New York (1963).
9. D. Jerome et al., *J. Phys. Lett.* 41, L-95 (1980).
10. J. Ferraro and J. Williams, *Introduction to synthetic electrical conductors*. Academic Press, New York (1987).
11. T. Ishiguro and K. Yamaji, *Organic superconductors*. Springer-Verlag, Berlin, 1990.
12. *Organic Superconductivity*. V. Kresin and W.A. Little, Eds., Plenum Press, New York, 1990.
13. Yu. Bychkov, L Gor'kov, and I. Dzyaloshinskii, *Sov. Phys. - JETP* 23, 489 (1966).
14. J. Hirsch and J. Scalapino, *Phys. Rev. B* 32, 5639 (1985).
15. a) V. Ginzburg, *Sov. Phys. - JETP* 20, 1549 (1965). b) *High temperature superconductivity*. V. Ginzburg and D. Kirzhnits, Eds., Cons. Bureau, New York, 1982.
16. D. Allender, J. Bray, and J. Bardeen, *Phys. Rev. B* 7, 1020 (1973).
17. a) B. Geilikman, *Sov. Phys. - JETP* 48, 1194 (1965). b) B. Geilikman, *Sov. Phys.- Vsp.* 8, 2032 (1966); 16, 17 (1973).
18. a) R. Greene, G. Street, and L. Suter, *Phys. Rev. Lett.* 34, 577 (1975). b)W. Gill et al., *Phys. Rev. Lett.* 35, 1732 (1975).
19. H. Gutfreund and W.A. Little, in *Highly conducting one-dimensional solids*, J. Devreese, R. Evrard, and Va. van Doren, Eds., p. 305. Plenum Press, New York, 1979.
20. Y. Browman and Yu. Kagan, *Sov. Phys. - JETP* 25, 365 (1967).
21. B. Geilikman, J. *Low Temp. Phys.* 4, 189 (1971); B. Geilikman, *Sov. Phys. - USP* 18, 190 (1975).
22. a) W. McMillan, *Phys. Rev.* 167, 331 (1968). b) R. Dynes, *Solid State Commun.* 10, 615 (1972).

23. P. Allen and R. Dynes, *Phys. Rev.* 312, 905 (1975).

24. V. Kresin, H. Gutfreund, and W.A. Little, *Solid State Commun.* 51, 339 (1984).

25. C. Owen and D. Scalapino, *Physica* 55, 691 (1971).

26. V. Kresin, *Phys. Lett A* 122, 434 (1987).

27. G. Bednorz and K.A. Müller, *Z. Phys.* B64, 189 (1986).

28. A.P. Mackenzie et al., *Phys. Rev. Lett.* 71, 1938 (1993).

29. M. Osofsky et al., *Phys. Rev. Lett.* 71, 2315 (1993).

30. A. Abrikosov and L. Gor'kov, *Sov. Phys. - JETP* 12, 1243 (1961).

31. a) D.Saint-James, G.Sarma,and E.Thomas, *Type II Superconductors,* Pergamon, Oxford (1969)

   b) K.Maki, in Superconductivity, Ed. by R.Parks ,p.1035, Marcel Dekker, New York, (1969)

32. A.Abrikosov, *Fundamentals of the Theory of Metals*, North-Holland,Amsterdam (1988)

33. a) C,Niedermayer et al., *Phys.Rev.Lett.*71,1764(1993);

   b) J.Tallon ,J.Loram, *J.of Supercond.*,7,15 (1994); J.Tallon et al. *Phys.Rev.Lett.* 74, 1008 (1995)

34. G.Williams,J.Tallon,and R.Meinhold,*Phys.Rev.*B52,7034 (1995)

35. V.Kresin and H.Morawitz, *Sold State Comm.*74,1203 (1990)

36. a) J.Schilling and S.Klotz, in *Physical Properties of High Temperature Superconductors,* v.3, D.Ginsberg,Ed.,World(Singapore,1982).

   b) P.Chu et al., in *Frontiers of High Pressure Research*, p.383, H.Hochheimer and R.Etters,Eds.,Plenum,NY (1991).

37. V.Kresin, S.Wolf, Yu.Ovchinnikov, *Phys.Rev.B* (in press)

38. a) G.Yu et al., *Solid State Comm.* 72, 345 (1989)

   b) V.Kudinov et al., *Phys.Lett.* A151,358 (1990)

   c) D.Lederman et al., *J.of Supercond.* 7,127 (1994)

39. a) L.Gor'kov, *Sov.Phys.-JETP* 10,593 (1969).

   b) E.Helfand and N.R..Werthamer, *Phys.Rev.Lett.*13, 686 (1964);

40. Yu.Ovchinnikov and V.Kresin, *Phys.Rev.B*52,3075 (1995)

41. a) Y.Dalichaouch et al., *Phys.Rev.Lett.* , 64, 599 (1990)

   b).M.Suzuki and M.Hikita, *Phys.Rev.B*44, 249 (1991)

   c)D. Walker et al., *Phys.Rev.* B51, 9375 (1995)

42. H.Homma et al., *Phys.Rev.B*33,3562 (1986)

43. S.Wolf , V.Kresin, Ref.12,p.31

44. a) W.A.Little, in *Novel Superconductivity*, S.Wolf and V.Kresin,Eds.,p.341,Plenum Press, NY(1987).

   b)G.Lee, J.Collman,and W.A.Little, J.Supercond.,3,197 (1990)

   c)W.A.Little, J.Collman, *J.of Supercond.*,7,175 (1994)

45. M.Holcomb, J.Collman, and W.A.Little, *Phys.Rev.Lett.* 74,3884 (1994)

46. W.A.Little,*Scientific American*, 212,21 (1965).

ARTIFICIAL MUSCLES OPERATING AT (NEARLY)

CONSTANT VOLUME

Pierre-Gilles de Gennes

Laboratoire de Physique de la Matière Condensée
U.R.A. - C.N.R.S. n° 792
Collège de France
11 place Marcelin-Berthelot
75231 Paris Cedex 05
FRANCE

ABSTRACT

Gels can be strongly swollen by external agents, but this mechanical response is not very practical for various reasons (a) the transition between compact and swollen states involves very strong local tensions and damages the sample (b) since solvent must diffuse in or out, the response time are limited by diffusion, and are very slow. We propose here a set up which would, to some extent, eliminate these difficulties. It is based on a network with a nematogenic backbone, swollen by a nematic solvent, with molecules very similar to the backbone. Then a slight drop in temperature (from the isotropic to the nematic regime) could induce a strong uniaxial deformation, with nearly 0 volume change.

MUSCLES BASED ON ISOTROPIC SWELLING

Since early days the construction of chemico-mechanical motors, transforming chemical signals into work, has been a challenge.[1,5] More recently, the swelling or deswelling of gels, with abrupt phase transitions induced by external agents, has been studied in great detail.[6,7] For instance many gels show a sharp transition from a swollen state to a collapsed state, induced by changes in temperature or pH.[8] These transitions imply a spectacular change in volume : thus, at intermediate stages the sample is swollen in some regions, not in others, and it experiences huge mechanical tensions ; a complicated set of fractures builds up.[9] This heals to some extent after completion (the gel returns to a transparent state) but for mechanical purposes this process is clearly bad.

A further complication exists if the agent initiating the transition is chemical :[2,3,10] for instance a change of pH displacing the ionization equilibrium of carboxylic functions

$$G - CO_2^- + H^+ \rightarrow G - CO_2H$$

where G represents the gel backbone. Very often the GCO_2^- form (in water) will be swollen by Coulomb repulsions, while the neutral form GCO_2H collapses whenever G is slightly hydrophobic.

From High Temperature Superconductivity to Microminiature Refrigeration
Edited by Cabrera *et al.*, Plenum Press, New York, 1996

221

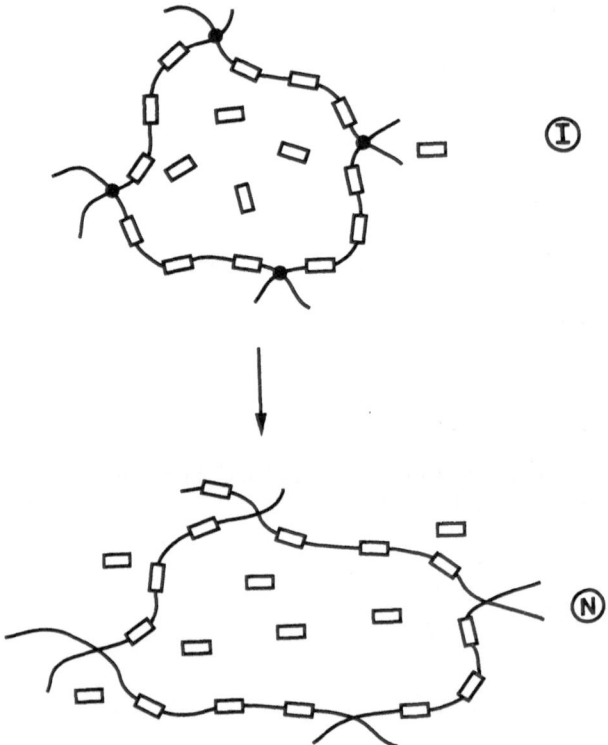

Figure 1. A nematogenic gel in an isotropic solvent (I) and in the nematic phase (N) of the same solvent. The crucial assumption behind this figure is that gel and solvent should not segregate in the N-phase.

The difficulty with these chemical action is that the reagent (e.g. H^+ ions) must reach all points in the gel. In most cases it moves in by diffusion, and the resulting response times are very slow. In some cases the ions can be pushed in by drift in an electric field -but here again the typical electrophoretic velocities are somewhat too slow.

One countermeasure could be of interest : it is based on a gel with two interpenetrated networks :[11] one which is elastic (e.g. the GCO_2^- gel mentioned above) and one which would be based on a backbone of conducting polymers : the latter plays the role of a very finely divided electrode, and allows for liberation (or absorption) of H^+ ions locally. At the moment, conducting polymers which are soluble in certain solvents begin to be available. But the whole idea is however very far fetched : even if one could achieve a swollen, conducting network, it would probably not resist many cycles of operation.

Finally, all systems based on isotropic swelling face a common limitation (which was pointed out to me recently by A.R. Khokhlov) : because solvent must diffuse in (for swelling) or out (for deswelling) the response times of the gel will always be limited by the (very long) diffusion times of the solvent : even the sophisticated interpenetrated network stretched above would not circumvent this difficulty.

Thus we should (probably) drop the idea of using isotropic swelling : an alternate solution is proposed in the next section.

NEMATIC GELS

The basic object is shown on fig. (1). We start from a backbone made of nematogenic molecules, separated by short flexible spacers. We swell this by a nematic solvent, which must mimic the backbone very precisely. This is a very stringent requirement. F. Brochard showed long ago[12] that a very slight mismatch between solvent and chain is enough to induce a collapse in the nematic phase (although the gel may be swollen in the isotropic phase).

For our purposes, we assume perfect compatibility : hopefully this can be achieved by a careful optimization of spacer lengths and other, similar, chemical parameters.

Let us now start from an isotropic solution, at a temperature T_0 just above the nematic clearing point T_n Wo then abruptly cool down to a temperature T_1 slightly below T_n. The system acquires a finite alignment, measured by the order parameter

$$S = < \frac{1}{2}(3\cos^2\theta - 1) >$$

where θ is the angle between one nematogenic block and the optical axis.

Clearly the gel deforms, and elongates along the optical axis.[13] Let us first estimate what is the volume change involved. We shall base this discussion on the Flory discussion of gel swelling.[14,15] One chain (between cross-links) in the network is assumed to have N monomers, and an unperturbed m. s. radius $R_0^2 = Na^2$ in the isotropic phase. In the deformed phase we have two sizes (along and normal to the optical axis)

$$R_{0z}^2 = \frac{1}{3}Na^2 <\cos^2\theta> = \frac{1}{3}Na^2(1+2S)$$

$$R_{0x}^2 = \frac{1}{3}Na^2 <\sin^2\theta> = \frac{1}{3}Na^2(1-S)$$

(1)

These eqs. assume that the order parameter is the same for solvent and gel -a reasonable idea for our case.

Let us now assume that the solvent is good, and that dissolved chains repel each other with an excluded-volume parameter v (we assume here that v is <u>independent of S</u>). In the deformed system the end-to-end vector of one chain is designed by X, Y, Z (Z being along the optical axis). The free energy F (per chain) has the scaling form

$$\frac{F}{kT} = \frac{3Z^2}{2a^2N(1+2S)} + \frac{3(X^2+Y^2)}{2a^2N(1-S)} + \frac{N^2v}{(X^2+Y^2)Z}$$

(2)

The first two terms are elastic. The third described interchain repulsions, with a chain of volume $Z(X^2+Y^2)$. Optimizing F versus Z, X, and Y we arrive at a swelling ratio

$$Q \equiv \frac{XYZ}{R_0^3} = (1+2S)^{1/5}(1-S)^{1/5}$$

(3)

Note that, at small S, the swelling ratio deviates from unity only at the level of an S^2 term

$$Q \to 1 - 3S^2/5 + O(S^3)$$

(4)

as expected from symmetry arguments. The main conclusion from eq. (3) is that $1 - Q$ is <u>small</u> when we use an S value corresponding to the transition point ($S \sim S_c \sim 0.44$) : namely $1 - Q \cong 10\%$.

DISCUSSION

1) The presence of a small volume contraction is encouraging from the point of view of mechanical strength : the danger of fracturing the gel is very much reduced.

2) The problem of solvent diffusion is in principle circumvented : if we have no volume change, the solvent can move in unison with the network, with zero relative velocity. Then the diffusion lags are not relevant. Of course, in actual fact, the gel wants to contract slightly. But we then expect a mechanical response with two components : one fast

(at constant volume) and one slow (with volume change). Further calculations on this point are under way.

3) In any case the response speeds are limited by the temperature change, i.e. by heat diffusion. Fortunately heat diffusion in liquids is much faster than matter diffusion (by a factor $> 10^2$).

4) A specific difficulty of nematics is the fact that when we cool down below the clearing point the local optical axis may start with different orientations in different parts of the sample (ultimately leading to complicated textures such as the "schlieren texture".[16]

Clearly the best way out of this problem is to keep a guiding field in the isotropic state : this could be a magnetic field, or an electric field, or a mechanical field (maintaining a weak uniaxial stress on the sample).[13]

Clearly we are still far from efficient artificial muscles, operating fast and with good durability. But the nematic systems of fig. (1) may provide an interesting approach.

ACKNOWLEDGMENT AND DEDICATION

This work greatly benefited from discussions with A. R. Khokhlov and F. Brochard.

I would like to conclude with some thoughts about Bill Little's professional career. If we consider the nature of inventions over the last century, we see that a number of them, like the Eiffel Tower, may be completely useless, but serve to keep our dreams alive. Other inventions are like the Brooklyn bridge; not only beautiful, but also very useful for mankind. I think that Bill Little is a person of our times who in his research has been building very useful inventions like Brooklyn bridges throughout his life, and I hope that he will continue to do so for many years to come.

REFERENCES

1. A. Katchalsky, *J. Polymer Sci.* 7:393-412 (1951).
2. A. Katchalsky, S. Lifson, H. Eisenberg, *J. Polymer Sci.* 7:571-574 (1951).
3. A. Katchalsky, M. Zwick, *J. Polymer Sci.* 16:221-233 (1955).
4. A. Katchalsky, A. Oplatka, *Handbook Sen. Physiol.* 1971:1-17.
5. A. Katchalsky, *Proc. Int. Congn. Pharmacol.* 4:245-248 (1970).
6. T. Tanaka, D. Fillmore, S.T. Sun, I. Nishio, G. Swislow, A. Shah, *Phys. Rev. Lett.* 45:1636-1639 (1980).
7. T. Tanaka, *Physica* 140A:261 (1986).
8. M. Shibayama, T. Tanaka, *Adv. Polymer Sci.* 109:1-62, Springer Verlag (1993).
9. T. Tanaka, S.T. Sun, Y. Hirokawa, S. Katayama, J. Kucera, Y. Hirose, T. Amiya, *Nature* 325:796-799 (1987).
10. See for example E. Kofukata, Y.Q. Zhang, T. Tanaka, *Nature* 351:302-304 (1991).
11. See for instance K. Frisch, D.K. Lempner, H. Frisch, *Polymer Engineering and Sci.* 22:1143-1152 (1982).
12. F. Brochard-Wyart, *J. Phys.* (Paris) 40:1049 (1979).
13. For a related discussion see P.-G. de Gennes, *C. R. Acad. Sci.* (Paris) 281:101-103 (1975) and P. Bladon, E. Terentiev, M. Warner, *J. Phys.* (Paris) 4:75-91 (1994).
14. P. Flory. *Principles of Polymer Chemistry*, Cornell University Press, Ithaca, New York (1971).
15. P.-G. de Gennes. *Scaling Concepts in Polymer Physics*, Cornell University Press, Ithaca, New York (1985).
16. P.-G. de Gennes, J. Prost. *The Physics of Liquid Crystals*, Oxford (1993).

HIGH T$_c$ ORGANIC SUPERCONDUCTIVITY REVISITED

H. Gutfreund

Racah Institute of Physics
The Hebrew University
Jerusalem, Israel

PREFACE

I was fortunate to spend my postdoctoral years at Stanford in the late sixties. After a year of work with Felix Bloch on the Tomonaga model of a one-dimensional electron system, he suggested to me, before leaving for a year of absence, that I join Bill Little to study with him the problems, implications and consequences of his ideas on organic superconductivity. I shall always remain grateful for this advice. It led to many years of collaboration and friendship between the two of us, it brought me back to Stanford many times, and it brought Bill Little several times to Jerusalem. It has enriched my life in general, and my life as a physicist, in particular. My brief contribution to this volume, honoring Bill Little on his 65th birthday, is a personal historical retrospect on our joint effort in exploring the prospects of high temperature organic superconductivity. I have omitted references to the original work of a large number of scientists on the topics and issues mentioned below. Most of these topics are discussed in detail, with proper references, in the relevant papers which are included in this volume.

THE IDEA

In 1964 W.A. Little published a paper entitled "Possibility of Synthesizing an Organic Superconductor." Little suggested that search for novel superconductors should be extended to organic conductors and conducting polymers, at a time when the understanding and phenomenology of organic conductivity were in their infancy, and when organic superconductivity had still to be demonstrated. He argued that in such compounds a new mechanism, which he named the 'excitonic mechanism' could induce superconductivity at room temperatures, at a time when it was generally believed that phonon exchange was the underlying mechanism in all the known superconductors and that, as a consequence, this phenomenon was in principle restricted to very low temperatures. Almost everything in this paper was out of line with the conventional wisdom about superconductivity in those days and it therefore met with skepticism, disbelief and sometimes with fierce criticism. The validity of the whole concept was questioned and objections to many of the notions contained in the paper were raised.

From High Temperature Superconductivity to Microminiature Refrigeration
Edited by Cabrera *et al.*, Plenum Press, New York, 1996

Little's argument was very simple. In conventional superconductors the electron-electron pairing is caused by the exchange of phonons and the basic parameters which determine the transition temperature are the Debye temperature Θ_D and the electron-phonon coupling constant λ_p. In the weak coupling limit the BCS formula for the transition temperature reads

$$T_C = 1.14\Theta_D \exp(-1/\lambda_p).$$

If the attractive interaction between the electrons involved in the superconducting state were mediated by the exchange of an exciton, or an electronic excitation in general, then the Debye energy, which corresponds to a temperature of several hundred degrees, would be replaced by a typical electron excitation. This would increase the prefactor in this formula by two-three orders of magnitude, setting a new scale for the transition temperature T_q. The simplicity of this argument is very misleading. The conclusion still depends on whether it is possible to achieve a reasonable strength of the electron-exciton coupling constant, which should now replace the parameter λ_p. In fact, much of the criticism and of our subsequent work focused on this issue.

Little's idea of an excitonic superconductor, requires two types of electrons-those which participate in the superconducting transition by forming Cooper pairs, and those which make up the exciton modes necessary to provide the attractive electron-electron interaction. These two electronic media are confined to two spatially separated regions in close contact with each other. The structure which was proposed to implement this idea consisted of a conducting one-dimensional spine of carbon atoms, with organic dye molecules chemically bound to this spine at regular intervals. An electron propagating along the spine polarizes the dye molecules by pushing the electronic charge away from the spine creating a positively charged region, which could attract another electron, thus forming an electron pair. The formation of Cooper pairs in a many electron system is a much more complicated process, but such a simple minded picture is frequently used to explain the role of ionic polarization in superconductivity. It is hoped that the ionic polarization may be replaced, under specific conditions, by the kind of electronic polarization described above.

The exciton mechanism is in principle not restricted to one-dimensional structures. The possibility of excitonic superconductivity in two-dimensional systems consisting of thin metallic films coated on one or both sides by highly polarizable layers has been proposed at about the same time. One can also think about three dimensional systems in which one group of electrons makes up a broad conduction band and the excitonic medium consists of electrons in localized states or in a narrow overlapping band. Little has argued for the advantage of the one-dimensional filamentary system, because such a geometry allows a dense packing of polarizable ligands in close contact with the conducting electrons. More detailed analysis, which came much later, showed that such a structure works to strengthen the coupling and seems to optimize the delicate balance between the different factors controlling the electron-electron interaction.

On the other hand, the one-dimensionality of the model poses a difficulty, which in the early days appeared to cast strong doubt on the whole idea. It was known that phase transitions involving a classical order parameter, such as magnetic ordering or a liquid-gas transition, cannot occur in one dimensional systems of particles with finite range interactions. It was shown that this conclusion could be extended to the superconducting phase transition. However, in real systems there is always an interchain interaction which makes the system quasi one-dimensional. Thus, such a system can undergo a phase transition at a finite temperature, although this temperature may be significantly lower than its mean field value. The problem of suppression of the transition temperature by fluctuations of the order parameter in a one-dimensional system occupied Bill Little for two to three years after the publication of his 1964 paper. He developed an ingenious topological argument to show that the fluctuations which destroy the off-diagonal long range order, which describes the superconducting state, hardly affect the persistent currents involved in superconductivity, so that states of extremely high conductivity can occur in filamentary materials despite their restricted dimensionality. Only when this point was sufficiently well understood, was Little ready to take the next step and only then did Felix Bloch resume his cautious support for the research effort which developed from Little's model.

BEFORE THE FIRST ORGANIC SUPERCONDUCTOR

In 1969 Bill Little organized in Honolulu the first "International Conference on Organic Superconductors" which brought together physicists and chemists to educate each other in the relevant aspects of their respective fields of inquiry. The close collaboration between these two scientific communities, which is so essential to this field and so typical of its development, started there. The physicists talked about the physics of superconductivity and the chemists explained the properties of several known organic conductors and discussed the problems of synthesizing the filamentary compounds proposed by Little. The physicists began to appreciate how much the chemists actually knew and understood about such 'physical' concepts as the Peierls instability and charge density waves. The chemists learned that the π-electrons in extended hydrocarbon molecules behave in some respects like the conduction electrons in metals. They may be studied by methods of solid state physics and they exhibit collective behavior like plasma oscillations. Bill Little and myself studied these aspects of the π-electron system and presented our work at this conference. The highlights of that conference were the presence of Felix Bloch, who gave a very inspiring talk on the history of superconductivity with a rich conceptual content, and of Vitaly Ginzburg, the founder and leader of the Russian School on High Temperature Superconductivity. Ginzburg articulated his confidence that superconductivity is not confined to low temperatures and argued that one-dimensional organic compounds are not the only option for the search for superconductors with high transition temperatures.

Five years later, in 1974, we organized a similar conference, but with a more cautious title, on the "Search for Higher Critical Temperature Superconductors" at Kefar Giladi, a kibbutz in the north of Israel near the Lebanese border. I shall say a few words about this conference, but first let me mention a discovery, which generated a broad and intensive interest in the field of organic conductivity and superconductivity.

In 1973 Alan Heeger and his group studied the conductivity of the charge transfer compound TTF-TCNQ and discovered a large conductivity peak around 60 K, followed by a sharp metal-insulator transition. The chain-like stacking of planar molecules in the TTF-TCNQ crystal renders a quasi one-dimensional metallic system, with high electrical conduction along the TCNQ stacks and low in the perpendicular direction. At first sight the structure of TTF-TCNQ, with its highly conducting TCNQ chains and polarizable TTF chains, bears a resemblance to the Little model. The discovery of the enhanced conductivity in this material raised several questions about its origin and about the nature of the metal to insulator transition. It was suggested that the conductivity peak was due to fluctuations towards a BCS superconducting state which does not materialize because a Peierls transition takes over. This was the beginning of the study of competing instabilities in one-dimensional systems with the idea that crossover from one type of behaviour to another can be induced by external means, like pressure or irradiation. This program did not succeed in TTF-TCNQ and related compounds but gave positive results in the organic superconductors which were discovered several years later.

The conference in Kefar Giladi was the first such international meeting after the discovery of the conductivity peak in TTF-TCNQ. Thus, the nature of this peak and the subsequent metal-to-insulator transition were discussed extensively, as was, in this context, the competition between the Peierls transition and the BCS superconducting transition. Among the participants at the conference were Ted Geballe and Berndt Mathias. Many of the then known superconducting compounds had been discovered by them through hard work, trial and error, and intuition based on extensive experience and a profound understanding of materials. They succeeded in pushing the transition temperatures in the ordinary superconductors to higher and higher values, which at that time was about 23 K. At the conference they argued that lattice instabilities would be the most severe obstacle towards achieving high transition temperatures. Incidentally, Mathias was one of the most ardent critics of Little's ideas, yet he came to the conference not only as a curious attendant but also as an active participant.

In 1975 I spent a Sabbatical year in Stanford. During that year I focused on two research projects. One was the theory of the possible instabilities in one-dimensional electron systems. The ground state of such systems may exhibit one of the following four types of order: charge-density- or spin-density-wave, and singlet- or triplet-superconductivity. Their occurrence depends on the strength of certain parameters, denoted by g; (i-1,2,3,4), which characterize the interaction between two electrons on the one-dimensional Fermi surface. The effect of interchain coupling, of pressure and of impurities was investigated in order to specify the conditions under which superconducting ordering prevails. This whole subject has frequently been referred to as the g-ology problem and it attracted much of attention in the course of development of the field.

The second project which took most of my time during that year was the beginning of a systematic study, together with Bill Little, of a variety of aspects of the excitonic mechanism of pairing. This led several years later, in 1979, to our review paper: "The Prospects of Excitonic Superconductivity" (reprinted in the present volume). This paper discusses the main differences between the exciton and phonon pairing mechanisms and addresses the various objections which had been raised. I shall return to this paper later.

SUPERCONDUCTIVITY IN CHARGE TRANSFER COMPOUNDS

The first superconducting transition in an organic compound was discovered in 1980 in the charge transfer compound $(TMTSF)_2PF_6$ This is one of the so-called Bechgaard salts $(TMTSF)_2X$, in which one electron is transferred from two TMTSF molecules to one acceptor molecule X, where X represents one of several molecules like PF_6, AsF_6, CIO and a few others. The conductivity in these compounds is predominately along the TMTSF columns, and the X ions stabilize the crystal structure by preserving charge neutrality and separate the conducting columns so that the structure is essentially one-dimensional. At ambient pressure the compound undergoes a metal to insulator transition at 12K, and the insulating phase exhibits magnetic ordering of the spin-density-wave type (SDW). At a critical pressure of 6.5kbar the SDW phase is suppressed and the system becomes insulating at 1.2 K.

Actually, the possibility of superconductivity in a polymeric structure had already been demonstrated in 195 with the discovery of a superconducting transition in the polymer $(SN)_x$ at 300mK. However, the discovery of superconductivity in the charge transfer salt $(TMTSF)_2PF_6$ even though the transition temperature is still very low, was the more significant breakthrough on the road towards high temperature organic superconductivity. This discovery demonstrated for the first time the possibility of superconductivity in a truly quasi one-dimensional organic compound. Moreover, it appeared as a triumph of a deliberate search for superconductivity by suppressing an insulating instability. It also opened a very intensive field of research leading to the discovery of more than thirty organic superconductors with transition temperatures up to .5 K. They may be classified into several groups of charge transfer salts based on a very small number organic molecules. Their properties have been studied and analyzed by all known experimental tools and theoretical methods. They exhibit a very rich phenomenology of behaviour and the nature and mechanisms involved in their superconductivity is still a controversial topic.

One may pose the question "why is the transition temperature in these organic compounds so low?" This questions has been asked in comparison with the cuprates, in which the transition temperatures are significantly higher, despite some striking similarities in their anisotropic structure and carrier concentration. I do not intend to address the question in this context, but rather in the context of the goal of Little's program. The structure of these compounds is very different from those proposed by Little. The charge transfer compounds, based on the donor molecule BBDT-TTF, which have the highest transition temperature around 10 K, exhibit a quasi two-dimensional layered structure. In the one-dimensional compounds of the (TMTSF)$_2$X family, the conducting electrons move along a very broad stack of TMTSF molecules. To achieve a reasonably strong excitonic interaction in such a structure it is essential for these electrons to be localized in a narrow spine of atomic dimensions. It seems that the basic conditions for the realization of Little's model are not fulfilled in these compounds.

BACK TO THE ROOTS

The consequence of the last paragraph is that the goal of high temperature organic superconductivity takes us back to Little's original proposal where the emphasis is on the three words organic, one-dimensional, excitonic. It is already clear that organic superconductivity exists and that one-dimensionality is not an obstacle. Let me now say a few words about the validity of the notion of the excitonic mechanism of pairing. The detailed discussion question is whether and under what circumstances can the electron-exciton interaction, which is assumed to replace the electron-phonon interaction, be strong enough.

The first difficulty that comes to mind is that of the exchange interaction between the conducting electrons and the electrons in the polarizable medium which make up the excitons. The exchange component of the interaction is of opposite sign to the direct component and this reduces the effective electron-exciton interaction. Another, apparent difficulty had to do with the effect of the virtual and thermally excited phonons on the exciton mechanism. Questions about the form of the equation for the transition temperature had to be resolved, in view of the apparent importance of vertex corrections and a consequent breakdown of certain approximations which are valid in ordinary superconductors. We have analyzed these and other issues to show that the objections which were raised were either invalid, or their effect could be minimized under proper conditions. The study of a detailed model showed that high transition temperatures can indeed be achieved in the proposed structure provided that the conducting electrons are concentrated along a narrow spine, surrounded by closely packed polarizable ligands at close distance to the spine.

In 1986 Bednorz and Mueller discovered superconductivity in cuprate ceramics with T_c around 35 K. Within a short time materials with transition temperatures exceeding 120 K were found. The controversy about the nature of superconductivity in this class of compounds persists until today. Every known mechanism of superconductivity in these compounds is of the BCS-type, but with an electronic contribution to the pairing interaction in addition to the ordinary phonon contribution. This is similar to the originally postulated excitonic mechanism, except that instead of the polarizable side chains an oxygen-to-copper charge transfer excitation at 1.6 eV was involved.

If this turns out to be true, as indicated by accumulating experimental and theoretical evidence provided by Little's group, then we would know that all the different elements of Little's original proposal can indeed be realized. In that case, the challenge which still remains would be to engineer and synthesize a system in which all these elements come in to play together. All our work together indicates that this is in principle possible though not easy to achieve.

SUMMARY

Much has happened in this field in the last thirty years. A large number of organic conductors have been synthesized and their experimental and theoretical investigation has progressed significantly. About thirty five organic superconductors with transition temperatures up to 12 K are known. They exhibit a variety of novel features related to their one-and two-dimensional structure. A very active field of research at the interface between physics and chemistry has emerged. Many of the early objections and doubts relating to organic superconductivity and to the destructive effect of one-dimensionality have disappeared, simply by the strength of existence proofs.

The search for high temperature superconductivity has achieved a partial success in the family of cuprates, diverting attention to that direction. It seems that many of the features and problems which characterized the development of the study of organics surface again in the discussion of cuprate compounds. In particular, issues such as the contribution of non-phonon mechanisms, the effect of restricted dimensionality and the role of magnetic ordering are at the heart of the debates on the nature of superconductivity in the cuprates. The comparison between these two classes of compounds have become a methodological tool for understanding and hopefully for drawing lesson which could be useful for future developments.

Many of the developments in the research on organic metals and superconductors and many aspects of the research of the cuprate ceramics can be traced back to Little's original and bold venture into almost unknown territory thirty years ago. Although the high-temperature organic superconductor has not yet been synthesized, this idea and the work which followed with its large variety of direct and indirect developments and implications, has had a distinct impact in science.

To conclude with a personal remark, I wish to say how grateful I am for having participated with Bill Little, during a certain stage in my life, in this exciting endeavor. I hope and wish that he will still celebrate the full realization of his vision.

RECENT ADVANCES IN CRYOGENICS

W. A. Little

Physics Department
Stanford University
Stanford, CA 94305

INTRODUCTION

I would like to thank the organizers of this Symposium for the opportunity to play at being "King for the Day" - to have my wish to bring together old friends, colleagues and co-workers for this Symposium. On this occasion I thought it would be appropriate to talk about Cryogenics, which has been my other life for the past 15 years - a mini-career in industry in our small company, MMR Technologies, Inc. in parallel with my University career. It is appropriate because when I came to Stanford in 1958 my charter was to set up the infrastructure for cryogenics on campus and to provide liquid helium for all users. Felix Bloch and George Pake had obtained funding from the Office of Naval Research for the acquisition of a Collins helium liquifier and I was to set it up and have it run. This I did for the next 15 years until I was able to pass the responsibility on to others. At first I was the only user, but in time others emerged in chemistry, material science, high energy physics, and when Bill Fairbank arrived on campus he became a major user. I have thus returned to my roots.

When I arrived at Stanford I had just completed 2 years at the University of British Columbia in Vancouver, Canada where I had build a cyclic adiabatic demagnetization refrigerator to maintain a bath at 0.10 K where I had hoped eventually to study the polarization of nuclei. This was before He^3 was generally available and, of course, before the advent of the dilution refrigerator, so the only way to achieve these temperatures was by adiabatic demagnetization of a paramagnetic salt. After my experience in Canada working on my own, I came to regard these experimental runs as heroic undertakings, lasting as they did, 5 or more days and nights, often with little to show for one's efforts. At Stanford, I planned to build a better machine, hopefully to reach 0.010 K, but found it even more demanding, for I was carrying at the same time a full teaching load on Solid State Physics, much of which was still new to me. This experience instilled in me a strong interest in simplifying the way cryogenics was to be done.

In 1958, I learned of some work that had indicated that superconductivity could persist in wires of niobium at 4.2 K in fields as high at 6,000 gauss, contrary to the popular held beliefs at that time. This led my post-doc, Marcel LeBlanc, and me to try to build a superconducting solenoid for this refrigerator using niobium. This we did, and it was one of the first such superconducting magnets ever built[1]. The other had been built by Stan Autler a few months earlier, and both preceeded by a couple of years the discovery of the really high field alloys and the subsequent explosion of interest in superconducting magnets. It was our first introduction to superconductivity. In the next several years I drifted away from the refrigerator project, distracted by superconductivity through the discovery of flux quantization by Fairbank and Deaver in the adjoining laboratory in the basement of the old Physics Department, and later through the related experiments I did with Ron Parks as a forerunner of the SQUID[2].

From High Temperature Superconductivity to Microminiature Refrigeration
Edited by Cabrera *et al.*, Plenum Press, New York, 1996

231

ORGANIC SUPERCONDUCTIVITY: POLITICAL FALLOUT

Shortly after our paper with Parks, I published a paper on the possibility of synthesizing an organic superconductor[3]. This paper seems to have inspired a number of young persons, but the older establishment in this country saw it more as a red flag to a bull and attacked it unmercifully. Bernd Matthias, in particular, then at Bell Labs and later at La Jolla took every opportunity to ridicule it and to deride our interest in the study of systems of limited dimensionality. I illustrate this with some quotes by Matthias from Comments on Solid State Physics[4].

"...Little who predicted superconductivity far above room temperature in some organic compounds of a more or less one-dimensional structure along with other science fiction dreams...", and,

" ...present theoretical attempts to raise the superconducting transition temperature are the opium in the real world of superconductivity where the highest T_c is, at present and at best, 21 K. Unless we accept this fact and submit to a dose of reality,... all that is left in this field will be these scientific opium addicts, dreaming and reading one another's absurdities in a blue haze".

Another nugget, again by Matthias, can be found in Physics Today[5] in regard to the discovery of a new superconductor, where he says, "It just goes to show, don't bother with two-dimensional structures. It won't work if you want high (transition) temperatures." In the light of recent discoveries of organic superconductors (quasi-one and two-dimensional structures), the high transition temperature cuprates (quasi-two dimensional, layered structures), and the fullerenes, these remarks illustrate all too well the folly of this thinking.

I took this kind of battery for a few years but eventually got back at Bernd in a response in Physics Today[6] in 1972, where I argued for the importance of studying a wider range of materials than the simple binary and ternary alloys, which was the limited arena within which the likes of Bernd Matthias had worked. But in addition, I said in effect, and this is relevant to our later work, that some of the money (in my opinion) being squandered on Matthias could be better spent on the improvement of cryogenic refrigerators by us! This is how I put it.

"The article by Bernd Matthias on "The search for high temperature superconductivity unfortunately gives a rather parochial view of the field. ...

...In view of the heroic efforts which have gone into raising the transition temperature of conventional superconductors to date, recognizing the rather modest return from this (less than 13% increase in 17 years), and the dismal prospects in continuing this approach, it would be far more worthwhile from the *technological point of view* to devote proportionately less effort to this dying approach and a much bigger effort to the goal of improving the efficiency of cryogenic refrigerators..."

I stressed the technological point of view because I felt that there was merit in trying to understand the basic science responsible for limitations on the transition temperature of superconductors, but that was a different problem from the practical problems of the use of superconductors.

In spite of these harsh words, in fact. Bernd and I had been quite good friends prior to this and nearly a decade later, after being stuck on a flight from Paris for twelve hours together, we settled our differences, and became good friends again.

GENESIS OF MICROMINIATURE REFRIGERATORS

Some years later I was taken up on my comment on refrigerators. I had suggested that if the Navy would give me a little money on this project, it would speed the day that we would have superconducting electronic devices in all our laboratories. The Navy took me up on this but gave me just a *very* little bit of money, but it was this that started what was to become MMR.

During those early days after the discovery of the Josephson effect it was realized that superconductivity would make possible a whole new class of supersensitive instrumentation - better voltage standards, more sensitive voltmeter, ammeters, fluxmeters, etc. But these would require refrigerators and at that time the simplest refrigerators in existence were Joule Thomson (JT) refrigerators, which consist of a simple counter-current heat exchanger with a throttle valve at the low temperature end. High pressure gas passing down the heat exchanger is pre-cooled by the returning, counter-flowing, low pressure gas, and at the end of the exchanger the high pressure gas is allowed to expand to low pressure and return up the exchanger. The JT effect leads to cooling and so the process continues until the gas liquefies. These "Miniature refrigerators", were the size of a short pencil but had a capacity of about 10 Watts and consumed a large mass of gas in operation. The SQUID devices we were trying to cool, on the other hand, needed at most 10 to 100 mW of refrigeration. So it seemed that *Micro-Miniature refrigerators* were needed - refrigerators with a capacity a hundred times less than that of conventional *Miniature* refrigerators. Simple scaling arguments[7] showed that to build such refrigerators, capillary tubing would be needed with the dimensions of a human hair! Brazing fins on such tubing was virtually impossible, as I soon discovered - and so the subject rested.

Some years later, however, in 1975, while I was working in the Integrated Circuits Laboratory at Stanford on a novel type of memory device for the neural network project we were then engaged in, I learned of the use of photolithography for the fabrication of integrated circuits and realized this would be a way to build the microminiature refrigerators. After a multitude of false starts we ultimately succeeded in doing so in the following way[8].

A thin glass plate, a little smaller than a microscope slide was coated with a thin solution of gelatine sensitized with ammonium bichromate and allowed to dry. It was then exposed to ultra-violet light through a photo-mask of the pattern of the desired heat exchanger configuration, then developed in warm water, and allowed to dry. The part exposed to light hardened and the unexposed part was washed away. The slide was then sand blasted with a jet of fine alumina powder, etching narrow channels in the glass. After the proper depth of etch was realized the slides were cleaned, inlet and outlet holes etched in them and then they were fused together to form the final refrigerator. This worked! In due course I rented some space in nearby Mountain View in the Silicon Valley and, with a colleague founded the company which is now MMR, Technologies, Inc. Here we began to develop means for the manufacture of these coolers and our first refrigerators were built there in December, 1980. Fig. 1.

APPLICATIONS OF THE REFRIGERATORS

The unique characteristics of these refrigerators[9] are that they are small, quiet, they respond fast, consume very little gas, and are inexpensive to operate. They can be

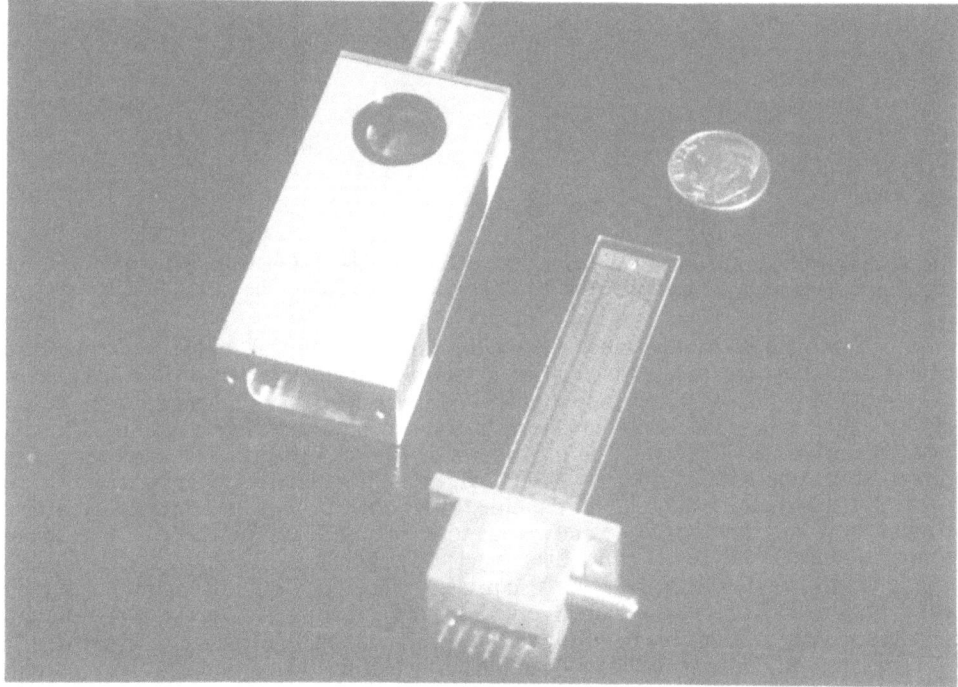

Figure 1. First microminiature refrigerator and vacuum enclosure built at MMR, December 28, 1980.

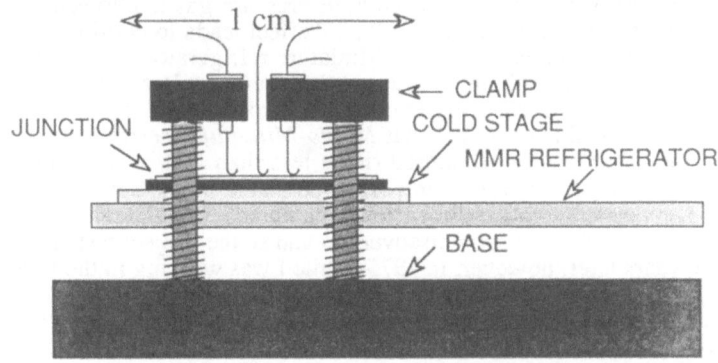

Figure 2. Schematic of microminiature refrigerator, sample, and tunneling spectrometer.

accurately set in temperature and rapidly shifted from one temperature to another. I will illustrate these characteristics as they applied to two experiments done in our laboratories in the past five years by Matthew Holcomb, originally a student with Collman and myself, then a postdoc, and now a colleague.

The first was an experiment[10,11] on the high T_c cuprate superconductor Tl-2223, which has a superconducting transition at about 120 K. Figure 2 shows a schematic of the refrigerator and of the sample configuration. Current is passed from one tinned copper wire to a layer of Ag deposited on top of the superconductor and out via a second tinned copper wire coated with Bi also resting on the Ag layer. The differential resistance of each junction is then measured as a function of the voltage across each junction. A simplified view of the junction is shown in Fig. 3.

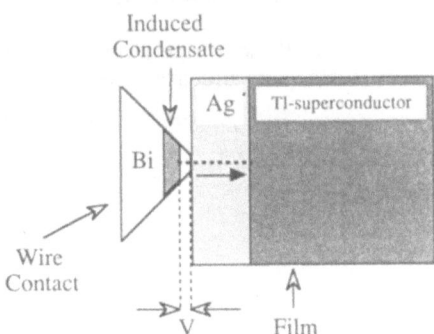

Figure 3. Simplified view of junction and location of region in the Bi-film believed to be driven into the superconducting state by the presence of the Tl-2223 superconductor across the 2,000Å thick Ag-interlayer.

The differential resistance across the junctions, Fig. 4, exhibits extraordinary beautiful interference fringes which show no thermal broadening even at 78 K. We found they could be explained if, and only if, the Bi became superconducting at these enormously high temperatures - induced into this state by the superconductivity of the Tl-2223, which was as much as 2000 Å away, on the other side of the Ag film. The discovery of this remarkable Long Range Proximity Effect, which we now recognize and understand, was made possible by the availability of the little refrigerator. Its temperature stability, absence of vibration, and rapid response allowed us to try many, many junctions in a reasonable time period and finally to determine exactly what was going on.

The second experiment[12] was also master-minded by Matt although we claim some contributions to its success. This involved again the study of the properties of the high temperature superconductors but using what we call a thermal difference spectrometer[13]. In this, the intensity of light reflected from the surface of the superconductor is measured at one temperature $T + \Delta T$ and then compared with its reflectance at $T - \Delta T$. By repeating this measurement many, many times as one scans the photon energy, small changes in the

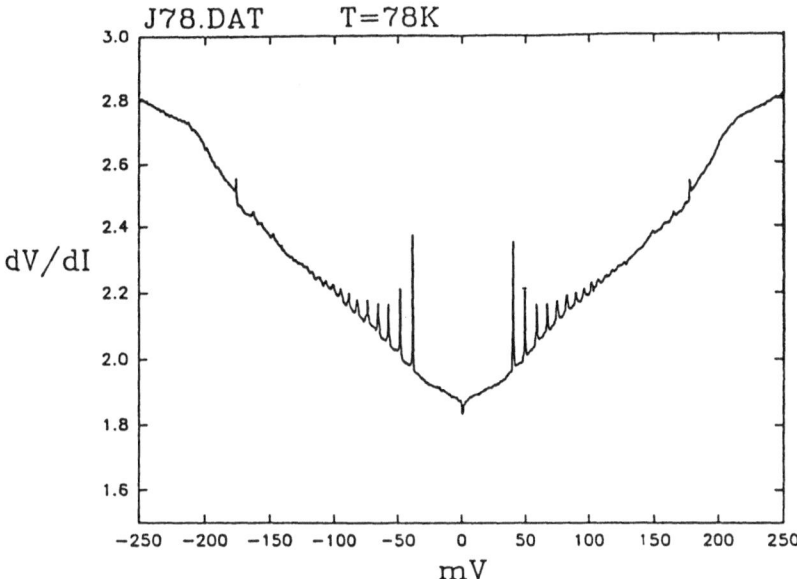

Figure 4. Differential resistance of Bi/Ag/Tl-2223 junction at 78 K as a function of applied voltage showing interference fringes between the Tl-2223 surface and the induced superconducting region in the Bi.

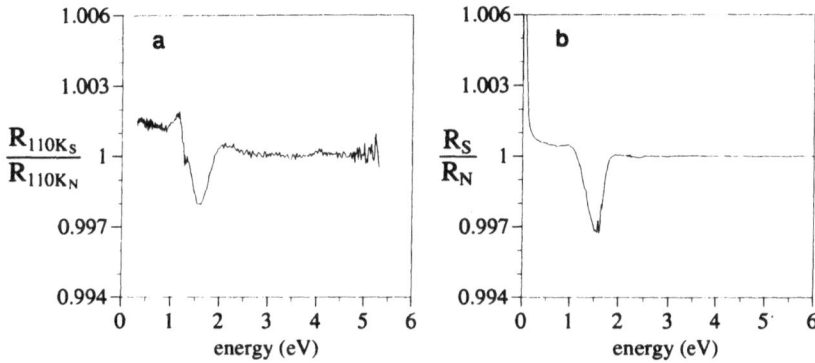

Figure 5. Reflectivity ratio vs photon energy for Tl-2223 in the superconducting state at 110 K to what it would be in the normal state at 110 K. (a) Observed. (b) Calculated assuming, in addition to the phonon contribution to the pairing, another contribution from an electronic excitation at 1.6 eV.

reflectivity can be measured with great accuracy - a few parts in a million! What the experiment showed was that, in addition to a contribution of the phonons to the pairing interaction in the superconductor, there was a contribution from another excitation, one at about 1.6 eV - which we identify as the oxygen-to-copper charge transfer excitation[14]. See Fig. 5. This model has succeeded in explaining both the wavelength dependence of the change of reflectance upon entering the superconducting state and its temperature dependence. Again what made this experiment possible was the absence of vibration of the sample on the refrigerator, the precise temperature control, and the rapid response of the refrigerator to the required changes in temperatures.

Technology Transfer

A great deal has been spoken recently on the subject of the transfer of technology from the Universities to Industry. Fig. 6 serves, perhaps, to illustrate some part of the problem for such transfers, in the particular case of our development of the microminiature refrigerators.

The MMR refrigerator was a cute idea, but could it be a business? As an illustration, there simply isn't a market for a hundred thousand microminiature refrigerators a month, no matter how cute they may be! But, that is almost what you have to sell to run a viable business, if that is *all* you have to sell. So developing a viable business is a more complex issue. One idea is not enough, much, much more is needed. A typical system that we now sell is shown in Fig. 6. This is a low temperature microprobe with temperature controller, vacuum control circuitry, flow gauges, a computer and software. They all add to the value delivered, and every one of these peripherals has to be designed and must be

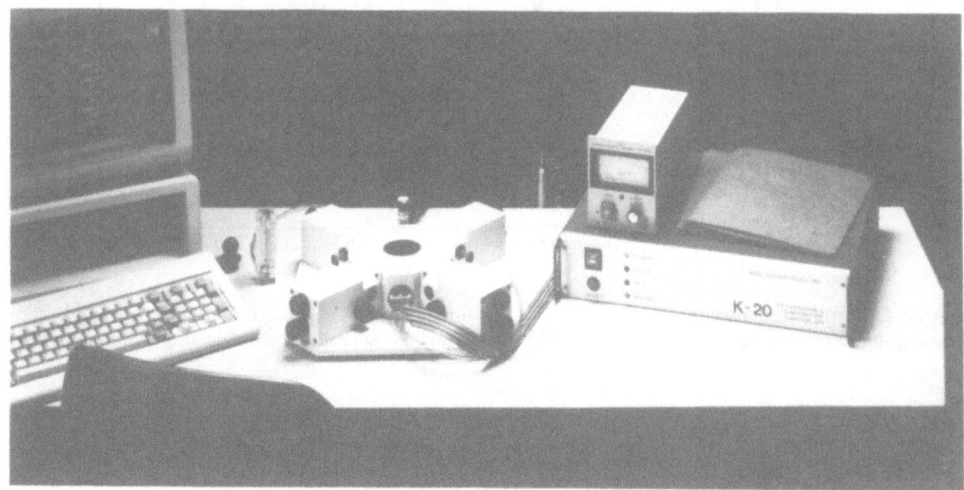

Figure 6. A typical MMR commercial product: Low temperature microprobe, vacuum chamber, micromanipulators, temperature controller, vacuum control circuitry, computer and software. The refrigerator is hidden from view, attached to the ribbon cable at the center of the photograph.

manufactured if the company is to remain viable - the cute little refrigerator is almost lost in the heap! My point is that it takes many others in engineering, documentation, software, marketing and sales to get the product to market. Likewise in most of our other products, the refrigerator plays a relatively minor role, and typically constitutes perhaps 10 % of the cost of the system. On the otherhand, in some cases, one unique component such as the refrigerator, can make it possible to solve a particular problem which might be virtually insoluble otherwise, and this then gives one an edge in the market.

COOLERS FOR CRYOELECTRONICS

Let me now turn to some new developments which can be expected to have a much wider impact on the way the world works! In 1982 I came across a remarkable piece of work by a group of physicists in Kiev in the former Soviet Union. This was a patent on the use of mixtures of nitrogen and low boiling point hydrocarbon gases in JT refrigerators[15]. They showed that some of these gas mixtures had a cooling effect that was 10 to 12 times greater than that of pure nitrogen and, in fact, gave a cooling effect in a JT cooler that appeared to be 30% to 50% of Carnot efficiency. This struck me as extraordinary as the JT expansion process is essentially irreversible. However, we found after testing these mixtures with our MMR refrigerators, that they did, indeed, have this remarkable cooling effect. It intrigued me enough that I began a study shortly thereafter of the phase diagrams of such mixtures in an attempt to understand how this could come about[16]. This work has continued to this day. We now are able to calculate the thermodynamic and transport properties of gas mixtures containing eight or nine components in a few seconds with acceptable accuracy. The understanding of this process has contributed to the development of a new class of cryogenic refrigerators of very low

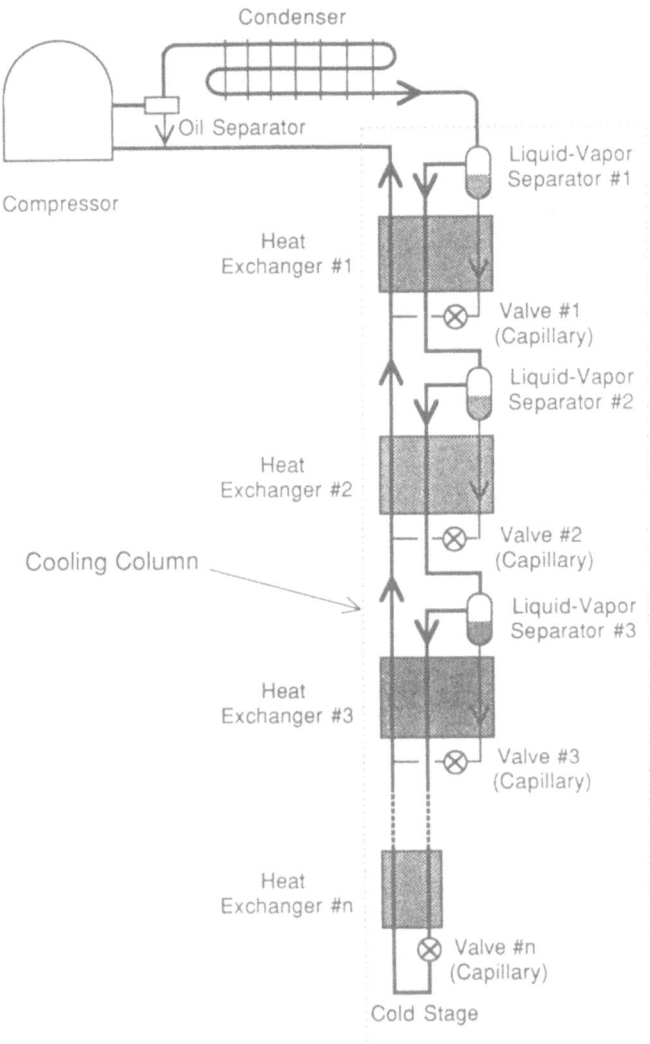

Figure 7. Schematic Diagram of refrigeration cycle proposed by Kleemenko. Compressor can be an oil lubricated compressor and one or more heat exchangers may be contained in the Cooling column.

cost, great durability and great simplicity that we believe will finally herald in the age of superconducting cryoelectronics.

Kleemenko Cycle Coolers

The essence of the idea can be understood from a brilliant paper by Kleemenko[17], also from Kiev, but published over 35 years ago. His concept was of a refrigerator illustrated in the figure. This uses an oil lubricated compressor to compress a gas mixture. The gas is then cooled and some of the high boiling point components condense. These are separated out in liquid-vapor separator and the liquid component is allowed to expand through a valve. In doing so it cools, and the cool returning gas precools the remaining high pressure gas, causing futher condensation. The process continues as shown and produces refrigeration at cryogenic temperatures with remarkable efficiency. Why is this so efficient and why does it works so well?

First, as pointed out by Kleemenko, whenever one expands a liquid, rather than a gas from high pressure, one does so with almost the same efficiency as for a reversible

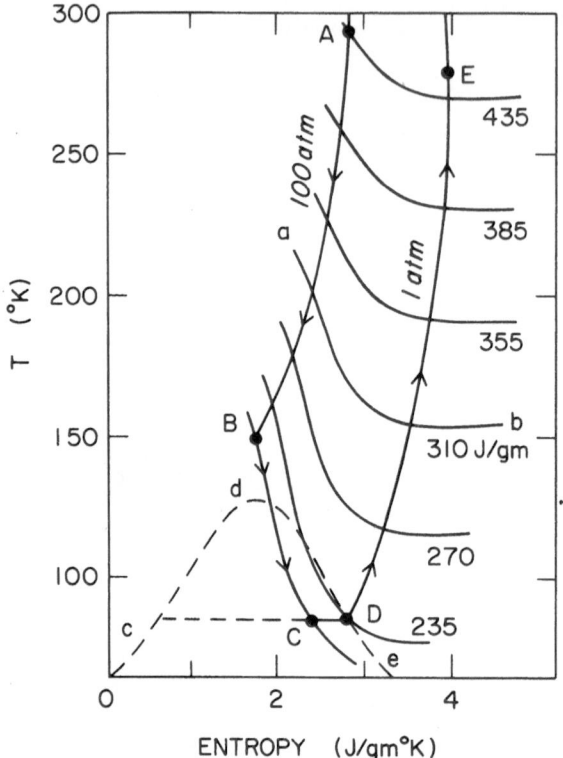

Figure 8. Plot of Temperature vs Entropy for nitrogen. Note that the lines of constant Enthalpy BC near the coexistence dome are almost vertical and hence almost parallel to the lines of constant Entropy. Therefor isenthalpic expansion in this region is almost reversible.

expansion. This can be understood from Fig. 8, where we show the phase diagram of nitrogen, and note that the isenthalps BC near the critical point d, *under* the co-existence dome are almost vertical in the T-S diagram, and thus are almost parallel to the lines of constant entropy. So isenthalpic expansion will occur here with high efficiency. The phase diagrams for mixtures, on the other hand, have distorted co-existence domes, but the above feature of the isenthalps is similar.

The second point is that in a counter-flow heat exchanger, in order for heat to flow from the warm stream to the cold, a temperature difference must, of course, exist between the two. This temperature difference leads to an increase in entropy and thus contributes to the inefficiency of the refrigerator. Unless the heat capacities of the high and low pressure fluids are the same, this temperature difference will diverge as one moves from one end of the exchanger to the other resulting in a large, nett increase in entropy in operation. To obtain the minimum increase of entropy the heat capacities of the two streams should be the same. This is never the case for a single component gas at these cryogenic temperatures, but can be so for mixtures. Thus by tailoring the composition of the mixture one can make the heat capacities of the two streams the same. This is illustrated in Fig. 9.

Here we plot the Enthalpy/unit mass, H for two mixtures, both at high, and at low pressures. Note that the slope of these curves reflects the effective heat capacity of the mixtures. In general, they are different. If the gas is compressed at 300 K and then allowed to expand at low temperatures then the difference in H at each temperature is the maximum amount of heat one can absorb at that temperature. In Fig. 9(a) H vs T is plotted for two pressures, 300 psia and 30 psia for a mixture used in a commercial refrigerator that operates at 150 K. Here the two isochores have slopes that are not the same and this results in a loss of efficiency.

Fig. 9(b) shows the corresponding plot for a mixture that we have developed. This gives essentially parallel isochores from ambient to about 100 K. Analysis of a refrigeration cycle with such a mixture shows that with a good heat exchanger and an off-the-shelf compressor, an efficiency of about 33% of Carnot should be possible. The first few coolers using this cycle have been built. Thus far the ability to refrigerate to 100 K and operate for thousands of hours without interuption has been demonstrated. The efficiency that has been observed is not as high as one would expect from the above analysis, however, special attention has to be paid to the design of the heat exchangers for the peculiar regime under which they must work at the lowest temperatures, and with improvements in these, higher efficiency can be obtained.

An example of a recent prototype is shown in Fig. 10. With this type of cooler, temperatures down to 76 K have recently been achieved and a capacity of 5 Watts at 90K

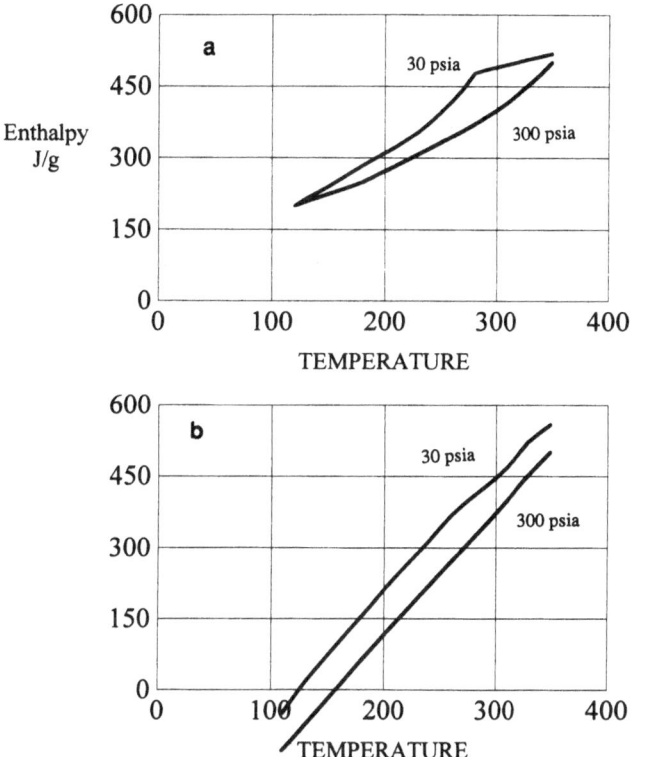

Figure 9. Plot of Enthalpy vs Temperature at two pressures for (a) Refrigerant mixture used in a commercial refrigerator operating at 150 K, and for (b) Refrigerant developed at MMR for operation at 100 K. The parallel isochores of 9(b) imply a high thermodynamic efficiency for the refrigerator.

have been realized with an input power of 180 Watts. The unique feature of these coolers is that the cost of the refrigerator, even in small quantities, is less than a fifth of any comparable cooler on the market today, and it is for this reason we feel that at last we will have a low-cost cooler that will make possible the Age of Cryoelectronics.

ACKNOWLEDGEMENTS

I am indebted to my colleagues at MMR and my former students for their many contributions toward the development of the microminiature refrigerators, in particular, Richard Hollman, whose thesis was devoted to the development of the refrigerator; Al Nash and Gideon Friedmann for work on the phase diagrams of mixtures; Steve Garvey, Mark DuBois, Carlos Fuentes, Herb Edman, Boris Reynov for the development of instrumentation and peripheral equipment; and Igor Sapozhnikov, Leonid Rappoport and

Figure 10. MMR's low-cost, cryocooler using the Kleemenko refrigeration cycle. The unit is designed to cool two Low Noise Amplifiers at 130 K or be used for biostorage at 115 K. The rule in the foreground is six inches in length.

Dimitriy Yam for contributions to the development of the Kleemenko coolers; and Robert Paugh for helping to put it all together as a business. We are indebted to the Office of Naval Research and the National Science Foundation for initial funding of the refrigerator development.

REFERENCES

1. M. A. R. LeBlanc and W. A. Little, Critical Current for superconducting niobium wire and ribbons in external magnetic fields, Proc. VII Int. Conf. Low Temp. Physics, Toronto, 362, (1960)

2. W. A. Little and R. D. Parks, Observation of quantum periodicity in the transition temperature of a superconducting cylinder, Phys. Rev. Lett, 9, 9 (1962)

3. W. A. Little, Possibility of synthesizing an organic superconductor, Phys. Rev. 134, A1416 (1964).

4. B. T. Matthias, High temperature superconductivity ??, Comments on Solid State Physics, 93 (1970); See also B. T. Matthias, The search for high temperature superconductors, Physics Today, August 1971, p.23.

5. B. T. Matthias, quoted in Physics Today, April 1972, p.18.

6. W. A. Little, Physics Today, Letters, April 1972, p.11.

7. W. A. Little, Scaling of miniature cryocoolers to microminiature size, Proc. NBS Cryocooler Conf., Ed. J. Zimmerman and T. M. Flynn; NBS Spec. Pub. No. 508, p.75 (1978).

8. R. Hollman and W. A. Little, Progress in the development of microminiature refrigerators using photolithographic fabrication techniques, Proc. NBS Conf. Refrigeration for Cryogenic Sensors and Electronic Systems, Ed. J. E. Zimmerman, D. B. Sullivan and S. E. McCarthy, Spec. Pub. No. 607,160 (1981).

9. W. A. Little, Microminiature refrigeration, Rev. Sci. Instrum. 55, 661 (1984).

10. M. J. Holcomb, J. P. Collman, and W. A. Little, New phenomena in proximity effect tunneling of high T_c Superconducting Cuprates, Physica C 185 - 189, 1747 (1991)

11. W. A. Little and M. J. Holcomb, The long range proximity effect: pair echoes, Journal of Superconductivity 7, 175 (1994)

12. M. J. Holcomb, J. P. Collman, and W. A. Little, Optical evidence of an electronic contribution to the pairing interaction in superconducting $Tl_2Ba_2Ca_2Cu_3O_{10}$, Phys. Rev. Lett. 73, 2360 (1994).

13. M. J. Holcomb, J. P. Collman, and W. A. Little, Thermal difference spectroscopy, Rev. Sci. Instrum. 64, 1867 (1993).

14. M. J. Holcomb, C. L. Perry, J. P. Collman, and W. A. Little, Thermal difference reflectance spectroscopy of the high temperature cuprate superconductors, Phys. Rev. B (in press 1996).

15. V. N. Alfeev et al., Great Britain patent No. 1,336,892 (1973)

16. W. A. Little, Advances in Joule-Thomson cooling, in Advances in Cryogenic Engineering 35, 1305 (1990).

17. A. P. Kleemenko, One flow cascade cycle, Proc. Xth Int. Congress of Refrigeration, Copenhagen, 1, 34 (1959) Pergamon Press, London.

RECOLLECTIONS OF THE LOW TEMPERATURE LAB

IN PHYSICS CORNER: 1958-1961

Bascom S. Deaver, Jr.

Physics Department
University of Virginia
Charlottesville, VA 22901

In fall 1958 Bill Little joined the Physics Department as an assistant professor and initiated research in low temperature physics at Stanford. He installed a helium liquifier in the basement of the High Energy Physics Laboratory, where the Mark III electron linear accelerator was located. At that time the Physics Department was located in Physics Corner, the northwest front corner of the main quadrangle. There Bill created the first lab for low temperature research in the basement in a large, mostly underground room. It had small windows up near the ceiling that looked out just above ground level on the two sides of the corner of the quad. There was a small enclosed space in the corner of the room in which Bill installed a large, high capacity vacuum pump used to lower the temperature in the liquid helium cryostats. Along the north side of the large open room Bill situated benches and equipment that he and his graduate students used for their initial experiments. Some space along the west side and three small partitioned rooms along the south wall he reserved to be used by William Fairbank, who would be joining the faculty in fall 1959. The people and their activities in that lab during the following few years constitute some of my fondest memories. My recollections of some of them are offered here as a tribute to Bill Little, who contributed so much to the excitement we all experienced there.

I first met Bill in the spring of 1959. I was just completing my second year as a part-time graduate student while working full time at Stanford Research Institute and would become a full-time student in fall 1959. It had already been arranged that I would be working with Fairbank when he arrived in the fall, however, I wanted to learn my way around in the Department - the shop, stockroom, the labs - and Bill Little kindly agreed to let me work part-time with his group during the summer of 1959.

When Fairbank arrived in the fall, he described to me a series of experiments he proposed to undertake. One that I found extremely appealing was to measure the flux trapped in a superconducting ring to test Fritz London's prediction of fluxoid quantization. He agreed that I could work on it and gave me one of the small partitioned rooms in the low temperature lab in which to begin work.

Little and his students already had several projects underway. I don't remember precisely the order of who started what and when, but during the next year or so Ronald Parks,

From High Temperature Superconductivity to Microminiature Refrigeration
Edited by Cabrera *et al.*, Plenum Press, New York, 1996

243

Robert C. Johnson, Nareshchandra Shah, Warren Sommer and Leonard Meyers were all working there with him.

Bob Johnson was carrying out a critical experimental study of the Kapitza effect, the abrupt jump in temperature at the interface between a solid and superfluid liquid helium. This was a topic on which Bill had done some theoretical work while he was at the University of British Columbia, where he was before coming to Stanford. It is interesting that Bob had originally started to work with Felix Bloch; however, with Bloch's gracious good wishes he went to work on these experiments with Bill. Bob relates that subsequently over a weekend Bloch did a beautiful calculation for his thesis of heat transfer from the conduction electrons in a metal to liquid helium phonons - a calculation that Bill characterized as one that would have taken him three months to do, if he could have managed to do it at all.

The major project in Bill's group was the development of a cyclic magnetic refrigerator for the polarization of nuclei. Later this led to an electron scattering experiment on the Mark III accelerator in collaboration with Robert Hofstadter. This was before the general availability of He^3 for refrigeration, so Bill had designed a cyclic adiabatic demagnetization refrigerator for cooling the target. Ron Parks and Naresh Shah were working on phases of this project as was John Gerig, an Electrical Engineering graduate student whose advisor was away for the year. John was working in Bill's lab designing and building electronics, including a new, highly sensitive, marginal oscillator for detecting NMR signals and some of the control circuits for cycling the refrigerator. He also designed a lock-in amplifier that several of us subsequently built for our own experiments. This was well before there were any commercial ones. It was also fun to talk to John about problems in our E & M and quantum mechanics courses because he had such good physical insight and often some helpful electrical analogies.

One critical requirement for the refrigerator was a high magnetic field. Bill had noticed reports in the literature that although Nb had a critical field of ~ 1800 gauss, superconductivity had been observed to much higher fields, so there was the potential for making a high field solenoid of Nb wire. In January 1959 Marcel LeBlanc came from the University of British Columbia to work with Bill as a post doctoral research associate, and they began constructing and testing Nb magnets. They achieved a 5 kG field at 4.2 K and 7 kG at 1.8 K in one of the first successful high field superconducting magnets for an actual application.

Marcel then set out to understand why Nb could carry such high currents in high fields. He discovered training effects and what is now known at the "knee effect" in superconducting magnets. He became one of the leaders in research on what subsequently would be known as Type II superconductivity, an area in which he continued to work as professor of physics at the University of Ontario.

In working on the polarized target experiment, Ron Parks had the idea of making highly paramagnetic metals that could be demagnetized for cooling. He then pursued a study of the interaction of rare earth atoms in dilute alloys.

Concurrently that year, members of the Fairbank group were assembling in the Physics Corner lab. John Goodkind and Dwight Adams, both of whom completed their doctoral research with Fairbank at Duke University, came to work as research associates and initiated NMR experiments to study liquid He^3. Their cryostats, magnets and gas handling equipment were situated in the bay along the west wall of the large room. Two more graduate students, George Hess and Pierce Webb, began work in the other two small partitioned spaces along side the one where I was working. George started experiments to measure the quantized rotational states of a small container of liquid helium, and Pierce initiated light scattering experiments to search for quantized vortices in rapidly rotating liquid helium.

Working in that lab in close proximity with so many interesting people engaged in research on so many fascinating topics was one of the most stimulating experiences I have ever had. Both Little and Fairbank were continually in the lab helping with experiments and talking about new ideas. There always seemed to be something going on in the lab at all hours of the day and night. Also the flow of famous visitors through the lab contributed to the excitement.

There were some minor frustrations in working together so closely too. For example, finding that just when you need it, your favorite screwdriver has been borrowed. I can remember deciding that rather than trying to retrieve screwdrivers I would just replace them until the lab was saturated. However after replacing a substantial number I realized that this was not a successful tactic.

On another occasion, I remember having Marcel come to my room to borrow my scissors, which were nearly always hanging in place on my compulsively organized tool board. I took him back to his work bench and showed him that there were three pairs of scissors within easy reach from his chair, although all were covered with some bit of clutter on the bench. Mine were easier to find.

There were small tragedies along the way that now make funny reminiscences. There was the time the fork of a lift being used to hoist a liquid helium container got caught under the edge of a lab bench and toppled someone's apparatus. When someone failed to vent his dewar after completing an experiment, there was a loud bang several hours later and his cryostat was launched out of the glass dewar upward toward the ceiling. There was the even louder expletive from the person who, while measuring the depth of a glass dewar, dropped the meter stick which then crashed through the bottom of the dewar.

During the years from 1959 through 1961 there were many new results emerging from the research in the basement lab. However, the time I remember most vividly began on May 3, 1961 when my experiments yielded the first definitive evidence for fluxoid quantization in superconductors through measurements of the flux trapped in 13 μm-diameter Sn cylinders. Fairbank and I then began six weeks of frantic activity and sleepless nights. By the middle of June we had convincing proof that the flux was quantized in units of h/2e, and we submitted a paper to Physical Review Letters.

Also in June there was a conference on superconductivity as part of the dedication ceremonies for the new IBM Research Center at Yorktown Heights which Bill Little was attending. The night before he left for this meeting I took copies of our data to Bill for him to discuss with people there. It was very fortunate that he presented our results there, because at that meeting Doll and Näbauer (about whose experiments we had not known) announced that they had similar results.

C. N. Yang was visiting Stanford during the course of these experiments. When the results were sufficiently convincing, together with Nina Byers, he sought a theoretical explanation. They showed that quite generally quantum mechanics requires that the free energy of a collection of electrons, confined to an annular region within which there is a magnetic flux, be a periodic function of the enclosed flux. For electrons constrained by the BCS pairing condition, there are very deep minima in the free energy for values of flux equal to integral multiples of h/2e.

After talking to Yang about this theory, Little realized that the periodic variation of the free energy of the superconducting cylinder with enclosed flux implied that the transition temperature is also a periodic function of the flux, and he used the BCS theory to calculate the variation. He and Ron Parks, who had by then completed his dissertation research, demonstrated this periodic dependence in a very direct and elegant way. They measured the variation of the resistance of Sn cylinders approximately 1 μm in diameter as a function of applied magnetic field along the axis while the temperature was held fixed at a value in the superconducting transition. The variation of the transition temperature produced corresponding periodic variations of the resistance with period h/2e.

It is amazing that these measurements from concept to submission of the paper to Physical Review Letters took only 30 days. There was an interesting bit of serendipity too. While Ron was seeking small fibers on which to deposit Sn for these experiments, he noticed that when mounting samples with dabs of GE 7031 varnish there was often a spider-web-like filament of varnish drawn out as the applicator was pulled away. He experimented with these filaments and found that they made excellent substrates onto which he could evaporate his Sn cylinders.

Subsequently Leonard Meyers, who was also a graduate student working with Bill, refined and extended the Little-Parks experiment with other materials including vanadium, niobium and indium.

Sometime during these early years Bill taught a course in what we would now call condensed matter theory in which most of the students in the two low temperature groups enrolled. I believe it was while he was teaching this course that he began thinking specifically about excitations in fermi liquids and decided to undertake with Naresh Shah experiments using soft x-ray scattering to observe zero sound in liquid He^3. Bill has mentioned discussions he had with Felix Bloch about the theory and some calculations Bloch made in the early 1930's concerned with oscillations in the shape of the fermi-surface.

Warren Sommer, whom I mentioned earlier as being one of Bill's graduate students, became interested in the interaction of free electrons with liquid helium. In some important early experiments in what was to become a very active research area, Warren measured the energy required to inject a free electron into the liquid and was the first to realize that the surface of liquid helium could be used as a barrier to support electrons.

Throughout those early years at Stanford, Fairbank talked extensively about Fritz London's concept of "long range order" and "macroscopic quantum effects" in superfluids, ideas that we have come to understand much more fully now, but that were largely new and unexplored then. London's prediction of fluxoid quantization was based on these ideas rather than any rigorous mathematical proof. He also investigated the unusually large diamagnetism of large aromatic molecules, and he speculated about the possible significance of long range order in biological molecules. Fairbank and Little talked a lot about these concepts.

Bill has commented on the significance of these discussions, together with his subsequent studies with Tad Day of some problems in biology, in shaping his interest in the problem of superconductivity in organic molecules. We all know now that his theoretical study of a model molecular system led to the idea of a new mechanism for superconductivity, and ultimately he was the leader in opening an entirely new field of research on organic conductors and superconductors.

In 1962 the Physics Department moved from the old Physics Corner into the new Varian Physics Building. At about this same time Bill embarked on research in the new areas in which he has worked so creatively since then. In essence these recollections are about activities that are a prelude to those encompassed by the title of this volume and that began after the move to the new building. They are about a time and place in which those of us who worked so closely together in the two low temperature research groups enjoyed a remarkably warm supportive environment and some incredibly diverse and exciting research.

I talked recently to several of the people who worked with Bill during that time: Marcel LeBlanc in the Physics Department at the University of Ontario, where he has been since leaving California; Bob Johnson, who has recently retired from a highly successful career in research and management at Dupont Central Research in Wilmington and is now involved in science education; Naresh Shah, who is a professor in the Physics Department at Boston University; Warren Sommer, who after some years at Wayne State University, has

246

been an independent inventor living in Milwaukee; John Gerig who continues a long career as an independent consultant.

In these conversations there were many recurring themes, because we all share some common perceptions of Bill Little: an incredibly creative and broadly knowledgeable physicist, an ingenious experimenter who also has a dogged determination to gain a deep theoretical understanding of experimental effects, a man of great energy with a charming and engaging personal style that generates inspiration and excitement, a person who is fun to work with and whom we all admire.

Ron Parks' early death deprived us of one of the most outstanding physicists and remarkable people to emerge from Bill's group of students. However among his legacies are two of his students who have particular pertinence to this discussion: Ron Groff, Ron's first Ph.D. student at Rochester University, was a member of Bob Johnson's exciton research group at Dupont. Another of Ron's students is Bill's colleague at Stanford, Steve Chu.

I am happy to have this opportunity to present these brief reminiscences as a tribute to Bill and to wish him all the best in his newest ventures.

ROUTES TO HIGH T_c MOLECULAR SUPERCONDUCTORS

T. Ishiguro and H. Ito

Department of Physics, Kyoto University
Kyoto 606-01, Japan

PROPOSAL OF HIGH Tc SUPERCONDUCTORS

The superconductivity discovered by Onnes in 1911 is a fascinating phenomenon appearing in metals. The observed diamagnetism and anomalous specific heat indicated that this behavior arises from the thermodynamic nature of the electron system. The list of known materials exhibiting superconductivity had expanded from elemental metals to alloys and intermetallic compounds. Concurrently the maximum observed values of superconductivity transition temperature (T_c) and the critical magnetic field strength had steadily risen. The mechanism of superconductivity was described by Bardeen, Cooper and Schrieffer in 1957. In their theory this mechanism is ascribed to a new electronic state caused by paired electrons through the mediation of lattice vibrations.

In 1964, by taking into account the possible mediating role of electronic polarization or excitons instead of phonons, an idea of the effective interaction between mobile electrons leading to the existence of high T_c material was introduced by Little [1]. This model describes what is known as an excitonic mechanism. Little's model stimulated not only debate on the validity of the theoretical concepts concerning the superconductivity but also the material developments required to realize the idea. In order to observe the excitonic mechanism, Little proposed studying a composite structure of conducting polymers like polyacetylene with polarizable side chains such as dye molecules, in which quasi one-dimensional superconductivity is sought. The one-dimensionality was criticized by physicists[2], because it contradicts the thermodynamical concept that here the transition to a long range ordered state is not stablized. Further, the strength of the coupling giving possiblity to realize the high T_c was criticized with respect to the spacing between the interacting molecules and the screening effect. The contradiction concerning one-dimensionality has been reconciled by taking into account weak but non-zero side interactions existing in a realistic system. These arguments simultaneously lead interests to one-dimensional electron systems, such as linear metals and magnets, not only from a theoritical

From High Temperature Superconductivity to Microminiature Refrigeration
Edited by Cabrera *et al.*, Plenum Press, New York, 1996

249

point of view but also from an experimental view. The efforts in the field of material development had begun to produce quasi one-dimensional conductors such as tetracyanoplatinates such as $K_2Pt(CN)_4Br_{0.3} \cdot 3H_2O$ (KCP) [3]. It may be asserted that the developed studies on low-dimensional systems, quasi two-dimensional as well as one-dimensional, have provided solid bases for investigations on novel functional materials, including the CuO_2 high T_c superconductors possessing pronounced two-dimensionality.

On the material side, the proposal has stimulated the interests not only of physicists but also chemists, particularly synthetic chemists, because probable molecular models were proposed. This is striking even as an example of pioneering work in the field of synthetic metals and superconductors. The first model based on a combination of metallic spine and polarizable side-chain with conducting polymers and dye molecules was revised in 1976, so as to enhance the interaction strength with a shorter distance between the conducting path and with polarization segments [4]. Further, an interaction via a back-bonding between transition metal d_{xz}-orbital and the π-orbitals of ligand which may work in KCP-like molecules was proposed to be useful for pairing in 1982 [5]. Experimentally superconductivity in polymeric material was first found in $(SN)_x$ [6], but this was not ascribed to the proposed excitonic mechanism and T_c remained below 0.6 K. The KCP salts with structure similar to that described by the Little's second model were studied intensively. Although the superconductivity has not been found in this kind of materials, strong interest has been focused on the phenomena related to significant one-dimensionality. As a result, the predicted Peierls transition characteristic to the one-dimensional metals has been observed in not a few materials including KCP [3].

SUPERCONDUCTIVITY IN CHARGE TRANSFER SALTS

Investigations aimed at producing electronically conductive materials in organic materials started in the 1940's. The first charge transfer salt exhibiting high electrical conductance was realized as perylene bromine complex in 1954 [7]. Following this discovery, a system of degenerate electrons in molecular materials, like in traditional metals, was reported to exist in TCNQ-salt [8]. (See Fig. 1 concerning the chemical structure of molecules). In particular, (TTF)(TCNQ) was found to exhibit genuine metallic characteristics [9]. The discovery of the giant metallic conductance in (TTF)(TCNQ), yielding a pronounced conductivity peak near 60 K [10], strongly stimulated physicists as well as chemists. In this case the high electrical conductance was interpreted as a precursor to high T_c superconductivity, although it is ascribed to conductance due to charge density waves (CDW), through later investigations [11].

The first superconductivity in molecular salts was found in $(TMTSF)_2PF_6$ under a pressure of 1.2 GPa with a T_c of 0.9 K [12]. The TMTSF salts are represented by $(TMTSF)_2X$, where X is a counter anion such as AsF_6^-, SbF_6^-, ClO_4^-, and FSO_3^-. Their crystal structures are isomorphous for various anions, although the patterns of the superstructures appearing at low temperature differ with respect to the anion shape. The similarity in the structure of these salts, however, enabled the elucidation of the roles of intermolecular spacings and constituent molecules. Among the TMTSF salts, $(TMTSF)_2ClO_4$ possessing the shortest intermolecular distance at ambient pressure, was the first ambient pressure organic superconductor to be found [13]. The TMTSF salts have attracted the interest of physicists not only due

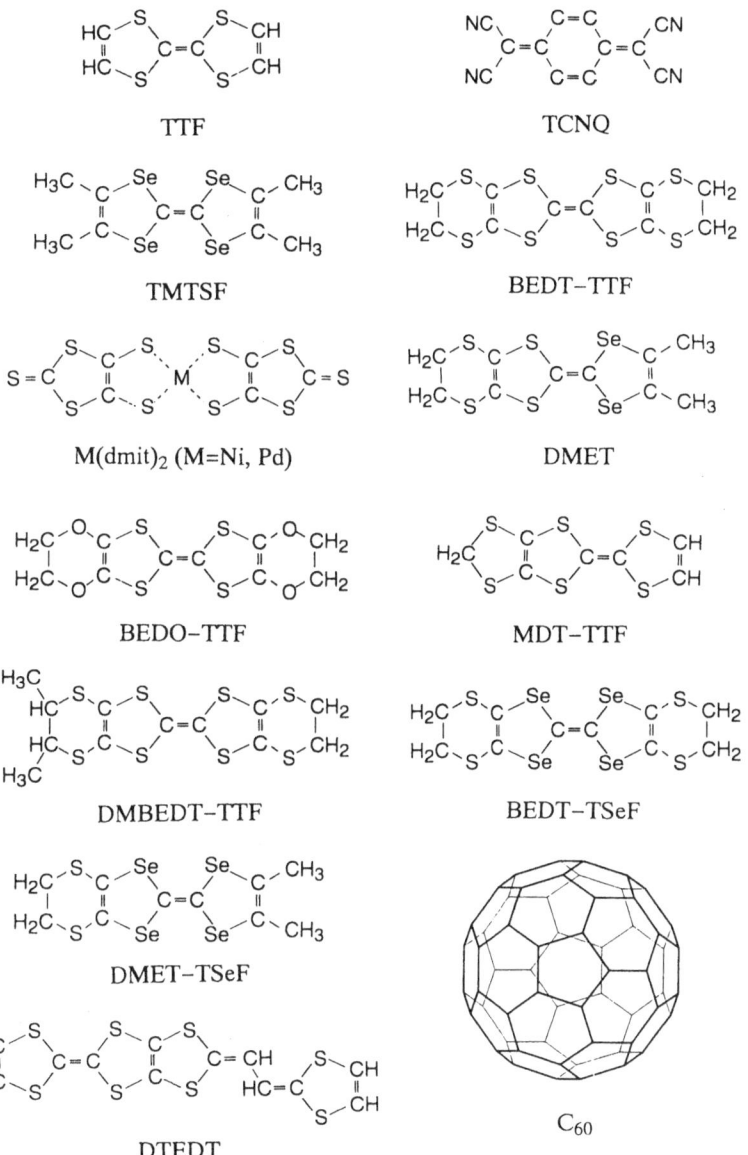

Figure 1. Chemical structures of TTF, TCNQ and principal molecules constructing charge transfer salts exhibiting superconductivity.

Table 1 Development of molecular superconductors.

Complex	P_c(kbar)	T_c(K)		Year of report
(SN)x	0	0.28		1975
$(TMTSF)_2PF_6$	12	0.9		1980
$(TMTSF)_2SbF_6$	10.5	0.38		1981
$(TMTSF)_2TaF_6$	11	1.35		1981
$(TMTSF)_2ClO_4$	0	1.4		1981
$(TMTSF)_2AsF_6$	9.5	1.4		1982
$(TMTSF)_2ReO_4$	9.5	1.3		1982
$(TMTSF)_2FSO_3$	5	~ 3		1983
$(BEDT\text{-}TTF)_2ReO_4$	4.0	2.0		1983
$\beta_L\text{-}(BEDT\text{-}TTF)_2I_3$	0	1.5		1984
$\beta\text{-}(BEDT\text{-}TTF)_2IBr_2$	0	2.7		1984
$\beta_H\text{-}(BEDT\text{-}TTF)_2I_3$	0	8.1		1985
$\beta\text{-}(BEDT\text{-}TTF)_2AuI_2$	0	4.9		1985
$\gamma\text{-}(BEDT\text{-}TTF)_3I_{2.5}$	0	2.5		1985
$\kappa\text{-}(BEDT\text{-}TTF)_4Hg_{2.89}Cl_8$	12	1.8		1985
$\theta\text{-}(BEDT\text{-}TTF)_2I_3$	0	3.6		1986
$\kappa\text{-}(BEDT\text{-}TTF)_2I_3$	0	3.6		1987
$\kappa\text{-}(BEDT\text{-}TTF)_4Hg_{2.89}Br_8$	0	4.3		1987
$(BEDT\text{-}TTF)_3Cl_2 \cdot (H_2O)_2$	16	2		1987
$\kappa\text{-}(BEDT\text{-}TTF)_2Cu(NCS)_2$	0	10.4	(8.7)*	1988
$\kappa\text{-}(BEDT\text{-}TTF)_2Cu(NCS)_2$ deuterated	0	11.2	(9.0)*	1988
$\alpha\text{-}(BEDT\text{-}TTF)_2NH_4Hg(SCN)_4$	0	0.8 - 1.7		1990
$\kappa\text{-}(BEDT\text{-}TTF)_2Cu[N(CN)_2]Br$	0	11.8	(10.9)*	1990
$\kappa\text{-}(BEDT\text{-}TTF)_2Cu[N(CN)_2]Cl$	0.3	12.8		1990
$\kappa\text{-}(BEDT\text{-}TTF)_2Ag(CN)_2H_2O$	0	5.0		1990
$\kappa\text{-}(BEDT\text{-}TTF)_2Cu[N(CN)_2]Br$ deuterated	0	11.2	(10.6)*	1991
$\kappa\text{-}(BEDT\text{-}TTF)_2Cu_2(CN)_3$	1.5	2.8		1991
$\kappa'\text{-}(BEDT\text{-}TTF)_2Cu_2(CN)_3$	0	4.1		1991
$(BEDT\text{-}TTF)_4Pt(CN)_4H_2O$	6.5	2		1991
$\kappa\text{-}(BEDT\text{-}TTF)_2Cu(CN)[N(CN)_2]$	0	11.2		1991
$\kappa\text{-}(BEDT\text{-}TTF)_2Cu(CN)[N(CN)_2]$ deuterated	0	12.3		1992
$(BEDT\text{-}TTF)_4Pd(CN)_4H_2O$	7	1.2		1992
$\alpha\text{-}(BEDT\text{-}TTF)_2KHg(SCN)_4$	0	0.3		1993
	1.2(uniaxial)	1.2		1994
$\alpha\text{-}(BEDT\text{-}TTF)_2RbHg(SCN)_4$	0	0.5		1994
$\alpha\text{-}(BEDT\text{-}TTF)_2TlHg(SCN)_4$	0	0.1		1994
$\kappa\text{-}(BEDT\text{-}TTF)_2Cu(CF_3)_4(TCE)_x$	0	9.2		1994
$\kappa\text{-}(BEDT\text{-}TTF)_2Ag(CF_3)_4TCE$	0	11.4		1994
$\kappa\text{-}(BEDT\text{-}TTF)_2Cu(CF_3)_4TBE$	0	5.2		1995
$\kappa\text{-}(BEDT\text{-}TTF)_2Ag(CF_3)_4112DCBE$	0	10.2		1995
$\beta''\text{-}(BEDT\text{-}TTF)_4Fe(C_2O_4)_3 \cdot H_2O \cdot PhCN$	0	6 - 7		1995

continued

continued

Complex	P_c(kbar)	T_c(K)	Year of report
TTF[Ni(dmit)$_2$]$_2$	7	1.6	1986
N(Me)$_4$[Ni(dmit)$_2$]$_2$	7	5.0	1987
α-(EDT-TTF)[Ni(dmit)$_2$]	0	1.3	1993
α-TTF[Pd(dmit)$_2$]$_2$	20	6.5	1988
β-N(Me)$_4$[Pd(dmit)$_2$]$_2$	6.5	6.2	1991
α-Me$_2$Et$_2$N[Pd(dmit)$_2$]$_2$	2.4	4	1992
(DMET)$_2$Au(CN)$_2$	3.5	0.8	1987
(DMET)$_2$AuCl$_2$	0	0.83	1987
(DMET)$_2$AuBr$_2$	1.5	1.6	1987
(DMET)$_2$AuI$_2$	5	0.55	1987
(DMET)$_2$I$_3$	0	0.47	1987
(DMET)$_2$IBr$_2$	0	0.58	1987
κ-(DMET)$_2$AuBr$_2$	0	1.9	1988
κ-(MDT-TTF)$_2$AuI$_2$	0	4.0	1988
(BEDO-TTF)$_3$Cu$_2$(NCS)$_3$	0	1.06	1990
(BEDO-TTF)$_2$ReO$_4$·H$_2$O	0	1.5	1991
κ-(DMBEDT-TTF)$_2$ClO$_4$	5.8	2.6	1991
λ-(BEDT-TSeF)$_2$GaCl$_4$	0	8	1993
λ-(BEDT-TSeF)$_2$GaBr$_x$Cl$_y$	0	7 - 8	1995
λ-(BEDT-TSeF)$_2$GaCl$_3$F	0	3.5	1995
(DMET-TSeF)$_2$AuI$_2$	0	0.58	1993
(DMET-TSeF)$_2$I$_3$	0	0.4	1994
(DTEDT)$_3$Au(CN)$_2$	0	4	1995
K$_3$C$_{60}$	0	19.3	1991
Rb$_3$C$_{60}$	0	29.6	1991
Cs$_2$RbC$_{60}$	0	33	1991
Ca$_5$C$_{60}$	0	8.4	1992
Ba$_6$C$_{60}$	0	7	1992
Na$_3$N$_3$C$_{60}$	0	15	1993

* The parethesized values are given through an authodox way of analysis, taking account of dimensionality and thermal fluctuation.

to a highly anisotropic nature but also due to an interplay among electronic phases related to the low-dimensionality, such as spin density waves (SDW) [11, 14]. In fact, the existence of the SDW phase adjacent to the superconductivity stimulated to revisit the relationship of the magnetism to the superconductivity, since they have been thought to dislike with each other in the traditional metals. This may imply appearance of a new mechanism for the superconductivity: involvement of a possible mediating role of the magnetic interaction [11]. The debate around this subject has not been settled to date.

STABILIZATION OF HIGHER T_c IN CHARGE TRANSFER SALTS

From the point of view of developing superconductors stabilized at higher temperature, it is thought necessary to suppress other electronic phases such as SDW and CDW appearing in high temperature region. For this purpose it has been preferred to raise the dimensionality of the material [15]. The BEDT-TTF molecule (Fig. 1), possessing rings at the outer ends of the TTF molecule with planar shape and thereby increases side-by-side interactions among aligned molecules to the same order of strength as face-to-face interactions, has enabled the production of two-dimensional conductors. It is notable, however, that the charge transfer complexes consisting of BEDT-TTF molecules are specific due to its variety in crystal structure: they have freedom with respect to molecular arrangement [14, 16]. For example, the salt (BEDT-TTF)$_2$I$_3$ has at least 5 types of crystal structure distinguished with greek letters (Table 1). Among these, the crystal structures with β-type and κ-type stacking arrangements [17, 18] shown in Fig. 2 possess a superconductivity phase giving

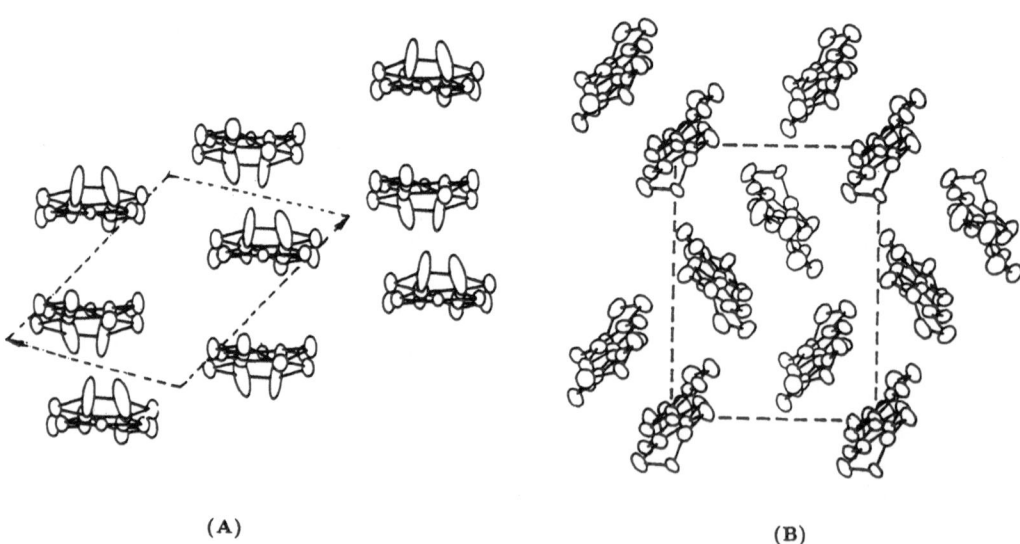

(A)   (B)

Figure 2.   Molecular arrangement patterns of BEDT-TTF in (BEDT-TTF)$_2$X for β-type (a) and κ-type (b). Intra-plane molecular arrangements are seen along the direction of the longer axis of molecule. (Courtesy of T. Mori).

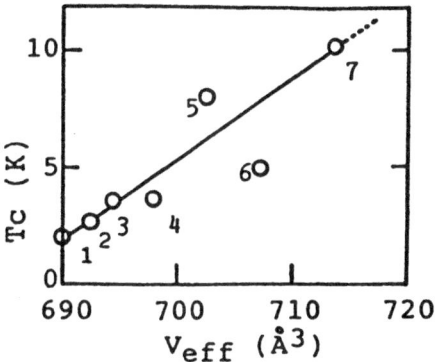

Figure 3. Relation of T_c to effective volume v_{eff} allotted to BEDT-TTF molecule in (BEDT-TTF)$_2$X superconductors.

1 : (X=)ReO$_4$ salt, 2 : β-IBr$_2$ salt, 3 : κ-I$_3$ salt, 4 : θ-I$_3$ salt, 5 : β_H-I$_3$ salt,
6 : β-AuI$_2$ salt, 7 : κ-Cu(NCS)$_2$ salt.

[Reprinted by permission from Ref. 22, copyright Elsevier Science S.A.]

higher T_c. The β-(BEDT-TTF)$_2$I$_3$ crystal was found to exhibit ambient-pressure superconductivity with a T_c of 1.4 K. Following this discovery, it was found that the T_c of this salt is raised discontinuously to $\sim$ 8 K by pressure of $\sim$ 1 kbar [19]. It has been revealed that an incommensurate superstructure appearing below 175 K at ambient pressure [20] and the related disorder in ethylene conformation decreases T_c: the higher T_c is realized when the superstructure is suppressed, although the mechanism of the descent has not yet been fully understood. The salts with κ-type arrangement of BEDT-TTF molecules have provided not a few stable superconductors, among which the highest T_c charge transfer salt, κ-(BEDT-TTF)$_2$Cu[N(CN)$_2$]Cl [21], is included. Table 1 lists the charge transfer salts exhibiting superconductivity classified by dominating molecules, together with (SN)$_x$ and alkali metal fullerides.

It is common to the superconductivity in TMTSF and BEDT-TTF salts that the pressure reduces T_c, except for the case involving phase transitions to other electronic phases, such as the SDW phase [14]. The change in T_c due to the replacement of counter anions of different size can be interpreted as the change in the intermolecular spacing similar to the pressure, provided that the crystal types are kept isomorphous: with decrease of the resultant spacing, T_c is reduced. Furthermore, even if the crystal types are not isomorphous, one can find a correlation between T_c and the effective space allotted to each BEDT-TTF molecule; larger spacings between molecules tend to accompany higher values of T_c, as shown in Fig. 3 [22]. On this basis, increase in the intermolecular spacing has been adopted as a guideline to develop high T_c molecular superconductors. The κ-(BEDT-TTF)$_2$Cu(NCS)$_2$ exhibiting a T_c close to 10 K [23] is located on this line. Incidentally, the salts with T_c higher than 10 K, such as κ-(BEDT-TTF)$_2$Cu[N(CN)]$_2$Br [24], κ-(BEDT-TTF)$_2$Cu(CN)[N(CN)$_2$] [25] and κ-(BEDT-TTF)$_2$Cu[N(CN)$_2$]Cl [21], can be located in the graph relating T_c with the intermolecular spacing. This tendency of the higher T_c with larger intermolecular spacing can be understood in terms of the higher density of states at the Fermi surface, which is caused by a decrease in the intermolecular electron transfer, provided that the BCS-like dependence of T_c on the density of states dominates the superconductivity.

The arguments concerning the mechanism of superconductivity in the charge transfer complex have not yet been settled. The NMR studies showing the absence of the Hebel-Slichter peak in the relaxation below T_c support the possibility of the exotic mechanism resulting in the pairing of d-type symmetry [26]. The controversy with respect to the types of the electron pairing, whether it is of s-like symmetry or not, exists even among the experimental results on the temperature depedences of the magnetic penetration depth, through measurements of the magnetic susceptibilities [27] and the muon-spin relaxation [28]. However it is understood that the model of the excitonic interaction does not fit these systems, since the polarizability of counter anions is not so strong as to mediate the pairing. Taking into account the roles of the intramolecular interaction in TTF-type molecules, a probable model was proposed [29] by extending the BCS mechanism. This was motivated by the fact that the principal molecules in superconducting charge transfer salts possess TTF-type backbone, as represented in Fig. 1. Spectroscopic data on the molecular vibration and the electron-molecule coupling strength have provided experimental background to evaluate T_c. The model enables to make semiquantitative discussion, since it can be combined with experimental molecular data: it provides a possibility to check its validity by the substitution of isotope constitute atoms, through varying the vibrational frequency and the strength of molecule-electron coupling.

The effect of isotope substitution is interpreted also in terms of the significance of the chemical roles of anions in superconductivity, because they bridge conducting layers through hydrogen-bond-like coupling between ethylene bases of BEDT-TTF molecules and anions [30]. The role of hydrogen bonding to induce superconductivity is not clarified, but one cannot rule out unknown mechanisms if there exist some meaningful experimental results. The electronic structures of the TTF-type donor stacks are regarded to be rather simple in the sense that the highest occupied molecular orbital and the lowest unoccupied molecular orbital are distinctively separated. However, for metal(dmit)$_2$ molecules which work as acceptors in a charge transfer salt, hybridized orbitals form the conduction bands result in a two-band, where an interband pair-transfer interaction between the conductive and a completely filled second band can occur. The interband process enhances the effect of the pair-transfer interaction, through reduction of the effective on-site Coulomb energy [31].

It is evident that there is a bound for increasing T_c by increasing the intermolecular spacing, because increase of the spacing by a sufficiently large amount restricts the electron transfer and, simultaneously, softens the crystal. In fact, κ-(BEDT-TTF)$_2$Cu[N(CN)$_2$]Cl with the highest T_c among the charge transfer salts faces an insulating state, whose boundary is affected dramatically even moderate pressure less than 0.5 kbar [32].

To find a new route to raise T_c, different types of molecules and their arrangements are required. Although great effort has been made in exploring new style of molecules outside of TTF-type, very few exceptions have been useful in synthesizing superconductors. Metal(dmit)$_2$-type molecules are examples of non TTF-type molecules, but still there is a similarity in the backbone structure, although there is a possibility to induce a new mechanism due to the difference in the electronic structure as stated above. As a new way to expand the list of TTF-type superconductors, fused molecules such as DTEDT (Fig. 1), in which two TTF molecules are bridged including a vinyl base, have recently been synthesized. Since it has an elongated structure, enhancement of the polarizability within molecule and thereby the superconductivity coupling strength is expected to be enhanced via excitonic process. The observed T_c to date, however, remained low: e.g., 4 K for (DTEDT)$_3$Au(CN)$_2$ [33].

FULLERIDE SUPERCONDUCTORS

A new fashion in molecular supreconductors was initiated by the discovery of superconductivity in alkali-metal fulleride, A_3C_{60}, where A represents K and Rb, in which case T_c is 18 K and 29 K, respectively [34, 35]. The highest value of T_c, 33 K, is found for Cs_2RbC_{60} [36]. The constituent molecule C_{60} has a so-called soccer ball shape (Fig. 1), and the A_3C_{60} crystal exhibiting the high T_c has an fcc structure. The type of the molecular structure for fullerene C_{60} belongs to quite different category from the TTF-type molecules with a planar structure: the C_{60} molecules are assembled to form three-dimensional crystal. In this case, however, there are some similarities with the planar shaped TTF-type molecules in the fact that the principal conduction carriers are π electrons: the carbon atoms constitute five- and six-member rings with combination of sp^2 and sp^3 orbitals, which dominate the electronic structure of A_3C_{60}. The mobile carriers are produced by the addition of extra electrons to the filled band system formed by C_{60}, through the charge transfer from alkali metal atoms to C_{60} molecules. Further, concerning the physical prosperities, T_c is decreased by pressure, indicating that T_c increases with intermolecular spacing as with the charge-transfer-salt superconductors. This dependence, implying that increasing the intermolecular spacing is favorable to obtain higher T_c, is supported by the substitution of alkali metals with different size and with different combination of alkali metals as shown in Fig. 4 [37].

Generally speaking, a good correlation between a lattice parameter and T_c is found. However, stable fcc phases of Cs_3C_{60} and Li_3C_{60} are not found. Secondly, although the lattice parameters seem to decrease monotonically as the diameter of alkali metals decreases, a significant deviation in T_c is observed for Na_3C_{60} [38]. The fcc C_{60} crystal has two types of interstitial sites which can accomodate alkali metals,

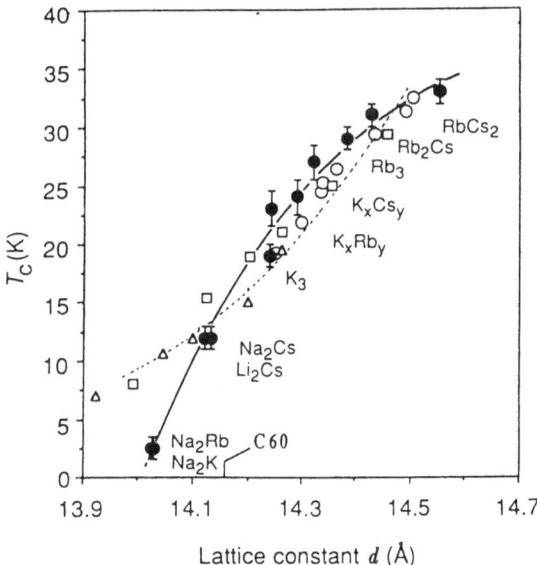

Figure 4. Relation of T_c to lattice constants for A_3C_{60} superconductors. Open circles are by F. M. Fleming et al. (*Nature* **352** : 787 1991)) and open triangles and squares are from pressure experiments for K_3C_{60} and Rb_3C_{60}, respectively, by O. Zhou et al. (*Science* **255** : 833 (1992)). [Reprinted by permission from Ref. 37, copyright Macmillan Magazines Ltd.]

Table 2. GL coherence lengths $\xi(0)$ of high T_c molecular superconductors, derived based on the fluctuation renormalization theory from the magnetization measurement. $\xi_\perp(0)$ and $\xi_\parallel(0)$ are the intra-plane and inter-plane coherence lengths at 0 K, respectivily. $\xi_\perp(0) = \xi_\perp(0)$ for A_3C_{60}. d is the inter-plane distance for κ-type BEDT-TTF salts and the unit cell length for the fcc structured A_3C_{60}.

	T_c(K)	$\xi_\perp$ (0) (nm)	$\xi_\parallel$ (0) (nm)	d (nm)
κ-(BEDT-TTF)$_2$Cu(NCS)$_2$	8.7 ± 0.2	0.31 ± 0.05	2.9 ± 0.5	1.52
κ-(BEDT-TTF)$_2$Cu[N(CN)$_2$]Br	10.9 ± 0.2	0.58 ± 0.1	2.3 ± 0.4	1.48
K_3C_{60}	19.3	2.1 ± 0.1		1.41
Rb_3C_{60}	29.6	1.3 ± 0.1		1.42

one being the larger octahedral site and the other being the two smaller tetrahedral sites. The larger cations are accomodated in the octahdral sites and the smaller ones in the tetrahedral sites. This implies that the absence of stable compounds in Cs_3C_{60} and Li_3C_{60} is due to incongruence of dopant atoms to the interstitial sites. Thus the T_c boundary of the fcc A_3C_{60} system is related to the crystal structure determined by C_{60} molecules.

COHERENCE LENGTHS OF HIGH T_c SUPERCONDUCTORS

It is a general trend that Pippard coherence length ξ_o representing the spread of superconductivity wave function decreases with increase of T_c, as given by $\xi_o \simeq \hbar v_F/(k_B T_c)$, where v_F is Fermi velocity. This value is close to the Ginzburg-Landau (GL) coherence length in the zero Kelvin limit $\xi(0)$, which can be evaluated from the superconductivity transition characteristics near T_c. For the molecular superconductors, the GL coherence lengths have been evaluated mostly from the temperature dependence of the upper critical field found in the resistive measurement in the transition region, with the aid of the formulae derived in the meanfield approximation. However, the recent investigations on the high T_c superconductors with low-dimensional structure have shown that the effect of thermal fluctuation cannot be neglected and, as a result, the upper critical field cannot be regarded as a phase transition point for long-range ordering. Experimentally, the resistive transition chartacteristics of the highly anistropic system becomes too broad to determine the transition point. For κ-type BEDT-TTF superconductors, it is not appropriate to regard them as a highly anisotropic system but to be treated as a stack of two-dimensional superconducting sheets with Josephson coupling between them. With the aid of a theoretical treatment based on the Lawrence-Doniach model, taking account of thermal fluctuation, the GL coherence lengths of the high T_c molecular superconductors have been evaluated from the magnetization data in the mixed states [39, 40] as listed in Table 2. In the evaluation procedure the κ-type BEDT-TTF salts are assumed to be two-dimensional but isotropic within the layer, whose GL coherence lengths are represented by the in-plane coherence length $\xi_\parallel(0)$ and the normal-to-plane one $\xi_\perp(0)$.

It is remarkable that the $\xi_\perp(0)$ of κ-type BEDT-TTF salt becomes shorter than the inter-layer distance, although $\xi(0)_\parallel$ is of considerable length covering a number of molecular units, being consistent with the Lawrence-Doniach model. For Rb_3C_{60}, the $\xi(0)$ becomes shorter than the fcc unit cell parameter d. The shortness of $\xi(0)$ may weaken the tolelance against thermal effect, resulting in the saturation behavior of T_c in larger d region, as illustrated in Fig. 4. The shortening of $\xi(0)$ with increase of T_c provides the following message. In order to stabilize the superconductivity by realizing long $\xi(0)$ in materials, it is preferable to adopt low-dimensionality, e.g., quasi two-dimensionality rather than three-dimensionality, since it is possible to unify different functions to realize superconductivity together with other roles such as binding to materialize in a bulk, by segregating relevant directions. This increases the freedom of material design leading to produce high T_c materials in a rational way.

FUTURE OF MOLECULAR SUPERCONDUCOR STUDY

Since the disceovey of the superconductivity in $(TMTSF)_2PF_6$, T_c has been raised but the highest T_c still remains below 40 K. On the other hand, the category of superconducting materials has been expanded steadily and the nature of existing superconductors reveals their exotic features with progress of research. The debate with respect to the types of superconductivity wave function is related to one of them. The photoelectron spectroscopy data show that a metallic Fermi edge is not observed in the density of states [41]. This has concern with the low-dimensionality of system and/or the effect of Coulomb interaction. The interplay with magnetism existing in the close neighbourhood of the superconductivity is another intriguing aspect of the molecular superconductors [42]. The competition and coexistence have been discussed in TMTSF salt with respect to one-dimensionality, in particular, in relation to SDW [11]. For BEDT-TTF salts, with quasi two-dimensional nature, the presence of antiferromagnetism and weak ferromagnetism has been reported [43, 44]. The magnetism stems not from magnetic ions but from mobile charge carriers. The Fermi surface nesting can be regarded as a cause of the magnetism, since the Fermi surface of BEDT-TTF salts has in-plane anisotropy resulting in a flat-shaped components which bring about a divergence in response function to a Fermi wave vector. The effect of the on-site Coulomb energy resulting in the Mott-Hubbard interaction is considered to explain not only the exotic nature of the superconductivity but also abnormal aspect of the metallic phase [45]. The roles of the magnetic ions in λ-$(BEDT\text{-}TSF)MCl_4$ are intiriguing, where the salt turns to be superconducting with M=Ga at 8 K, but it becomes insulating for M=Fe at the same temperature [46].

The superconductivity in polymer has not been observed except for $(SN)_x$. In doped polyacetylene $(CH)_x$, very high electrical conductivity has been realized [47]. It is a highly disordered system but shows metallic phase even at very low temperature, exhibiting Fermi-liquid-like behavior. However, superconductivity has not been found. Recently, a strategy for finding superconductivity in such conducting polymers is proposed as follows [48]. An insulating phase appearing at zero doping is a commensurate CDW state. Upon doping, the long range order in CDW is disturbed, but the short range order remains commensurate. At a critical dopant concentration, where the system undergoes a phase transition to a metallic state, the polymer may possess a highly correlated electronic state which resembles to that in the high T_c cuprate superconductors.

We remember that the Little model induced ideas to find pairon-condensation or superconductive-type enhanced conductivity in double-stranded DNA (Watson-Click helix) [49, 50]. It will be difficult to find a bulk material exhibiting the phenomena. However, the recent development of methods to investigate mesoscopic or atomic scale characteristics may provide chances to visit new types of materials where such novel superconductive-type phenomena can be found.

REFERENCES

[1] W. A. Little, *Phys. Rev.* 134 : A1416 (1964).

[2] R. A. Ferrell, *Phys. Rev. Lett.* 13 : 330 (1964),
T. M. Rice, *Phys. Rev.* 140 : A1889 (1964),
C. G. Kuper, *Phys. Rev.* 150 : 189 (1966), P. C. Hohenberg, *Phys. Rev.* 158 : 383 (1967).

[3] H. J. Keller (ed.), *Chemistry and Physics of One-Dimensional Metals*, Plenum Press, N.Y., (1976).

[4] D. Davis, H. Gutfreund and W. A. Little, *Phys. Rev.* B 13 : 4776 (1976).

[5] W. A. Little, *J. Phys. Paris* 44 C3 : 819 (1983).

[6] R. L. Greene, G. B. Street and L. J. Suter, *Phys. Rev. Lett.* 34 : 577 (1975).

[7] H. Akamatsu, H. Inokuchi and Y. Matsunaga, *Nature* 173 : 168 (1954).

[8] R. G. Kepler, P. E. Bierstedt and R. E. Merrifield, *Phys. Rev. Lett.* 5 : 503 (1960).

[9] J. Ferraris, D. O. Cowan, V. Walatka and J. H. Perlstein, *J. Am. Chem. Soc.* 95 : 498 (1973).

[10] L. B. Coleman, M. J. Cohen, D. J . Sandman, F. G. Yamagishi, A. F. Garito and A. J. Heeger, *Solid State Commun.* 12 : 1125 (1973).

[11] D. Jérome and H. J. Schultz, *Adv. Phys.* 31 : 299 (1982).

[12] D. Jérome A. Mazaud, M. Ribault and K. Bechgaard, *J. Phys. Lett.* 41 : L-95 (1980).

[13] K. Bechgaard, K. Carneiro, M. Olsen, F. B. Rasmussen and C. S. Jacobsen, *Phys. Rev. Lett.* 46 : 852 (1981).

[14] T. Ishiguro and K. Yamaji, *Organic Superconductors*, Springer Verlag, Heidelberg (1990).

[15] G. Saito, T. Enoki, H. Inokuchi and H. Kobayashi, *J. Phys. Paris* 44 C3 : 1215 (1983).

[16] J. R. Ferraro and J. M. Williams, *Introduction to Synthetic Electrical Conductors*, Academic Press, N.Y. (1987).

[17] E. B. Yagubskii, I. F. Shchegolev, V. N. Laukhin, P. A. Kononovich, M. V. Kartsovnik, A. V. Zvarykina and L. I. Buravov, *JETP Lett.* 39 : 12 (1984).

[18] A. Kobayashi, R. Kato, H. Kobayashi, S. Moriyama, Y. Nishio, K. Kajita and W. Sasaki, *Chem. Lett.* 1987 : 459.

[19] V. N. Laukhin, E. E. Kostyuchenko, Yu. V. Sushko, I. F. Shchegolev and E. B. Yagubskii, *JETP Lett.* 41 : 81 (1985),
K. Murata, M. Tokumoto, H. Anzai, H. Bando, G. Saito, K. Kajimura and T. Ishiguro, *J. Phys. Soc. Jpn.* 54 : 2084 (1985).

[20] P. C. W. Leung, T. J. Emge, M. A. Beno, H. H. Wang, J. M. Williams, V. Petricek and P. Coppens, *J. Am. Chem. Soc.* 107 : 6184 (1985).

[21] U. Geiser, A. J. Schultz, H. H. Wang, D. M. Watkins, D. L. Stuyka, J. M. Williams, J. E. Schirber, D. L. Overmer, D. Jung, J. J. Nova and M. H. Whangbo, *Physica* C 174 : 475 (1991).

[22] G. Saito, H. Urayama, H. Yamochi and K. Oshima, *Synth. Metals* 27 : A331 (1988).

[23] H. Urayama, H. Yamochi, G. Saito, K. Nozawa, T. Sugano, M. Kinoshita, S. Sato, K. Oshima, A. Kawamoto and J. Tanaka, *Chem. Lett.* : 55 (1988).

[24] A. M. Kini, U. Geiser, H. H. Wang, K. D. Carlson, J. M. Williams, W. K. Kwok, K. G. Vandervoort, J. E. Thompson, D. L. Stupka, D. Jung and M. H. Whangbo, *Inorg. Chem.* 29 : 2555 (1990).

[25] T. Komatsu, T. Nakamura, N. Matsukawa, H. Yamochi, G. Saito, H. Ito, T. Ishiguro, M. Kusunoki and K. Sakaguchi, *Solid State Commun.* 82 : 843 (1991).

[26] M. Takigawa, H. Yasuoka and G. Saito, *J. Phys. Soc. Jpn.* 56 : 873 (1987),
Y. Hasegawa and H. Fukuyama, *J. Phys. Soc. Jpn.* 56 : 877 (1987).

[27] K. Kanoda, K. Akiba, K. Suzuki, T. Takahashi and G. Saito, *Phys. Rev. Lett.* 65 : 1271 (1990),
M. Lang, N. Toyota, T. Sasaki and H. Sato, *Phys. Rev. Lett.* 69 : 5822 (1992),
M. Dressel, O. Klein, G. Griiner, K. D. Carlson, H. H. Wang and J. M. Williams, *Phys. Rev.* B 50 : 13603 (1994).

[28] D. R. Harshman, R. N. Kleinman, R. C. Haddon, S. V. Chichester-Hicks, M. L. Kaplan, L. W. Rupp, T. Pfis, L. D. Williams and D. B. Mitzi, *Phys. Rev. Lett.* 64 : 1293 (1990),
Y. J. Uemura, L. P. Lee, L. M. Luke, B. J. Sternlieb, W. D. Wu, J. H. Brewer, T. N. Reiseman, C. L. Seaman, M. B. Maple, M. Ishikawa, H. G. Hinks, J. D. Jorgensen, G. Saito and H. Yamochi, *Phys. Rev. Lett.* 66 : 2665 (1991).

[29] K. Yamaji, *Solid State Commun.* 61 : 413 (1987).

[30] J. M. Williams, H. H. Wang, T. J. Emge, V. Geiser, M. A. Beno, P. C. W. Leung, K. D. Carlson, R. J. Thorn, A. J. Schultz and M. H. Whangbo, *Progress Inorg. Chem.* 35 : 51 (1987),
J. M. Williams, A. J. Schultz, U. Greiser, K. D. Carlson, A. M. Kini, H. H. Wang, W. K. Kwok, M. H. Whangbo and J. E. Schirber, *Science* 252 : 1501 (1991).

[31] Y. Shimoi, K. Yamaji and T. Yanagisawa, *Synth. Metals* 70 : 1017 (1995).

[32] Yu. V. Sushko, H. Ito, T. Ishiguro, S. Horiuchi and G. Saito, *Solid State Commun.* 87 : 997 (1993).

[33] Y. Misaki, N. Higuchi, H. Fujiwara, T. Yamabe, T. Mori, H. Mori and S. Tanaka, *Angrew. Chem. Int. Ed. Engl.* 34 : 1222 (1995).

[34] A. F. Hebard, M. J. Rosseinsky, R. C. Haddon, D. W. Murphy, S. H. Glarum, T. T. M. Palstra, A. P. Ramirez and A. R. Kortan, *Nature* 350 : 600 (1991).

[35] M. J. Rosseinsky, A. P. Ramirez, S. H. Glarrum, D. W. Murphy, R. C. Haddon, A. F. Hebard, T. T. M. Palstra, A. R. Kortan, S. M. Zahurak and A. V. Makhija, *Phys. Rev. Lett.* 66 : 2830 (1991).

[36] K. Tanigaki, T. W. Effesen, S. Saito, J. Mizuki, J. S. Tsai, Y. Kubo and S. Kuroshima, *Nature* 352 : 222 (1991).

[37] K. Tanigaki, I. Hirosawa, T. W. Ebbessen, J. Mizuki, Y. Shimakawa, Y. Kubo, J. S. Tsai and S. Kuroshima, *Nature* 356 : 419 (1992).

[38] M. J. Rosseinsky, D. W. Muphy, R. M. Fleming, R. Tycko, A. P. Ramirez, T. Siegrist, G. Dabbagh and S. E. Barrent, *Nature* 356 : 416 (1992).

[39] H. Ito, M. Watanabe, Y. Nogami, T. Ishiguro, T. Komatsu, G. Saito and N. Hosoito, *J. Phys. Soc. Jpn.* 60 : 3230 (1991).

[40] A. Otsuka, T. Ban, G. Saito, H. Ito, T. Ishiguro, N. Hosoito and T. Shinjo, *Synth. Metals* 55-57 : 3148 (1993).

[41] R. Liu, H. Ding, J. C. Campuzano, H. H. Wang, J. M. Williams and K. D. Carlson, *Phys. Rev.* B 51 : 13000 (1995).

[42] T. Ishiguro, Yu. V. Sushko, H. Ito and G. Saito, *J. of Supercond.* 7 : 657 (1994).

[43] K. Miyagawa, A. Kawamoto, Y. Nakazawa and K. Kanoda, *Phys. Rev. Lett.* 75 : 1174 (1995).

[44] U. Welp, S. Fleshler, W. K. Kwok, G. W. Crabtree, K. D. Carlson, H. H. Wang, U. Geiser, J. M. Williams and V. M. Hitsman, *Phys. Rev. Lett.* 69 : 840 (1992).

[45] H. Kino and H. Fukuyama, *J. Phys. Soc. Jpn.* 64 : 2726 (1995).

[46] H. Kobayashi, T. Udagawa, H. Tomita, K. Bunk, T. Naito and A. Kobayashi, *Chem. Lett.* : 1559 (1993),
A. Kobayashi, T. Udagawa, H. Tomita, T. Naito and H. Kobayashi, *ibid* : 2179.

[47] J. Tsukamoto, *Adv. Phys.* 41 : 509 (1992),
T. Ishiguro, H. Kaneko, J. P. Pouget and J. Tsukamoto, *Synth. Metals* 69 : 37 (1995).

[48] S. A. Kivelson and V. Emery, *Synth. Metals* 65 : 249 (1994).

[49] R. M. Pearlstein, *Phys. Rev. Lett.* 20 : 594 (1968).

[50] J. Ladik, G. Bicźo and J. Rédly, *Phys. Rev.* 188 : 710 (1969).

NON-FERMI LIQUID BEHAVIOR OF KINETIC

CHARACTERISTICS IN QUASI-ONE-DIMENSIONAL

ORGANIC CONDUCTORS

Lev P. Gor`kov

National High Magnetic Field Laboratory
Florida State University
Tallahassee,Fl 32306

INTRODUCTION

Years ago W. Little has suggested a mechanism[1] by which, he hoped, it may be possible to dramaticaly increase the transition temperatures for superconductivity in the new generation of materials based on linear organic components. This publication stimulated enourmous experimental activity in syntheses and study of the absolutely new class of conducting materials-the organic conductors. Now, as the result of collective international efforts, we have the fascinating new world of organic compounds, and some of them are superconductors with a remarkably high (~10K) transition temperature. Recall that the vast majority of these new metals do not even contain any metallic element at all!

In what follows, we concentrate our attention on properties of the so-called "Bechgaard salts ", $(TMTSF)_2 X$, where superconductivity among the organic materials has been discovered for the first time . Since that, however, this class of materials has acquired much of attention, both of experimentalists and theorists. There are a few reasons for that. On the theoretical side, these compounds realize the best known so far example of the so-called "quasi-one dimensional" (Q1D) materials with a pronounced chain structure. The theory of electron-electron interactions in a one-dimensional (1D) system has been elaborated in great details[2,3] The strictly one-dimensional system would possess numerous interesting properties, dramatically different from properties of electrons in ordinary metals, described in frameworks of the familiar Fermi liquid theory. Therefore, unusual features of the gas of interacting one-dimensional electrons are very often considered as an illustrative example, a "prototype" of a nontrivial new physics, which may be pertinent to some other systems, such as the 2D elec-

From High Temperature Superconductivity to Microminiature Refrigeration
Edited by Cabrera *et al.*, Plenum Press, New York, 1996

263

tron gas, heavy fermions, high T_c oxides and so forth, where electronic correlation are strong enough to produce considerable deviations from the Fermi liquid theory predictions and, hence, result in distinctive new properties (see,e.g. in [4]) .

Experimentally, the detailed information is now available concerning thermodynamic, transport and galvanomagnetic properties of these materials, especially, for the two of the most studied compounds, $(TMTSF)_2PF_6$ and $(TMTSF)_2ClO_4$ (for rewiews see [5]). Taken as a whole, numerous existing low temperature experimental data are consistent with the idea of a strongly anisotropic three-dimensional metallic behavior in these systems . On the other hand, there are some basic facts which from the very beginning would need to be treated differently. Namely, among many striking features characteristic of these materials, there is a proximity to some transition, a structural transition, like the anion ordering (AO), or a spin density wave (SDW) instability, which occur at rather low temperatures (10-20K). Theoretically, it is well known that electron-electron interactions in the one-dimensional conductors reveal a tendency to produce various instabilities [2,3]; the instabilities then may end up as a thermodynamical phase transition [6], if additional three-dimensional features, such as tunneling, or interactions between electrons on different chains, were introduced into the purely one-dimensional analysis. It seems, at least, at first sight, that for the selenium-based compounds, $(TMTSF)_2X$, there is no evident need to invoke the one-dimensional physics at all. All transitions in the materials mentioned above, are rather well defined mean field transitions where critical fluctuations are not important, with the exception of a narrow enough vicinity of the transition temperature. The mechanism for the SDW state itself, its suppression by external pressure and the peculiar phenomenon of restoration of the SDW-like states in high magnetic fields, in particular, can be quantitatively described in a simple and plausible model of weekly interacting electrons with the metallic Fermi surfaces possessing an approximate "nesting" property [7].

Some *sulfur* analogs of the Bechgaard salts, such as $(TMTTF)_2PF_6$, however, display somewhat more pronounced one-dimensional features (see discussion in [5d]). A number of them show instabilities at considerable higher temperatures, which are comparable to estimated values of the transverse tunneling integrals. Therefore in these materials one may anticipate a more important role of the on-chain interactions, although it seems that low temperature properties of both the sulfur- and selenium-based materials qualitatively are rather similar and can be mapped on the top of each other by applying external pressure.

The fact that the most of the Bechgaard salts may exist in a non-metallic ground state at T=0, implies that, strictly speaking, their low temperature properties are not properties of a Fermi liquid theory. On the other hand, the SDW-phase can be removed by an external pressure, and the same material will now behave metallically down to lowest temperatures. The proximity to an insulating state, hence, casts some reasonable questions regarding the Fermi liquid description in the adjacent metallic phase. There is a growing amount of evidences that, at least, some properties of these materials somehow differ from simple expectations of the theory of metals. Among these properties are: unusual temperature dependence of resistivity, large magnetoresistance in weak enough fields (~10 Tesla) and temperatures of order of 10K, a non-Korringa temperature dependence of the NMR-relaxation rate (see in [5d]). Note that in all three cases new features occur in the kinetic properties.

We show below that the aforementioned simple model of two Fermi surface sheets with an approximate "nesting" properties, could explain also some key new features in kinetics of the Q1D-organic conductors. This picture, however, does not rule out an involvement of different physics [4].

MODEL

Although the Bechgaard salts belong to the monoclinic system, their properties are usually well approximated in the framework of an orthorombic symmetry. For the lattice parameters, (a, b, c), lets a are the lattice period along the chain direction, while b and c correspond to periodicity in directions transverse to the chain. For $(TMTSF)_2PF_6$ values of the three main tunneling integrals along three corresponding axes are estimated as:

$$t_a : t_b : t_C = 0.2 eV : 200K : 10K \qquad (1)$$

This anisotropy is consistent with data on the anisotropy of conductivity, as obtained at room temperatures. Strong anisotropy of conducting properties suggests that the main of the features of the electronic spectrum can adequately be described in terms of a tight binding model with only a few tunneling (hopping) matrix elements involved. Usually in such a model one consider the hopping integrals between the nearest --neighboring chains (for the b-- and c-- directions) and tunneling to the next-nearest neighbor chain in the (a, b)--plane. (In view of the exceedingly large anisotropy between b--and c--directions, the materials are sometimes called quasi-two-dimensional (Q2D) conductors). In this approximation the tight binding spectrum of non-interacting electrons acquires the following form:

$$E(p) = v(\pm p_a - p_F) - 2t_b \cos p_b b - 2t_c \cos p_c c - 2t\hat{}_b \cos 2p_b b \qquad (2)$$

The first contribution in this expression is of the purely one-dimensional origin, i.e., corresponds to an electron moving along the chain. It is linearized in vicinity of each of the two open sheets of the Fermi surface, for its right and left sides, respectively. Two other terms, t_b and t_c, describe hopping between nearest neighboring chains along b and along c; as for the $t\hat{}_b$--term, it measures hopping between the next nearest neighboring chains, as explained above. In absence of the latter term ($t\hat{}_b = 0$) the spectrum in eq.(2) possesses the so-called "nesting" degeneracy:

$$E(p + Q) = -E(p) \qquad (3)$$

Here

$$Q = Q_0 \equiv \left(2p_F; \frac{\pi}{b}; \frac{\pi}{c}\right) \tag{2'}$$

At $t`_b \neq 0$ eq.(3) takes place only approximately, when $t`_b$ is small enough compared to t_b. (In what follows the tunneling integral t_c introduces no new physics and will be omitted for brevity).

The Bechgaard salts, $(TMTSF)_2 X$, are the charge transfer compounds with one electron transferred to each anion, X. To say it in a different manner, these salts are hole conductors with one hole per unit cell in the conducting TMTSF-network. The conduction band is half-filled; in the one-dimensional approximation, i.e., for a single metallic chain, $4p_F = \frac{2\pi}{a}$. This is a so- called "commensurate" value for the Fermi momentum.

For the sake of simplicity of the analysis, electron-electron interactions are usually taken in a form of a short ranged (Coulomb) interaction. The effective screening of the Coulomb potential in real materials is due to the fact that all chains are packed together into the lattice, so that even if there were no interchain hopping at all, charges on neighboring chains may adjust themselves to completely screen a charge added to a given chain. Such a screening may still be rather anisotropic. For electrons confined to one metallic chain all interactions at the Fermi surface are reduced to the following three interaction constants :

$g_1 = g(+-;-+)$--the amplitude for the backward scattering of one electron by another ;

$g_2 = g(+-;+-)$--the forward scattering amplitude (each electron remains on its side of the Fermi surface after scattering);

$g_3 = g(++;--)$--the Umklapp processes;the total momentum is not conserved.

(For the Hubbard model all three constants are equal: $g_1 = g_2 = g_3$. The dimensionless inter-action constants above are formed by the interaction strenght itself, mutiplied by the bare density of states at the Fermi level). The matrix elements for the three scattering processes are shown schematically in Figure 1.

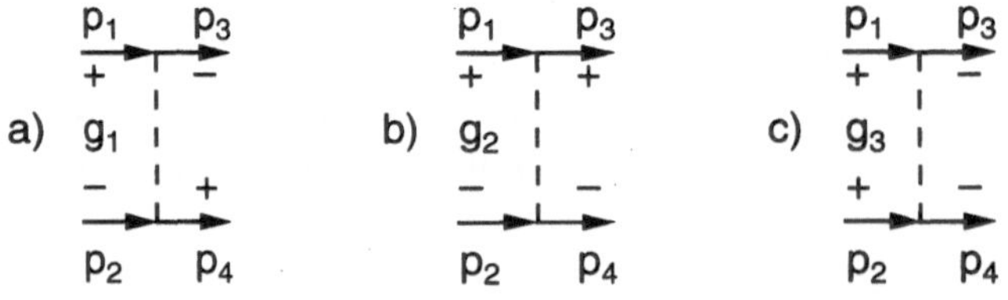

Figure 1. Three matrix elements for interactions between one-dimensional electrons: a) the backward scatter-ing amplitude; b) the forward scattering process; c) matrix element for the Umklapp scattering.

As it was shown in [2,3], interactions between electrons on a metallic one-dimensional chain may lead to a few instabilities which develop simultaneously and compete with each other for opening of a gap at the Fermi level. There are three channels for these instabilities: superconducting channel (SC); formation of charge density wave (CDW); the spin density wave channel (SDW). Of course, no thermodynamic phase transition is possible in one dimension.

To separate the above channels and to fix one of the above instabilities as the thermodynamical phase transition, it is necessary to account for some three-dimensional features of the real materials. There are only two distinctly different three-dimensional effects [6]: 1) interactions between electrons on different chains and 2) interchain electron hopping. The type of the ground state will essentially depend on which of these two effects prevails. We recall that according to [6], a finite tunneling is always necessary for a ground state to have a broken time reversal symmetry (for instance, the superconducting, or the spin density wave state). In that sense the model spectrum of eq. (2) meets this criterion. Another fact, important for our considerations, is that for weak enough interaction the onset of the new state, i.e., the phase transition, happens at some finite temperature, T_c. The transition itself turns out to be a mean field transitions. The provision that the temperature interval for critical fluctuations is narrow, coincides with the assumption of smallness of all the interaction constants, g_i , i.e., $g_i \ll 1$ (in addition, it is also assumed that $T_c \ll t_b$).

The characteristic energy scale for $(TMTSF)_2 X$ defined by the temperature of the SDW- transition, for $X=PF_6$ is about 10K. It seems therefore that properties of these salts fit well the model of weekly interacting electrons. On the other hand, transverse tunneling integrals seems to be considerable less in others, the sulfur based Bechgaard salts, $(TMTTF)_2 PF_6$. Here, the "resistive transition" starts at T~ 150-200K, in other words, at temperatures comparable with estimated value of the transverse hopping integrals (see in [5d]).

SDW-INSTABILITY

Since nesting instabilities in different realizations have been extensively studied by many authors (e.g., see [5] for review), we shall only outline below major steps at calculations. Thus, in case of the SDW -instability the analysis starts with calculation of the generalized linear magnetic response to an applied staggered magnetic field, $H(Q)$:

$$M(Q) = \chi(Q)H(Q)$$

The diagrammatic approach is the most helpful one for our purposes. The response function, $\chi(Q)$, is proportional to the matrix element constituted of two Green functions, shown in Figure 2 :

$$\chi(\mathbf{Q}) \sim T\sum \int d\mathbf{p} G(\omega_n;\mathbf{p}) G(\omega_n;\mathbf{p}-\mathbf{Q}) \Rightarrow$$

$$\Rightarrow \int d\mathbf{p}[E(\mathbf{p}) - E(\mathbf{p}-\mathbf{Q})]^{-1} \times [\tanh(E(\mathbf{p})/2T) - \tanh(E(\mathbf{p}-\mathbf{Q})/2T)] \tag{4}$$

Provided that $\mathbf{Q}$ is defined as in eq.(2'), $\mathbf{Q} = \mathbf{Q}_0$, the last integral becomes:

$$\nu(E_F)\int \frac{dE}{E}\tanh\frac{E}{2T} \sim 2\nu(E_F)\ln\left(\frac{\overline{E}}{T}\right) \tag{4'}$$

where $\nu(E_F)$ is the electronic density of states (per spin), $\overline{E}$ is a cut-off energy. Note that when the vector $\mathbf{Q}$ is commensurate, the integral over momentum, $\mathbf{p}$, comprises of the two equal logarithmic contibutions arrising each from two sides of the Fermi surface (+ or -), because $2\mathbf{Q}$ is the reciprocal lattice vector.

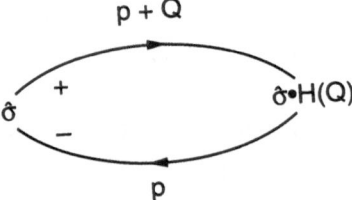

Figure 2. The generalized magnetic susceptibility for free electrons in its diagrammatic form: two lines correspond to the two Gree function in the matrix element of eq.(4).

With the above information concerning calculation of the single "bubble" diagram in Figure 2, consider corrections to $\chi(\mathbf{Q})$ due to electro-electron interactions. Some of those corrections are drawn schematically in Figure 3. In accordance with the above analysis, addition of each new internal block in Figure 3 introduces a factor, which actually is the integral of exactly the same form as calculated before, in eqs.(4,4'). This integral is logarithmically

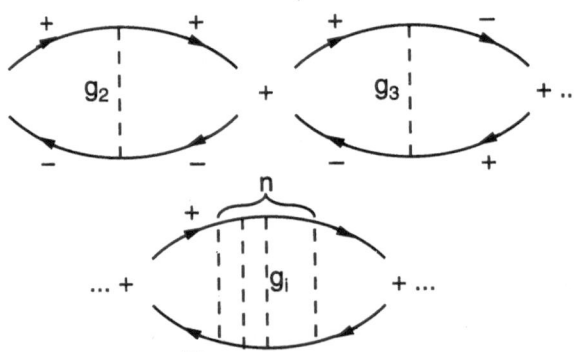

Figure 3. Electron-electron interactions(the dashed lines) form the "ladder" corrections to $\chi(\mathbf{Q})$; the two Green function inside each block produce the logarithmic term of the form in eqs.(4,4').

divergent at low temperatures. Each logarithmic contribution in Figure 3 appears together with the one of interaction constants, g_2 or g_3. Therefore, if the diagram contains some **n** interactions, the main contribution into the matrix element at low temperature is of order of

$$(\mathbf{g})^n \left[\ln\left(\overline{E}/\, \mathbf{T}\right) \right]^n$$

This is the familiar logarithmic problem: at low enough temperatures the large logarithmic factors may compensate weekness of interaction, and the whole expression has to be re-sumed to take this fact into account. For our purposes the most convenient approach to the problem is to find the renormalization of the effective interactions, Γ_2 and Γ_3. The diagra-matic equations for the two vertices are shown in Figure 4. Their form is self-explanatory.

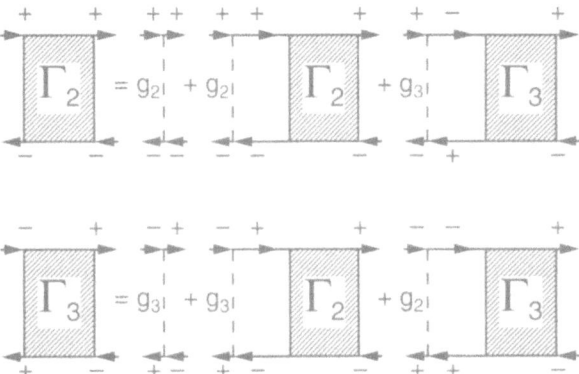

Figure 4. Equations for the renormalized verticies, Γ_2, Γ_3 ; their structure reflects the structure of diagram-matic corrections in Figure 3. Temperature dependence of Γ_2, Γ_3 determines the temperature behavior of $\chi(Q_0)$ in (4) and, hence, the onset of the SDW-phase.

With the bare interactions, $\mathbf{g}_2$ and $\mathbf{g}_3$, independent on momentum , the two equations become a system of the two algebraic equations. At low temperatures each vertex has a form:

$$\Gamma^* \equiv \Gamma_2 = \Gamma_3$$

(5)

$$\Gamma^* = \mathbf{g}^* \left[1 - 2\mathbf{g}^* \ln\left(\overline{E}/\, \mathbf{T}\right) \right]^{-1} \equiv 1 \Big/ 2 \ln\left(\frac{\mathbf{T}}{\mathbf{T}_0}\right)$$

(at the simple choice of $\mathbf{g}_2 \equiv \mathbf{g}_3$, denoted above schematically as $\mathbf{g}^*$). Here $\mathbf{T}_0$, as defined by the standard relation:

$$1 = 2\mathbf{g}^* \ln\left(\overline{E}/\, \mathbf{T}_0\right)$$

(6)

has the meanning of the temperature for the onset of the SDW-phase. It is easy to verify that the mean field character of the transition is guaranteed if

$$T_0 \ll t_b \ll E_F$$

To summarize, the SDW- phase transition, as described in framework of the nesting model (eqs.(2,3)), carries the mathematical features of a typical logarithmic scheme. The main contribution into eq.(4), which provides the pole in Γ`s , is due to a large logarithmic contribution from the extended interval $T_0 \ll E \ll \bar{E} \sim E_F$. Therefore deviations from the exact nesting condition, for instance, due to non-zero t_b in eq.(2), which only weakly change the contribution from that main integration interval, may not destroy conditions for a low temperature SDW-phase. This comment will be essential below at discussion of origin of the non-Fermi liquid features in kinetics of the Bechgaard salts.

PRECURSOR EFFECTS IN KINETICS

As it was already emphasized above, the nesting model leads to the mean field thermodynamical transition, i.e., everywhere outside a very narrow interval of temperatures around T_0, fluctuations introduce only negligable corrections to all physical quantities. Therefore the effect of critical fluctuations would not be able to account for considerable deviations from, say, the familiar linear Korringa-like temperature dependence in the NMR relaxation rate, seen for Bechgaard salts in a wide temperature interval (10-40K). Similar unexpected features have also been observed in the temperature dependence of resistivity (with and without magnetic filed) for $(TMTFT)_2 X$, where X= PF_6 and ClO_4 (see in [5d,8]).

The reason why fluctuations play only minor role in thermodynamics of the SDW-transition and, in particular, do not manifest themselves at temperatures even slightly above the transition temperature, T_0, is clear from the concluding remarks at the end of the previous Section. The onset of the SDW-instability is defined by the temperature at which the generalized magnetic susceptibility, $\chi(Q)$, in (4,5) diverges for the first time, and namely for that specific value of the wave vector, Q_0 in (2`). On the other hand, the vertecies $\Gamma_{2,3}(Q)$ in ( 5 ) show fast decrease with deviations of $Q$ from this specific value. Above $T_0$, corrections to the thermodynamic potential of the system due to interactions come from any $Q$`s, because electrons and holes are thermally excited along the whole Fermi surface and may have arbitrary momenta, whereas the verticies $\Gamma_{2,3}(Q)$ are large at $T \sim T_0$ only in a narrow range of Q around Q_0 (i.e., $Q = Q_0 + q$):

$$vq_a \sim 2t_b q_b b \sin(p_b b) \sim T_0 \tag{7}$$

We show now that at calculation of characteristics of the kinetic origin, such as some relaxation times, the above interval in the momentum space *is singled out by the conservation laws*, which have to be fullfiled for real processes of electron-electron scattering. This property is an interesting peculiarity of the nesting model [9].

In this paper we only illustrate the idea by considering a simple example. Thus, consider in some more details the temperature dependence of resistivity, $\rho(T)$. When electron-electron scattering is the dominating mechanism of resistivity, as is, probably, the case for these rather clean materials in the temperature range under consideration, the dissipation rate for momentum, $1/\tau_U$, is to be proportional to the matrix element of the Umklapp scattering, g_3, shown in Figure 1c. Omitting all nonessential factors, one can write $1/\tau_U$ in the following familiar form:

$$\frac{1}{\tau_U} \propto |g_3|^2 \frac{1}{T} \int n_{p_1} n_{p_2} \left(1 - n_{p_3}\right)\left(1 - n_{p_4}\right) \cdot$$

$$\delta\left(p_1 + p_2 - p_3 - p_4 - K\right)\delta\left(E_1 + E_2 - E_3 - E_4\right)dp_1...dp_4 \tag{8}$$

Here the Fermi functions, n_p, together with the δ-functions expressing conservation laws of energy and momentum, would lead to the well-known T^2-dependence for $\rho(T)$. Making use of the expression (2) for the total electron energy spectrum, as a sum of two components for the logitudinal and transverse disperssions, respectively,:

$$E(p) = \xi(p_a) + \varepsilon(p_b)$$

and choosing the Umklapp vector as $K = (4p_F; 0; 0)$, the conservation laws in (8) may be identically re-written in the following manner:

$$\delta\left(p_{1b} + p_{2b} - p_{3b} - p_{4b}\right) \cdot \delta\left(\xi_1 + \xi_2 + \frac{\Omega}{2}\right) \cdot \delta\left(\xi_3 + \xi_4 - \frac{\Omega}{2}\right) \tag{9}$$

(Here $\Omega = \sum \varepsilon(p_{bi})$). The resulting expression for (8) can be easely integrated over all the longitudinal components of momentum (energy), ξ_i, and for the relaxation rate $1/\tau_U$ one obtains:

$$\frac{1}{\tau_U} \propto \frac{|g_3|^2}{T} \int \Omega^2 \sinh^{-2}\left(\frac{\Omega}{2T}\right) \times \delta\left(p_{1b} + p_{2b} - p_{3b} - p_{4b}\right)dp_{1b}...dp_{4b} \tag{10}$$

The merit of this transfomation is that now one can explicitly see which part of the Fermi surface makes the main contribution into resistivity. In fact, at low enough temperatures, $T \sim T_0$, such a contribution comes from the vicinity of the phase space where $\Omega \sim T$ at $T \ll t_b$, that is Ω in (10) is small. Without the $t`_b$-term, for the brevity sake, it is possible to present $\Omega = \sum \varepsilon(p_{bi})$ in the explicit form:

$$\Omega = -8t_b \cos\left(\frac{(p_{1b} + p_{2b})b}{2}\right) \cdot \cos\left(\frac{(p_{3b} - p_{1b})b}{2}\right) \cdot \cos\left(\frac{(p_{3b} - p_{2b})b}{2}\right) \tag{11}$$

There are three planes in the $(p_{1b}; p_{2b}; p_{3b})$ --phase space on which $\Omega = 0$; in the discussion of the SDW-transition we need $p_{3b} - p_{2b} = \pi/b$.

In fact, as shown in Figure 5, instead of the bare matrix element g_3, one should consider all perturbative corrections of the "ladder"-type, similar to the diagrams summed already in Figure 3, because the matrix elements for each "bubble" again contain the same large logarithmic contribution, as diagrams in Figure 3, which have led to the expressions (5) and (6). The plane $p_{3b} - p_{2b} = \pi/b$ together with the longitudinal component, $2p_F$, is exactly the choice of the nesting vector Q_0, eq.(2`), for the SDW-instability. Finite temperature in the integral (10) will spread the essential Q's over interval (7). Therefore, the set of diagrams shown in Figure 5, provides considerable enhancement of the bare matrix element g_3.

It is straightforward to write down equations for these diagrams, which will have the same diagrammatic form, as shown in Figure 4. Solving these equations, however, becomes somewhat more compicated and tedious, and would demand numerical calculations to obtain the result for the effective mean free time, τ_U. (Partially, complications are due to that matrix elements also explicitly depend on energies of electrons participating in the scattering process, as seen from the integral of eq.(8); all energies are of order of temperature due to presence of the Fermi functions). We will not come into any further details and will describe the physical implications from the above considerations in a somewhat oversimplified fashion[9].

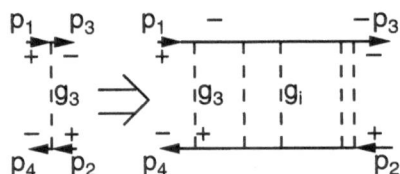

Figure 5. The bare Umklapp matrix element, g_3, renormalized by interactions along the SDW-channel: $P_3 - P_2 = Q_0$.

The physical importance of the corrections to bare matrix element g_3 (and g_2), shown in Figure 5, is that relaxation rates for various scattering processes acquire an additional temperature dependence due to proximity of the metallic phase in the Bechgaard salts under pressure to a SDW-state. This dependence does not exactly coincide with the temperature dependence of Γ_2, Γ_3, say, in eq.(5), which would determine the temperature behavior of the generalized magnetic susceptibility, $\chi(Q_0)$. The scattering processes between electrons depend on energies and momenta of interacting electrons, as compared to $\chi(Q_0)$. However, all the complicating features result in a minor modification of the effective matrix element [8]:

$$g_3 \Rightarrow \frac{1}{2}\left[\ln\left(\frac{T}{\alpha T_0}\right)\right]^{-1} \tag{12}$$

(Here $\alpha < 1$ is a constant which can be determined only from the full solution).

The new effectve interaction is *of order of unity* in the temperature range $T \sim T_0$, even if the bare interactions were small .The dominating contribution into resistivity in this temperature range is now of the form :

$$\rho(T) \Rightarrow \frac{T^2}{\left[\ln\left(\frac{T}{\alpha T_0}\right)\right]^2} \tag{13}$$

Temperature dependence in eq.(13) displays a new peculiarity in that it is not monotonous and has a minimum at

$$T_{min} = e\alpha T_0 ; \ (e = 2.71..) \tag{14}$$

The resistivity in (13), however, remains finite at the transition temperature. The fast increase of resistivity will only start below the transition, together with the development of the SDW-gap. (Close enough to transition all quantities, including resistivity, are slightly corrected by critical fluctuations in the very vicinity of the transition).

So far we were omitting the so-called "anti-nesting" term in eq.(2), $t`_b$. This term introduces additional complications at calculations of $\chi(Q)$ and all other relevant quantities. From the physical point of view, however, the role of this term is to suppress transiton temperature into the SDW-phase and with the further increase above some critical value, $t`_{b,cr}$, finally to stabilize the metallic state. (The term $t`_b$, therefore, mimics the role of the external pressure). Although no caltulations have been done yet to incorporate $t`_b$ rigorousely into the above analysis, one may speculate that, qualitatively, taking it into account would change eq.(13) in the following manner:

$$\rho(T) \Rightarrow \frac{T^2}{\left[\ln\left(\frac{T + \beta t`_b}{\alpha T_0}\right)\right]^2} \tag{15}$$

Here β is another constant. To find T_{SDW} for the actual SDW-transition as a function of pressure ($t`_b$ ), one have to return back to equations in Figure 4 in presence of $t`_b \neq 0$.

The temperature at which $\rho(T)$ may have a minimum, would now be determined from equation:

$$\ln\left(\frac{T + \beta t`_b}{\alpha T_0}\right) = \frac{T}{T + \beta t`_b} \tag{16}$$

Depending on the $t`_b$-value, the resistivity minimum may, or may not be seen in $\rho(T)$.

The temperature dependence of resistivity in the metallic phase of $(TMTSF)_2PF_6$ still remains a somewhat controversial issue. Metallic state itself in this material shows up only under external pressure. However, the main peculiarities in the resistivity behavior are known for a long time. In fact, resistivity displays rather unusual features [5a].The most remarkable are 1) considerable deviations of $\rho(T)$ at lower temperatures (10-30K) from the T^2-dependence (resistivity *decreases faster* !) and 2) large positive magnetorestance in the same temperature range for rather low fields of order of 10 Tesla. It was first suggested in [5a] that these peculiarities may be a manifestation of strong superconducting fluctuations, which develop in these chain materials due to strong electron-electron correlation effects, specific for the onedimensional physics. In view that the superconducting transition temperature in $(TMTSF)_2PF_6$ is about 1K, and supreconductivity itself has a well-defined three-dimensional character, the idea, probably, is somewhat too exotic.

A more detailed recent study[8] of magnetoresistance effects performed on $(TMTSF)_2ClO_4$, has revealed that *there is* a minimum in resistivity plotted as a function of temperature at a given magnetic field (in the field range 12-27 Tesla). The corresponding temperature, $T_{min}(H)$, increases considerable with the field increase.The two materials, $(TMTSF)_2PF_6$ and $(TMTSF)_2ClO_4$, are known for similarity in many of their properties, and we will discuss whether the theoretical results above can help with better understanding (at least , on a qualitative level) of the experimental findings [5a,8] .

As for different temperature dependences of resistivity of $(TMTSF)_2PF_6$ above and below, say, 30K, note first that no minimum has been observed at zero field [10]. In accordance with eq.(16), this fact would impose some restrictions (from below) on the actual value of $t`_b$. Assuming again the similar behavior for both materials, $ClO_4$ and $PF_6$ (at least, in higher fields), we conclude that for the minimum, $T_{min}(H)$, to appear with the field increase, magnetic field itself must effectivelly diminish the effect of $t`_b$.Such a mechanism is well known [7a]. Taking the scale for the effect of the magnetic field on the "anti-nesting" term $t`_b$ from [7a], we will *assume* in eqs.(15,16):

$$t`_b \Rightarrow t`_b \cdot \frac{\lambda}{1+\lambda} \;\; ; \text{ where } \;\; \lambda = \frac{2ct`_b}{veHb} \equiv \frac{2t`_b}{\omega_c}$$

With the furhter increase of the cyclotron frequency, ω_c, $T_{min}(H)$ in large fields will tend to its value from eq.(14). If the interpretation suggested above, is correct, a plot of $\rho(T) \cdot T^{-2}$ *vs.* **T** would merely show the temperature dependence of the effective relaxation time.

CONCLUSIONS

The model of two open Fermi sheets possessing an approximate nesting property, seems to be sufficient in explaining essential deviations from familiar Fermi-liduid temperature dependencies in number of kinetic characteristics among conductors of the Bechgaard salts family. However, the analysis, as it is described above, so far remains on a qualitative level, and a more elaborated comparison with expimental data is, of course, highly desirable. Nevertheless, effects of the temperature and the magnetic field dependence for relaxation times and some other characteristics are based on a rather transparent physics, specific for a subtle mathemathical structure of the theory. In this connection it is worth of emphasizing once again that this model looks to be rather well justified from the point of view of high anisotropy of all properties in these conductors. At least for the sellenium based Bechgaard salts, the picture of weak bare interactions is consistent with a low energy scale typical for the most of phenomena .

Acknowledgements

The author is very pleased to see his paper published in the volume commemorating the scientific contributions of W.A.Little. The stimulating discussions of many years with D.Jerome are acknowledged with the gratitude.

This work was supported by NHMFL through NSF cooperative agreement #DMR-9016241 and the State of Florida.

REFERENCES

1. W.A.Little, *Phys.Rev.* 134A: 1416 (1964).
2. Yu.A.Bychkov, L.P.Gor`kov and I.E.Dzyaloshinskii, *Soviet Phys. JETP* 23:489 (1966).
3. I.E.Dzyaloshinskii and A.I.Larkin, *Soviet Phys. JETP* 34:422 (1972).
4. D.G.Clarke, S.P.Strong and P.W.Anderson, *Phys. Rev.Lett.* 72:3218 (1994).
5. a) D.Jerome and H.J.Schulz, *Adv.Phys.* 31:229 (1982); b) L.P.Gor`kov, *Soviet Phys.Usp.* 27:809 (1984); c) T.Ishiguro and K.Yamaji. *OrganicSuperconductors,* Springer-Verlag, Berlin (1990); d) D.Jerome, in *Organic Conductors* , J.P.Farges, ed., M.Dekker, New York (1994).
6. L.P.Gor`kov and I.E.Dzyaloshinskii, *Soviet Phys. JETP* 40:198 (1975).
7. a) L.P.Gor`kov and A.G.Lebed, *J. Phys.(Paris) Lett.* 45:L-443 (1984); b) P.M.Chaikin, *Phys. Rev.* 31:4770 (1985); c) M.Heritier, G.Montanbaux andP.Lederer, *J.Phys. (Paris) Lett.* 45:L-943 (1984).
8. L.P.Gor`kov, *Europhys. Lett.* 31:49 (1995).
9. K.Behnia et al., *Phys.Rev. Lett.* 74:5272 (1995).
10. L.Balicas and D.Jerome (unpublished).

BILL LITTLE AND HIGH TEMPERATURE SUPERCONDUCTIVITY

V.L. Ginzburg

P.N. Lebedev Physical Institute,
Russian Academy of Sciences,
117924GSP Moscow, Russia

For a full 65 years, the science of superconductivity was a part of physics of low temperatures, i.e., those of liquid helium (and in some cases liquid hydrogen). Thus, for example, the critical temperature of the first known superconductor, mercury, discovered in 1911, is equal to $T_c=4.1K$; that of lead, whose superconductivity was discovered in 1913, is $T_c=7.2K$. If I am not mistaken, higher values of T_c were not achieved until 1930, although it was, of course, understood that higher T_c's were very desirable. The next important step came in 1954 when the compound Nb_3Sn with $T_c=18.1K$ was synthesized. Despite extensive efforts, it was not until 1973 that the compound Nb_3Ge with $T_c=23.2-24K$ was synthesized. Subsequent attempts to raise T_c were unsuccessful up until 1986, which saw the first indications (soon confirmed) of superconductivity in the La-Be-Cu-O system with $T_c\sim35K$ [1]. Finally, in early 1987 a truly high temperature superconductor, $YBa_2Cu_3O_7$ with $T_c=80$-$90K$ was created [2].*

The discovery of high temperature superconductors (HTSC) created a sensation and gave rise to a real boom. One of the indicators of this boom is the number of publications. For example, in the period 1989-1991, about 15,000 papers devoted to HTSC appeared, that is, on the average, approximately 15 papers a day. For comparison, one of the reference books states that in the 60 years from 1911 to 1970, about 7,000 papers total were devoted to superconductivity. Another indicator is the scale of conferences devoted to HTSC. Thus, at

---

* This statement reflects my opinion that the term "high-temperature" is appropriate only for superconductors with $Tc > T_{b;N_2}=77.4K$, where $T_{b;N_2}$ is the boiling temperature of liquid nitrogen at atmospheric pressure.

From High Temperature Superconductivity to Microminiature Refrigeration
Edited by Cabrera *et al.*, Plenum Press, New York, 1996

the conference M^{2}HTSC III in Kanazawa (Japan, July 1991) there were approximately 1,500 presentations and the conference proceedings occupied four volumes with a total size of over 2700 pages (see below, reference [20]). Undoubtedly, such a scale of research is to a large degree explained by the high expectations for HTSC applications in technology. These expectations, by the way, from the very beginning have appeared to me to be somewhat exaggerated, as has been confirmed in practice. But, of course, the potential importance of HTSC for technology, medicine (MRI imaging), and physics itself leaves no doubts. Nevertheless, I still do not completely understand such a hyperactive reaction of the scientific community and of the general public to the discovery of HTSC; it is some sort of social phenomenon.

In any case, the problem of HTSC was born not in 1986, but at least 22 years earlier. This problem, in its current form, was first stated by Bill Little in 1964 [3,4]. First of all, Little posed the question (and to pose a question is frequently equivalent to doing half the work): Why was the critical temperature of the superconductors known at the time not so high? Secondly, he pointed out a possible way of raising T_c to the level of room temperatures or even higher. Little proposed to replace the electron-phonon interaction, responsible for superconductivity in the model of Bardeen, Cooper, and Schrieffer (BCS) [5] by the interaction of conduction electrons with bound electrons, in other words, with excitons. In terms of the well-known BCS formula for the critical temperature

$$T_c = \theta exp(-1/\lambda_{eff}) , \qquad (1)$$

the meaning of the exciton mechanism is that the region of attraction between conduction electrons q can be set to q~q_{ex}, where $k_B q_{ex}$ is the characteristic exciton energy (here k_B is the Boltzmann constant). On the other hand, for the electron-phonon mechanism of attraction in (1), q~q_D, where q_D is the Debye temperature of the metal. Since a situation in which $q_{ex} >> q_D$ is quite possible and even typical, then for the same value of the effective interaction parameter λ_{eff}, one gains a factor of q_{ex}/q_D in T_c. Little proposed to create an "excitonic superconductor" on the basis of organic compounds, specifically by designing a long conducting (metallic) organic molecule ("spine") surrounded by side polarizers - other organic molecules [3,4].

It is not appropriate to go into great detail here. Let me just point out that Little's work did not remain unnoticed. Just the opposite, it attracted a lot of attention and gave rise to a large number of other papers. In particular, I also followed up Little's work by suggesting a somewhat different version: roughly speaking, replacing the quasi-one-dimensional conducting thread in Little's model with a quasi-two-dimensional structure ("sandwich"), i.e., with a conducting thin film placed between two polarizers (dielectric plates) [6]. This model was quite naturally related to two-dimensional superconductivity which had been considered somewhat earlier [7]. In fact, I believe even now that regardless of the problem of HTSC, it would be of great interest to synthesize and study a two-dimensional superconductor on the surface of a bulk dielectic (this refers not to the deposition of a very thin superconducting film, but to superconductivity based on surface states) [7,8].

Since 1964 the problem of HTSC based on the excitonic mechanism has been discussed a number of times (see, e.g., [9,10]). Furthermore, in 1977 in Russia and 1982 in the English translation, there appeared a monograph [11] devoted to the problem of HTSC in its different versions. Particular attention was paid to my "favorites", layered compounds and "sandwiches" of the type dielectric-metal-dielectric. But even earlier than that, in 1969, Bill Little organized a special conference devoted to organic and high temperature superconductors [12]. Back then, in 1969, neither layered nor organic superconductors had been synthesized. As a result there were people who declared this conference essentially absurd. But soon thereafter, in 1971, layered superconductors appeared, and in 1980 organic superconductors were synthesized.

The ideas of high temperature superconductivity were subjected to particular ridicule. For example, B.T. Matthias declared the excitonic mechanism "mysterious" [13] and disproven [14]. It was even claimed that raising T_c by many tens of degrees was impossible in principle [13,14].

Multiple attempts to find a reliable and reproducible high temperature superconductor were, indeed, unsuccessful for a long time. As a result, after a certain upswing, the level of activity decayed measurably. Based on these circumstances, I characterized the situation in a popular article published in 1984 [15] as follows:

"It somehow happened that research in high-temperature superconductivity became unfashionable (there is good reason to speak of fashion in this context, since fashion sometimes plays a significant part in research work and in the scientific community). It is hard to achieve anything by making admonitions. Typically it is some obvious success (or reports of success, even if erroneous) that can radically and rapidly reverse attitudes. When they sense a 'rich strike' the former doubters, and even dedicated critics, are capable of turning coat and become ardent supporters of the new work. But this subject belongs to the psychology and sociology of science and technology, and I shall not dwell on it here. In short, the search for high-temperature superconductivity can readily lead to unexpected results and discoveries, especially since the prediction of the existing theory are rather vague".

The events of 1986 which followed the quoted article by just two years have, I believe, fully confirmed the above.

As I mentioned earlier, in the whirlwind of activity which followed the discovery of HTSC in 1986-7 earlier work in the field was typically completely ignored. For example, at the 18th International Conference on Low-Temperature Physics (LT-18, Kyoto, August 1987) not a single plenary talk (accompanied by a large number of references) mentioned anything about Little's work or any other work on HTSC besides that devoted to the oxides [16]. To be fair, not everyone reflected the history of HTSC in the manner of LT-18. For example, E.A.Edelsack, D.U. Gubser, and S.A.Wolf in their Introduction to the proceeding [17] of the first large international conference on HTSC (June 1987) gave a very objective history of the problem.

I think that the HTSC-psychosis is already behind us, but I dwelled on it so that the scientific community can draw some lessons for the future. However, as is well known, the only thing that history teaches us is that history teaches us nothing.

Here, it is important to address another, more important, side of the issue. Namely, is it appropriate to think that the discovery of oxide HTSC has confirmed the expectations of Little, of myself, and others relating to the excitonic mechanism of superconductivity? I think that, unfortunately, this question has to be answered in the negative. If the oxide superconductors were described by the weak-coupling BCS theory (which we can refer to as the BCS model), when the coupling parameter

$$\lambda_{eff} \ll 1 \qquad (2)$$

and, as a result, Eq.(1) were valid, then high values of T_c could arise only for $\theta \gg \theta_D$, as in the case of the excitonic mechanism. But in the BCS model, the ratio $2\Delta(0)/k_BT_c=3.52$ [$\Delta(0)$ is the energy gap at T=0] and a number of other relations and features follow. At the early stage of the work on oxide superconductors, an analysis carried out by Little indeed gave rise to a hope that these superconductors could be described by the BCS model. I thought that Little's arguments and suggestions [18] were quite convincing but of course they needed to be verified [18,19]. In the end, it was discovered that the BCS model, by and large, does not describe the oxide superconductors. As a result, the pendulum sharply swung in the opposite direction and the prevalent opinion became that in the known HTSC's the electron-phonon interaction does not play a significant role and the BCS theory is not applicable at all. More than that, a popular point of view arose that in the superconducting oxides, even in the normal state, the Fermi-liquid picture does not apply. P.W. Anderson even went so far [20] as to state that HTSC in cuprates is undoubtedly described by the so-called RVB (resonating valence bond) theory, which is radically different from BCS.

Meanwhile, it had been known for a long time that it is possible to raise T_C even within the framework of the BCS theory by increasing the force of attraction between the pairing electrons, as characterized by the parameter λ_{eff}. Indeed, while the idea behind the excitonic mechanism is, roughly speaking, to increase the pre-exponential factor θ in the BCS equation (1), the same goal can be achieved by increasing λ_{eff}. There was, however, some kind of prejudice against this route. This prejudice was rooted in the wide success of the BCS model in which the coupling is weak. Indeed, for "usual" superconductors in many cases the inequality (2) was known to be satisfied. Furthermore, there was an opinion (not without foundation) that in the case of strong coupling, when

$$\lambda_{eff} > 1 \qquad (3)$$

the lattice becomes unstable. In fact however, in a number of superconductors, for example in lead, the coupling is strong but the critical temperature is not large due to the smallness of the Debye temperature θ_D (for lead, $\theta_D=96K$ and $\lambda=1.55$, see, e.g., [11], p.171[1]). Thus the slogan of the search for HTSC immediately after the appearance of the BCS theory in 1957 could be formulated as follows: Look for superconductors with the strongest coupling and, at the same time, a high Debye temperature. I should note, in passing, that Bednorz and Müller, to a large extent, followed this very ideology in their research (see [16], p. 1781 and [21]).

The issue of lattice stability in the presence of strong coupling, was a very important one and, to a certain extent, remains so. In paper [22] it was stated that high temperature superconductivity was impossible in principle precisely because of the lattice stability requirement. The fact of the matter is that the parameter λ_{eff} can be written in the following form:

$$\lambda_{eff} = \lambda - \mu^* = \lambda - \mu[1+\mu ln\ (\theta_F/\theta)]^{-1}. \tag{4}$$

Here λ and μ are the dimensionless coupling constants for the phonon or exciton attraction and Coulomb repulsion, respectively and $k_B\theta_F$ is the Fermi energy. At the same time, in the simplest approximation (homogeneity and isotropy of the material, weak coupling):

$$\mu-\lambda=<4\pi e^2 N(0)/q^2\varepsilon(0,q)>, \tag{5}$$

where $\varepsilon(\omega,q)$ is the longitudinal dielectric function for frequency ω and wave number q and the factor $[q^2\varepsilon(0,q)]^{-1}$ understood as a certain average with respect to q; $N(0)$ is the density of states on the Fermi edge of the metal in the normal state. If, as assumed in [22], the stability condition has the form

$$\varepsilon(0,q) > 0, \tag{6}$$

then it follows from (5) that

$$\mu > \lambda. \tag{7}$$

It follows from this and from Eq.(4) that superconductivity (for which, of course, λ_{eff} >0) is possible at all only thanks to the deviation of $\mu*$ from μ . In addition, the magnitude of

---

[1]I ask the reader's forgiveness for the large number of references to my own papers and to the book [11]. This is due not to a desire to emphasize their importance, but to purely practical considerations. The amount of published literature has become so huge that I am starting to literally sink in it. It is easier to find certain items in your own compositions and to make use of them.

T_C is not high. However, even empirically it was already known that $\mu<0.5$ and sometimes $\lambda>1$ and therefore the inequality (7) is violated. In addition to this and some other arguments which were put forth already at the early stage [10], later on it was rigorously shown (see [11,23] and references therein) that the stability condition (6) is not valid and in fact should be written as follows:

$$1/\varepsilon(0,q) \leq 1, \tag{8}$$

i.e., is satisfied if one of the following two inequalities hold:

$$\varepsilon(0,q) \geq 1, \quad \varepsilon(0,q) < 0, \tag{9}$$

The second inequality makes it clear that the parameter λ may exceed $\mu*$. On the basis of this fact, our group came to the conclusion even before 1977 (i.e., prior to the Russian edition of the book [11]) that the general stability requirement does not limit T_C and it is quite possible, for instance, to have $T_C <300K$.

In the case of strong coupling (3) the BCS equation (1) is clearly inapplicable, and a number of other expressions for Tc have been proposed (see, e.g., [11,17,24,25] and references therein). The simplest of them is as follows:

$$T_C=\theta exp\{-(1+\lambda)/(\lambda-\mu*)]\tag{10}$$

Of course, in the weak-coupling limit (2), taking into account (4) or more precisely the condition $\lambda<<1$, Eq.(10) goes over into (1), as it should. If we set the parameter $\mu*=0.1$ in Eq.(10), then, for example, for $\lambda=3$, $T_C=0.25\theta$. As a result, even for $\theta=\theta_D=400K$, quite reasonable for a phonon mechanism, we already have $T_C=100K$. More precise formulas also lead to the conclusion that a phonon mechanism with strong coupling can easily lead to temperatures $T_C\approx100K$ and even $T_C\approx200K$. Furthermore, as pointed out above, strong coupling is quite realistic: indeed, λ is measured in tunneling experiments which for HTSC cuprates have yielded, for instance, $\lambda\approx2$ and $T_C\approx100-125K$ (see references cited in [25-27], see also [28]).

Thus it is possible to obtain the observed values of T_C in HTSC cuprates already within the framework of the phonon mechanism. However, the properties of these HTSC are characterized not just by the value of T_C, but by a range of other characteristics. In our institute, E.G. Maximov, O.V. Dolgov and their colleagues state that the phonon mechanism with strong coupling allows to explain the bulk of the experimental data on HTSC cuprates, both in the normal and in the superconducting state near T_C (see the references in talks and reviews [25-27]).

This includes the issue of infrared and microwave spectra [29,30]. However, the situation at low temperatures is not clear and the behavior of HTSC cuprates does not fall within the framework of the standard isotropic BCS theory with strong electron-phonon coupling. The final word in the arguments about the mechanism of superconductivity in the cuprates has not yet been spoken. I do not have the authority or the intent to play the judge in this issue. However, to the extent I can judge, the aforementioned opinion about the role of the phonon mechanism appears to me at the very least possible and well founded. Of course it is not so easy to distinguish the interaction of electrons with phonons from their interaction with other bosons, including excitons. Therefore, it cannot be excluded that excitons play a role. Bill Little does discuss this possibility in reference [31].

Since I have not attended international conferences on HTSC for almost three years now (since January 1993 when I gave the talk [25]), I do not have a sufficient impression of the opinions and the spirit which are prevalent at present (in October 1995) in the community of HTSC researchers. It appears that the almost universal rejection of the role of the electron-phonon mechanism in HTSC cuprates characteristic of the recent past [20] is already behind us. The current hot issue of the character (symmetry) of pairing in the cuprates is closely related to their strong anisotropy, but as far as I can tell, is not critical from the point of view of the mechanism of superconductivity itself [32-34].

Let us allow for the sake of argument that in the presently known HTSC's the excitonic mechanism plays no role. This is important and interesting, but does not in any way discredit the very possibility of an excitonic mechanism. As I remarked earlier, we are not aware of any general rules forbidding the existence of such mechanism. On the other hand, it is clearly not easy for the excitonic mechanism to reveal itself. This requires some sort of special conditions which are not yet sufficiently understood (see, in particular, [26]).

The highest presently known critical temperature (for $HgBa_2Ca_2Cu_3O_{8+d}$ under pressure) is 164K. This value is still realistic for a phonon mechanism but if higher temperatures $T_c > 200K$ are discovered, then the phonon mechanism will not likely be sufficient (for $\lambda = 2$ a critical temperature of 200K would require $\theta_D \cong 1000K$). However, for the excitonic mechanism T_c is not limited even to room temperatures. The search for HTSC with the highest possible critical temperatures are of course continuing. I still believe that the compounds most promising in this regard are layered structures and "sandwiches" of the type dielectric-metal-dielectric.* These structures would naturally be constructed by the technique of atomic layer-by-layer synthesis [35]. The role of the dielectric in such sandwiches could be played by organic compounds as well. In fact, the possibilities here are truly endless. This makes it even more sensible to call upon qualitative considerations for guidance (see, e.g., [11], p. 49).

Bill Little has made a number of other outstanding contributions in addition to the papers that I mentioned above. Let me point out, for example, the papers [36] and [37] also devoted to superconductivity. Undeniably, Little's name is well known in the scientific world and on the whole, his work has been given just credit. I believe that all members of the physics community should be aware of his remarkable contributions.

---

* In addition to intuitive considerations in favor of such quasi-two-dimensional structures (see [6,10,11,12]) there exist certain concrete arguments as well [26].

The world scientific community (in particular, that of physicists) is nowadays a democratic one. There exists, so to speak, a democratic republic of physicists. Unfortunately, however, democracy does not at all imply full brotherhood and justice. The shortcomings of democracy and that of other social systems (for instance, absolute monarchy) gave rise to communist and socialist ideas. However, all known attempts to realize these ideas have failed completely and/or turned into despotism, arbitrary rule, vileness, and blood. As someone who lived under "socialism" almost all of his already long life, I can personally testify to the above. Sad examples from the history of Soviet physics include that of L.V. Shubnikov (the actual discoverer of superconductivity of the second kind) who was arrested and executed, L.D. Landau who spent an entire year in jail and was saved only by a miracle, and many other remarkable scientists who were either executed or died in prison or in internal exile. I am not even mentioning such "small details" as the great difficulty (and frequently impossibility) of attending international scientific conferences or submitting one's papers to foreign journals for publication.

Thus one has to agree with Winston Churchill who stated that democracy is a very poor way of government but all other known ways are even worse.

By analogy with the above, we can say that the universal democratic republic of scientists is, on the whole, the best possible arrangement. Undoubtedly, much can be improved, but my impression is that this applies mostly to details.

I am now 79 years old and I do not have a high chance to witness a new large milestone in the physics of superconductivity. As an atheist, I cannot hope to find out more in the next world, either. But Bill Little is only 65 and has a long life ahead of him. I wish him to see the appearance of materials that superconduct at room temperature.

REFERENCES

1. Bednorz J.G. and Muller K.A. Z. Phys. B64, 189 (1986)

2. Wu M.K. et al. Phys.Rev.Lett., 58, 908 (1987)

3. Little W.A. Phys. Rev. 134, A1416 (1964)

4. Little W.A. Scientific American 212, No 2, 21 (1965)

5. Bardeen J., Cooper L.N. and Schrieffer J.R. Phys.Rev. 108, 1175 (1957)

6. Ginzburg V.L. Phys.Lett. 13, 101 (1964); Sov.Phys.-JETP 20, 1549 1964); Contemp.Phys. 9, 355 (1968)

7. Ginzburg V.L. and Kirzhnits D.A. Sov. Phys.-JETP 19, 269 (1964)

8. Ginzburg V.L. Physica Scripts, T27, 76 (1989)

9. Allender D., Bray J. and Bardeen J. Phys.Rev. B7, 1020, 4433 (1973)

10. Ginzburg V.L. Ann.Rev. Materials Sci. 2, 663 (1972)

11. High Temperature Superconductivity. Ed. Ginzburg V.L. and Kirzhnits D.A. Consultants Bureau, New York (1982)

12. Proc.Int.Conf. on Organic Superconductors. Ed. W.A. Little Interscience Publishers (1970)

13. Matthias B.T. Comments on solid state phys. 3, 93 (1970)

14. Matthias B.T. Physics Today, 24, No.8, 21 (1971)

15. Ginzburg V.L. Energia (in Russian) 9, 2 (1984)

16. Proc.18th Intern.Conf.on Low Temperature Physics, part 3. Japanese Journ.Appl.Phys. 26, Suppl.26-3 (1987)

17. E. Edelsack, D. Gubser, and S. Wolf in Novel Superconductivity. Ed. S.A. Wolf and V.Z. Kresin. Plenum Press, New York (1987), p.1

18. Little W.A. (a) Ref.17, p.341;b) Science 242, 1390 (1989)

19. Ginzburg V.L. Physics Today 42, No.3, 9 (1989)

20. P.W. Anderson, in Proc.Inter.Conf. on Materials and Mechanisms of Superconductivity. High Termperatue Superconductors III (conference M2HTSC III), Physica C, 185, 11(1991)

21. Bednorz J.G. and Muller K.A. Rev. Mod. Phys. 60, 585 (1988)

22. Cohen M.L. and Anderson P.W. Superconductivity in d and f band metals. AIP conference proceedings. Ed. D.H. Douglass. New York, AIP, p.17 (1972)

23. Ginzburg V.L. Contemp.Phys. 33, 15 (1992)

24. Ginzburg V.L., Progress in Low Temperature Physics, 12, 1 (1989)

25. Ginzburg V.L. Physica C, 209, 1 (1993)

26. Ginzburg V.L. and Maksimov E.G. Superconductivity: physics, chemsitry, technique, 5, 1505 (1992)

27. Maksimov E.G. J.of Supercond. 8, 433 (1995)

28. D. Shimada et al., Phys.Rev. B51, 16495 (1995)

29. Rieck C.T., Little W.A., Ruvalds J. and Virosztek A. Phys.Rev. B51, 3772 (1995)

30. Dolgov O.V., Maksimov E.G. and Shulga S.V Phys.Rev.B (in press)

31. Holcom M.J., Collman J.R. and Little W.A. Phys.Rev.Lett. 73, 2360 (1994)

32. Abrahams E. et al. Phys.Rev. B52, 1271 (1995)

33. Fehrenbacher R. and Norman M.R. Phys.Rev.Lett. 74, 3884 (1995)

34. O'Donovan C. and Carbotte J.R., Physica C, 252, 87 (1995)

35. Bozovic I. et al. Journ.Supercond. 7, 187 (1994)

36. Little, W.A., Phys.Rev. 156, 396 (1967)

37. Little, W.A. and Holcomb M., J.of.Supercond. 7, 175 (1994)

COMPUTATION BY SYMMETRY OPERATIONS IN A STRUCTURED NEURAL MODEL OF THE BRAIN: MUSIC AND ABSTRACT REASONING

Gordon L. Shaw

Department of Physics and Center for the Neurobiology of
Learning and Memory   University of California, Irvine, CA 92717

INTRODUCTION

Predictions [1,2] from our structured neural model [3-7] of the cortex led us to the hypothesis that music could causally enhance spatial-temporal reasoning. We have shown [8,9]: a) College students scored significantly higher on a spatial-temporal reasoning task after listening to a Mozart Sonata, but not after listening to silence or to minimalist music. b) Preschool children who received private keyboard lessons for 6 months improved dramatically on a spatial-temporal reasoning task while appropriate control groups did not improve significantly [10]. Enhancement a) lasted roughly 10 minutes and established the causal effect, while enhancement b) lasts long enough to have major educational implications. Here we review the model, in particular, the "built-in" ability of the cortex to recognize symmetry relations [7] among the inherent spatial-temporal firing patterns, which we suggest is a crucial feature of the cortical relationship between music and spatial-temporal reasoning. Then, we summarize the striking behavioral results [8-10], and make further predictions relevant to education.

Historical Note

It is an honor and pleasure to write this article in tribute to W.A. Little. I take this opportunity to acknowledge the crucial role that Bill Little played in the entire field of neural networks, and in my career

From High Temperature Superconductivity to Microminiature Refrigeration
Edited by Cabrera *et al.*, Plenum Press, New York, 1996

in brain theory. As an example of the former, it is important to emphasize and remember that the quite famous Hopfield model [11] is a modification of the Little model [12]. In 1973, I received a preprint from Bill which proved to be the **seminal** paper [12] in bringing the powerful techniques in statistical mechanics to neural networks and brain theory. Little introduced the explicit role of temperature T and interaction energy E in a Boltzmann probability distribution, proportional to exp(-E/kT), into the neuronal firing dynamics; the interaction energy depended on the firing states of the neurons with firing or not firing being analogous to up or down spin states. I had recently heard a lecture by J.G. Taylor about the statistical fluctuations in the release of neurochemicals at the synaptic junction between incoming axon and target neuron demonstrated by Katz and collaborators [13]. Vasudevan and I [14] were then able to show that the physiological basis for the temperature T and the Boltzmann probability equations in Little's model followed from these statistical fluctuations in synaptic transmission. I then had the wonderful opportunity to collaborate with Bill on two papers [15,16]. Thus began my gradual shift in research interest from elementary particle theory to brain theory. Over the past 20 years, Bill has provided continued insight, inspiration and encouragement to me in my quest to understand how we think and reason. The main theme of my present research and this paper is that music is a window into higher brain function, and that music training can be used at an early age to enhance children's ability to reason.

Symmetries

Symmetries have long been recognized as a vital component of physical and biological systems. It is apparent that as neuroanatomical and neurophysiological techniques have improved in the past decade, more and more structure has been found in the cortex. We expect this trend to continue. We proposed [1,7] that symmetry operations performed by the brain are a crucial feature of higher brain function and result from this spatial and temporal structure of the cortex. This modular structure with symmetry among the connections introduces symmetries among the "inherent" spatial-temporal firing patterns in the cortex. The symmetries of these inherent firing patterns can then be "exploited" to perform higher level computations or symmetry operations. Learning introduces small breaking of the symmetries in the connectivities which enables a symmetry in the patterns to be recognized in the Monte Carlo evolution of the patterns [7,17]. Using the trion model of the cortex, we presented specific, simple examples of this in the recognition of rotational invariance and in the recognition of a time reversed pattern (see Figs. 9 and 10 below). This then led to our main theme of pattern development, below.

Music and Abstract Reasoning

A profound dilemma of historical [18] origin is the similarity among such higher brain functions as music, mathematics and chess. There are many correlational [19] and anecdotal reports [20,21] of such relationships.

Leng and Shaw in their model of higher brain function [1] proposed a causal link between music and spatial-temporal reasoning The model was developed from the trion model [3-7], a highly structured mathematical realization of the Mountcastle [22] organization principle with the column as the basic neuronal network in the cortex. According to the model, newborns possess a structured cortex which yields an inherent repertoire of spatial-temporal firing patterns at the columnar level which can be excited and strengthened by small changes in connectivity via a Hebb [15-16,23] learning rule. These firing patterns evolve over time in a probabilistic manner from one to another in natural sequences related by specific symmetries, see Fig. 11 below. *These inherent patterns form the common neural language of the cortex.* The results were striking when evolutions of the trion model firing patterns were mapped onto various pitches and instruments producing recognizable styles of music [2,24]. (A cassette tape of trion music is available upon request.) This gave the insight for us [1-2] to relate the neuronal processes involved in music and abstract spatial-temporal reasoning.

The key component of spatial-temporal reasoning may be the "built-in" ability of the columnar networks to recognize the symmetry relations [7] among cortical firing patterns in a sequential manner. *We refer to this sequential process of a temporal sequence of pattern recognition processes as pattern development* [1]. These pattern development mental processes may last some tens of seconds to minutes as compared to the pattern recognition process, such as face recognition which might be accomplished in some fraction of a second. Music clearly involves this pattern development concept as does spatial-temporal reasoning (for an example, see Fig. 12 below) as well as the ability to create, maintain, transform and relate complex mental images even in the absence of external sensory input or feedback [1,21,25-26].

Although cognitive abilities such as music and spatial-temporal reasoning crucially depend on specific, localized regions of the cortex, all higher cognitive abilities draw upon a wide range of cortical areas [27]. Recent studies have demonstrated that sophisticated cognitive abilities are present in children as young as 5 months [28-29]. Similarly, musical abilities are evident in infants [30-31]. Music, then may serve as a "pre-language" [1] (with centers [32] distinct from language centers in the cortex), available at an early age, which can access inherent cortical spatial-temporal firing patterns and enhance the cortex's ability to accomplish pattern development.

These ideas of Xiao Leng and myself [1] then led to our behavioral experiments to test the prediction that music training at an early age, when the child's cortex is very plastic, would enhance the ability to use pattern development in spatial-temporal reasoning. It became clear to Fran Rauscher and myself that these experiments we started with pre-school children in September 1992 would take us years at considerable financial cost [10]. Thus we came up with the idea for the "Mozart effect" experiments [8-9] which could do done relatively quickly: college students scored significantly higher on spatial-temporal reasoning (see, Fig. 12) after listening to a Mozart Sonata (K.448), but not after listening to silence or to minimalist music. We chose Mozart since he was composing at the age of four. Thus we expect that Mozart was exploiting the inherent repertoire of spatial-temporal firing patterns in the cortex. These experiments were the first to demonstrate a causal link for music enhancing spatial-temporal reasoning. However, this enhancement lasted roughly 10 minutes. The results of our study with a substantial number of preschool children showed that the group which received private keyboard lessons for 6 months improved dramatically on a spatial-temporal reasoning task while appropriate control groups did not improve significantly [10]. Clearly these findings have major scientific implications. Although *much more work needs to be done*, as described in the concluding section, we already know that the enhancements found for the preschool children last at least a day. *Thus , even at this early stage of the research, the preschool results have enormous educational implications.*

TRION MODEL

Mountcastle [22] proposed that the well-established [33] cortical column is the basic network in the cortex and is comprised of small irreducible processing units called minicolumns. A very simple pinwheel representation of the minicolumns in visual cortex had been suggested. The optical recording results by Bonhoeffer and Grinvald [34] not only show a strong similarity to these representations but find both helicities in the representation of the orientation minicolumns. We display this in a very highly idealized, structured and generalized scheme in Fig. 1. The column has the capability of being excited into complex spatial-temporal firing patterns. The assumption is that higher mammalian processes involve the creation and transformation of such complex spatial-temporal firing patterns (in contrast to a "code" which involves sets of neurons firing with high frequency). Evidence is accumulating in support of the viability of this spatial-temporal code for the "internal language" of the cortex.

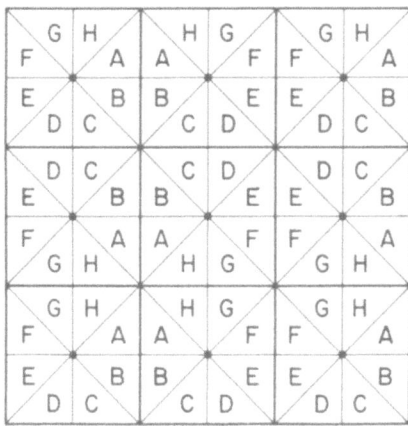

Figure 1. Highly schematic representation of the Mountcastle [22] principle of cortical organization. Each square represents a cortical column (horizontal dimension roughly 700 μ, comprising the six vertical layers of dimension roughly 2000 μ) while each triangle is a *distinguishable* minicolumn which encodes the relevant parameter in the stimuli such as line orientation in the visual cortex shown here by capitol letters. We note the optical recording results by Bonhoeffer and Grinvald [34] in secondary visual cortex show a strong similarity to the cartoon idealized cortex shown here.

The trion model of the cortical column [3-7] is a mathematical realization of Mountcastle's organizational principle· It was developed starting from Little's [12] neural network analogy to the Ising spin system, and modified in a direction inspired by the ANNNI (axial next nearest neighbor Ising) model results of Fisher and Selke [35]. A trion, Fig. 2, represents an idealized *distinguishable* minicolumn or roughly 100 neurons, and has three levels of firing activity, above average, average, and below average.

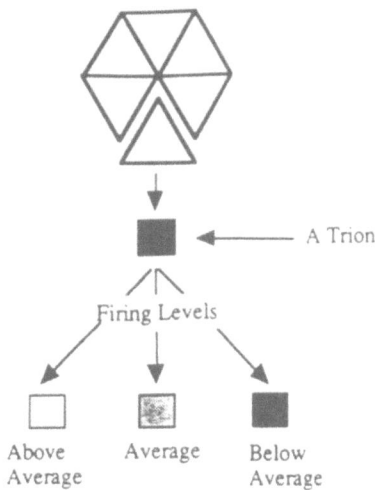

Figure 2. We identify a minicolumn with the idealized trion and the basic network of trions is the cortical column. As shown, the trion has three levels of firing activity.

A column with a small number of trions having structured connections yields a large repertoire of quasi-stable, periodic spatial-temporal firing patterns, defined as magic patterns or MPs which can be excited. These inherent patterns are called magic patterns or MPs because of their ability to be learned or enhanced via a Hebb learning rule to a large cycling probability. The repertoire of (periodic) MPs is found by evolving all possible initial states (of the first two time steps) by following the most probable or deterministic path. The symmetries relating the MPs in a repertoire are discussed below. In a full probabilistic (or Monte Carlo) evolution, the MPs evolve in natural sequences from one to another. The probability of each MP remaining in that pattern can be enhanced by even a small change in connection strengths using a Hebbian learning rule. We consider these symmetry operations in the dynamics of the cortical column to be the basic elements of higher brain function.

The interactions among the trions are taken to be localized, competing (between excitation and inhibition) and highly structured, and the firing state of the network (cortical column) of the distinguishable trions at time $n\tau$ is updated in a probabilistic way related to the states of the two previous discrete time steps $(n-1)\tau$ and $(n-2)\tau$ as in Fig. 3.

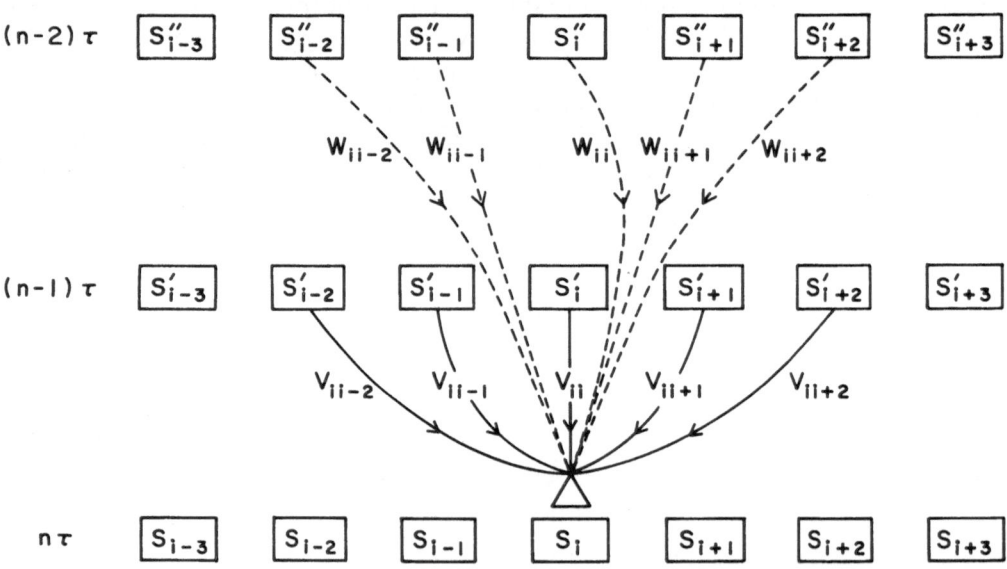

Figure 3. One network of trions at three time steps, showing the firing states S and the connections V and W, see Eq. (1). For N trions in the columnar network, we have the ring-like connections trion i = trion i + N as in Fig. 2.

We expect these time steps τ to be roughly 25-50ms. The probability $P(S_i)$ of the ith trion having a firing level or state S_i at time $n\tau$ is given by

$$P(S_i) = \frac{g(S_i)\exp[BM_iS_i]}{\sum\limits_{S} g(S)\exp[BM_iS]}$$

$$(1)$$

$$M_i = \sum_{j} [V_{ij}S'_j + W_{ij}S''_j] - V_i$$

where S'_j and S''_j are the states of jth trion at the two earlier times ((n-1)τ and (n-2)τ respectively. V_{ij} and W_{ij} are the interactions between trion i and j at time $n\tau$ from times (n-1)τ and (n-2)τ respectively; V_i is an effective firing threshold. The three possible firing states (of each trion) denoted by +, 0, - for S = 1, 0, -1 represent, respectively, a large "burst" of firing, an average burst, and a below average firing (see Fig. 2). The term g(S) with g(0) >> g(+/-) takes into account the number of equivalent firing configurations of the *distinguishable* trion's internal *indistinguishable* neuronal constituents [36]. (For example, in a trion representing a group of 90 neurons, firing levels of +, 0, - could correspond to 90-61, 60-31, 30-0 neurons firing, respectively. There are many more equivalent combinatorial ways of generating the 60-31 level from the *neurons*. This feature, g(0) >> g(+) = g(-), gives stability to the trion model firing patterns.) The fluctuation parameter B is inversely proportional to the noise and results [14] from the statistical nature of neurotransmitter release from the synapses [13]. Studies of the trion model for learning and memory and higher brain function have been reported. Basically, the success of these studies is due to fact that the localized, competing (between excitation and inhibition) with high symmetry yield a huge repertoire of inherent quasi-stable, periodic firing patterns, MPs, any of which can be readily learned or enhanced [3,6] with only small changes in the interaction strengths using the Hebb learning algorithm [14-15,23]:

$$\Delta V_{ij} = \varepsilon \sum_{n}^{pattern} S_i(n\tau)S_j((n-1)\tau)$$

$$(2)$$

$$\Delta W_{ij} = \varepsilon \sum_{n}^{pattern} S_i(n\tau)S_j((n-2)\tau)$$

$$\varepsilon > 0.$$

which allows for both increases and decreases of interaction strengths. (Simply extending this learning rule to a third time step using the correlation

$$\sum S_i(n\tau)S_j((n-3)\tau)$$

significantly enhances the effects of learning with a smaller change in the total connectivity [2,17].) Let us define the cycling probability $P_C(MP)$ that firing pattern for the columnar trion network remains in the MP for one cycle of the repeating MP. The $P_C(MP)$ is calculated by multiplying the probabilities $P(S_i)$, Eq. (1), of each trion i being in the state S at time $n\tau$, given by that MP for its whole cycle length:

$$P_C(MP) = \Pi_n \Pi_i P(S_i(n\tau)) \tag{3}$$

Then as a result of learning an MP using the Hebb [23] algorithm (2), the cycling probability $P_C(MP)$, (3) is increased. Further, after learning, many more initial states will go to the learned MP (and some related MPs). Note that these MPs evolve in natural sequences from one to another in a probabilistic Monte Carlo calculation.

It was shown in [5] that there exist a series of phase transitions at precise values B(n) giving new repertoires of firing patterns: Rewriting the statistical factors $g(S)/g(0) = \exp(-u^2S^2)$, then (1) becomes

$$P(S_i) \propto \exp(-u^2 S_i^2 + BS_i[\Sigma_j(W_{ij}S_j'' + V_{ij}S_j') - V_i]) \tag{4}$$

The cancellation of the u^2 and B terms in the exponential allow the $S_i = 0$ level to compete with the 1 and -1 firing levels. There then exist a series of "transition temperatures B^{-1}"

$$B(n) = u^2/n \tag{5}$$

for specific integers n related to the Vs and Ws. These B(n) separate regions with different logic and thus different repertoires of MPs. There are two especially interesting features near a "transition temperature" (5): i) The learning enhancement through the Hebb rule is striking as seen in Fig. 8, and ii) in a Monte Carlo evolution, the flow of the MPs changes in a manner illustrated in Fig. 11B which we propose is of behavioral importance.

The simulations of a trion columnar network are simply performed:
1) We specify the parameters of the trion network: the number of trions N, the degeneracy factors $g(S_i)$, the connectivities V_{ij} and W_{ij}, the firing thresholds V_i, and the fluctuation parameter B.
2) A choice for the firing states for the initial two times steps is made. Since each of the N trions in each time step has 3 possible firing levels S, there are 3^{2N} possible initial choices.
3) Given the firing states for each trion at the two earlier times $(n-1)\tau$ and $(n-2)\tau$, the probability $P(S_i)$ for the ith trion being in state S_i at time $n\tau$ is calculated from Eq. (1).

Having made the choice of parameters 1) above, the repertoire of MPs or inherent, quasi-stable, periodic firing patterns is found as follows: For a given initial firing state 2), follow the procedure of always choosing the S for each trion which has the largest probability (1) or most probable path, i.e., the largest exponent in (4) for P(S). Then the time evolution rapidly goes into a repeating spatial-temporal pattern or MP. Define the operator Γ which temporally evolves an MP according to its most probable path for its cycle length N_c. Then an MP is an eigenfunction of Γ with eigenvalue 1:

$$\Gamma MP = MP. \tag{6}$$

An explicit representation of Γ can be written down from (1) or (4). Going through all possible initial states 2) gives all the MPs (the repertoire of MPs) as well as the number of initial states recalling each MP and the average time to recall an MP. (See Fig. 4 below for an explicit example of a repertoire of MPs.) An MP has the property of being readily learned or enhanced using the Hebb learning rule in Eq. (2) with only a relatively small change in the connections V and W. After learning, more initial states will go to the learned MP (and some related MPs) and the cycling probability $P_C(MP)$, Eq. (3) will be increased. Furthermore, an arbitrary spatial-temporal pattern cannot be readily learned. Only an MP can be learned in a selective manner [6].

Symmetries of the MPs in a Repertoire

Consider a symmetry operator α acting on (6): $\alpha \Gamma MP = \alpha MP$. Then if α commutes with Γ, αMP is also an MP:

$$\alpha \Gamma MP = \alpha \Gamma \alpha^{-1} \alpha MP = \Gamma \alpha MP = \alpha MP \tag{7}$$

$$\text{for} \quad \alpha \Gamma \alpha^{-1} = \Gamma.$$

Thus, we expect for our structured connectivity in the trion model that there will be a number of symmetries α that will be useful to use to characterize a repertoire of MPs. To be explicit, let us examine a specific example of the repertoire of MPs for structured connections in an N = 6 network. Consider the connectivities and other parameters in Eq. (1) to be as follows:

$$V_{ii} = 2, \ V_{i,i+1} = V_{i,i-1} = 1, \ W_{ij} = -V_{ij}, \tag{8}$$

thresholds $V_i = 0$, all other V_{ij} equal to 0, $g(0)/g(+/-) = 500$ or $u^2 = 6.215$, and B above the first transition $B(1) = 6.215$ in (5). Then following the calculations in Section 2 above, each of the $3^{12} = 531,441$ possible choices for the initial states is followed to find the 155 MPs shown in Fig. 4.

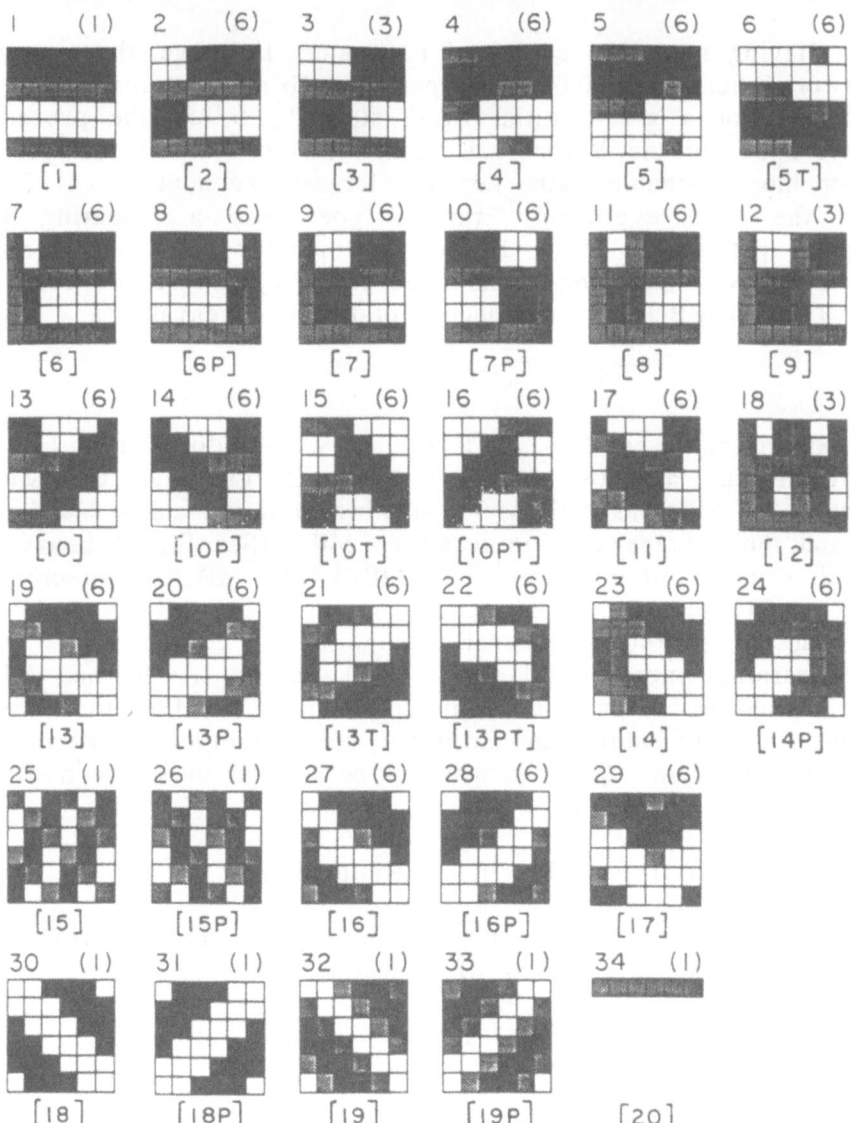

Figure 4. The initial repertoire [7] of MPs for a 6 trion network with the connectivity given by Eq. (8) are found by following the most probable path in evolving all possible 3^{12} initial states according to Eq. (1) until repeating patterns, the MPs, are obtained. Each square represents a trion with 3 levels of firing activity as in Fig. 2. Each horizontal row represents a ring of interconnected (as in Fig. 2) trions (so that the 6th square wraps around to the 1st) and time evolves downward. There are a total of 155 MPs which can be completely classified by their distinct spatial rotations into 34 groups of MPs shown here. The group number is listed on the top left corner while the number of MPs in each group is given on the top right by ( ) (cyclically rotate the MP so that the 1st column is the 2nd, etc.; if the MP is not a temporal rotation of any of the other elements in the group it is considered distinct). These 34 sets of MPs are further classified, below each MP, into 20 symmetry groups [ ] according to the additional symmetries of parity P (reflection of trion number about a line separating two trions), time reversal T and a combination PT. An explicit example is shown in Fig. 5. In a Monte Carlo calculation these MPs would flow from one to another.

These 155 MPs can be placed into 34 sets where the MPs in a set are related to each other by a spatial rotation among trions, R, among the (*distinguishable*) trions. It is evident from (8) and our ring boundary conditions that $\alpha = R$ commutes with Γ. Other symmetries among these MPs can be used to categorize groups of MPs. We see that a parity reflection, P, a time reversal operation, T, and the combination PT will relate different MPs as shown in Fig. 5. That T commutes with Γ is not obvious. In Fig. 4, the 34 sets of MPs are placed in 20 groups with MPs in each group related by the symmetry operations α equal to. P, T and PT (Fig. 5) in addition to rotation R.

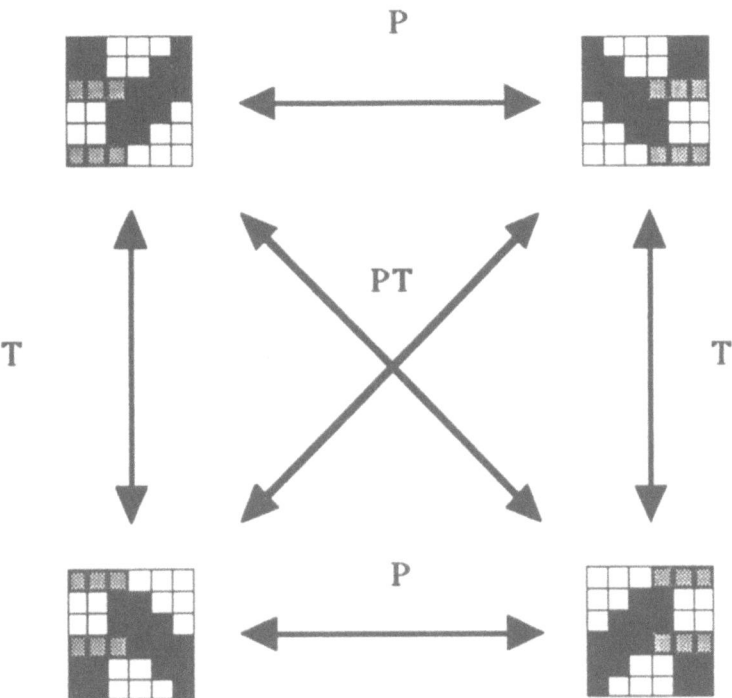

Figure 5. An explicit example [7] of the symmetry relations among the MPs in the symmetry group [10] in Fig. 4.

An additional symmetry operator changes firing level "spin" S to -S. In analogy with physical systems, let us define this as C, the "charge" conjugation operator. An example of a repertoire in which distinct MPs are related by C is seen in Table 5 of [4]

There are 1804 MPs in the repertoire given by the connections [3],

$$V_{i,i+1} = V_{i,i-1} = 1, \quad W_{i,i+2} = W_{i,i-2} = -1, \tag{9}$$

thresholds $V_i = 0$, all other V_{ij} equal to 0, $g(0)/g(+/-) = 500$, N = 6 trions and B > B(1) in Eq. (5). Using these symmetry operations R, P T and PT, in addition to an additional one, R_T rotating in space and time, present for these special set of connections and number of trions, these 1804 MPs can be placed in 73 symmetry groups. This repertoire proved to be especially interesting when mapped onto music [1-2,24] and onto robotic motion [37]. An example of the MPs in two of these symmetry groups is given in Fig. 6. In one of these groups, the MPs with respect to spatial (and temporal) rotations have been arranged to make the symmetry relationships among the MPs more transparent. We leave it as an exercise for the reader to see these relationships among the complex spatial-temporal patterns in Fig. 6. This helps illustrate the power of these networks that can readily recognize these relationships.

We suspect that there may be additional general symmetries to be discovered, especially when several columnar networks are coupled together. We suggest that these groups or categories of MPs defined by symmetry operations are not only useful in understanding aspects of pattern recognition such as rotational invariance, but will prove

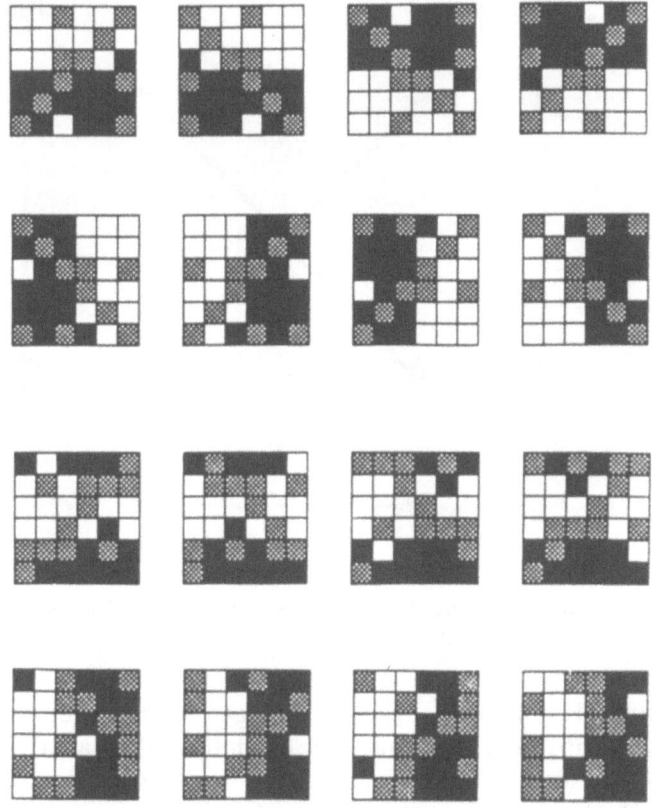

Figure 6. Example [7] of the MPs in two symmetry groups in the repertoire (see Figs. 4 and 5) for the connectivity (9). Each symmetry group consists of 8 MPs related by combinations of P, T and a symmetry operation specific to this repertoire corresponding to a spatial-temporal rotation of 90° about the center of the MP. In the first symmetry group, the MPs with respect to spatial (and temporal) rotations have been arranged to make the symmetry relationships among the MPs more transparent. We leave it as an exercise for the reader to see these relationships among these MPs. This exercise will help illustrate the power of these trion networks that can readily recognize these symmetry relationships.

invaluable in understanding the nature of the *sequences* of transitions of the MPs among themselves. A relevant example of an MP evolving into other MPs in Monte Carlo evolutions is shown in Fig. 7. This then forms the basis of inherent sequences of MPs.

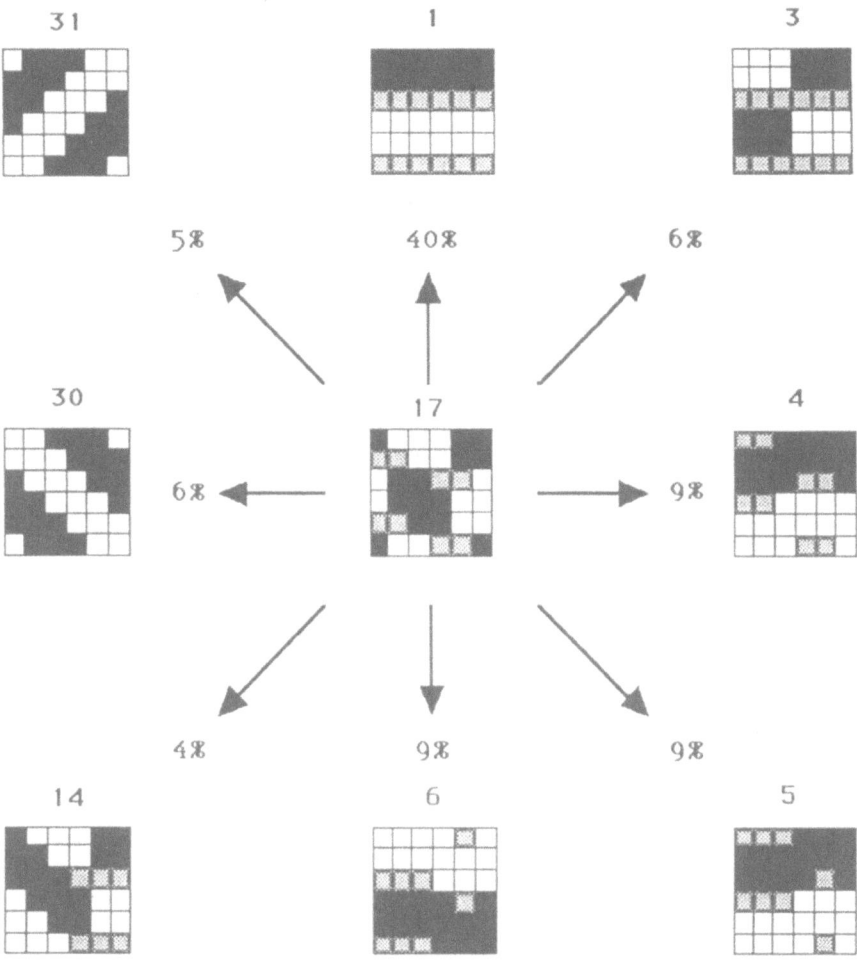

Figure 7. Monte Carlo calculations [7] for MP 17(0). The numbers at the end of the arrows give the percentage of 1000 Monte Carlo calculations that go to the eight other MPs (including all spatial rotations). Shown here are those MPs that are accessed with percentages greater than 3%. These relations can be substantially modified through learning.

We have called the above symmetries "global" in our columnar MPs, in contrast to "local" ones in which the temporal patterns for two specific trions might be interchanged (with a possible phase shift in time). For example, for the repertoire in Fig. 4, MP 2 is related to MP 1 by shifting the patterns for trions 1 and 2 by 3 time steps (or applying C to just these two trions). We note then that all the MPs in Fig. 4 consist "simply" of combinations of just three temporal patterns for individual trions:

299

$$a = (+, +, 0, -, -, 0),$$
$$b = (+, +, +, -, -, -), \tag{10}$$
$$c = (0, 0, 0, 0, 0, 0).$$

It would be of strong interest to analytically determine the repertoire starting from this "alphabet" (10). (Note that this alphabet (10) holds for the repertoire from the connectivity (8) with any number of trions > 3.) We see from Table 6 of [6] that for connectivity $V_{ii} = 1$, $V_{ii+1} = V_{ii-1} = 1$, $W_{ij} = - V_{ij}$, the repertoire has 246 MPs and the alphabet consists of (11) plus $d = (-, +, -, +, -, +)$, $f = (+, -, 0, -, +, 0)$, $g = (+, 0, +, -, 0, -)$ and $h = (+, 0, 0, -, 0, 0)$. We are thinking of these trion temporal firing patterns (dependent on the connectivity) as letters, the columnar MPs as words. The extension to a higher level architecture has been discussed [1-2,7].

Learning an MP

Consider now learning an MP from Fig. 4 using the Hebb learning algorithm, Eq. (2). The example of learning MP 6(2) {where the (2) denotes the MP obtained from the MP 6 specifically shown in Fig. 4, by operating twice with spatial rotation operator R} is shown in Fig. 8 where we plot the probability of cycling P_C (as defined in Eq. 3) versus ε.
The B value was 6.3 for this striking example of learning is particularly enhanced near this "transition" B(1), see Eq. (5). Note as a result of learning this MP 6(2), the connections Eq. (8) will be modified according to Eq. (2). For example, here, $V_{5,6} = 1$ and $V_{5,4} = 1 + 2\varepsilon$, so

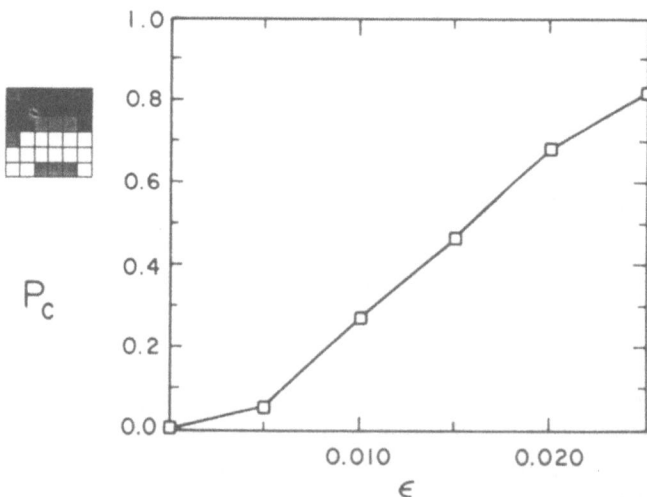

Figure 8. The cycling probability, Eq. (3), P_C to remain in MP 6(2) (where (2) denotes a rotation of two trion sites from the MP 6 pictured in Fig. 4) versus the learning coefficient ε, Eq. (2) for B = 6.3. This striking example [7] of learning is particularly enhanced near the "transition" B(1), Eq. (5).

that we say that the precise symmetry, $V_{i,i+1} = V_{i,i-1} = 1$, is slightly broken by learning for small ε. This small symmetry breaking has been shown [6] to form the basis for a temporally rapid selectional [38] learning in contrast to a much slower fine tuning of the parameters necessary for instructional learning [39]. (Both are probably necessary to understand behavior) Below, we show that this symmetry breaking in the Hebb learning forms the basis for the recognition of the symmetries among the MPs present in a repertoire.

Recognition of Rotational Invariance and Time Reversal

Here we discuss the recognition of spatially rotated and time reversed objects in the trion model. This recognition of spatial rotational invariance is built into the highly structured trion model due to its natural symmetry relations. Consider a simplified example in which a specific visual object, VO, is represented in the cortex by one of the MPs. Rotated VOs are represented by rotations of the MP of the standing VO. The VO seen in a normal standing position is learned with the Hebb rule and the symmetry among the connections is broken by a small amount. When a rotated VO is seen, the rotated VO MP evolves in a Monte Carlo calculation into the MP for the standing VO, thereby identifying it as a VO; the number of time steps to evolve is linearly related to the amount of rotation, in agreement with experiment [40]. A similar scenario is considered for recognition of time reversed MPs. Here we do not give an explicit physical representation for the abstract MPs, although the mappings onto music [2,24] and robotic motion [37] are relevant.

Rotational invariance is shown in the following example. Initially, when the subject sees an object, an MP will be selected out, e.g., MP 4(0) where the ( ) indicate the rotation of MP 4 in Fig. 4, so that 4(0) is the unrotated MP shown in Fig. 4. Before learning (B = 7.0) this MP will then evolve into other MPs.

If the network is presented with a rotated object then a rotated MP will be selected out. Since the network has 6 trions each rotated MP corresponds to an angle of 30 degrees, with MP 4(3) being a rotation of 90 degrees. When the various rotated MPs were run in Monte Carlo calculations they evolved to MP 4(0) with a percentage inversely dependent on the rotation as shown in Fig. 9.

In addition to being able to identify rotations the network can also identify a time-reversal operation. Besides the rotational symmetries for group 6, the time reversal T yield distinct MPs forming their own rotational group 5, so that MP 6(2) is transformed into MP 5(2). The rotation group for MP 5 and MP 6 are related through a time-reversal operation T as shown in Fig. 4. The Monte Carlo evolution results identifying both rotations and time reversal after learning MP 6(2) with an ε = .025 at B = 7.0 are shown in Fig. 10

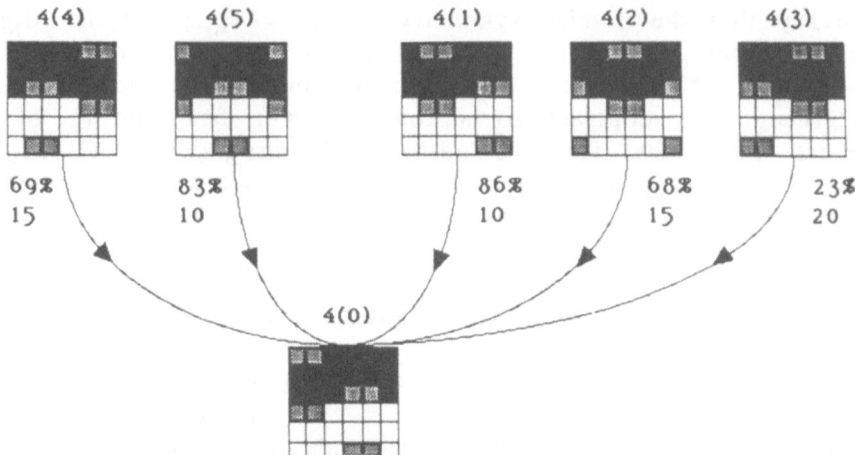

Figure 9. Example [7] of recognition of rotation invariance. Monte Carlo (1000 runs for 50 time steps) results from starting in each of the five rotated MPs (and then searching for MP 4(0) within the 50 time steps) after learning MP 4(0) with an $\varepsilon = 0.025$ at $B = 7.0$. Both the percentage of runs evolving to MP 4(0) and the average number of time steps taken $<t>$ are shown.

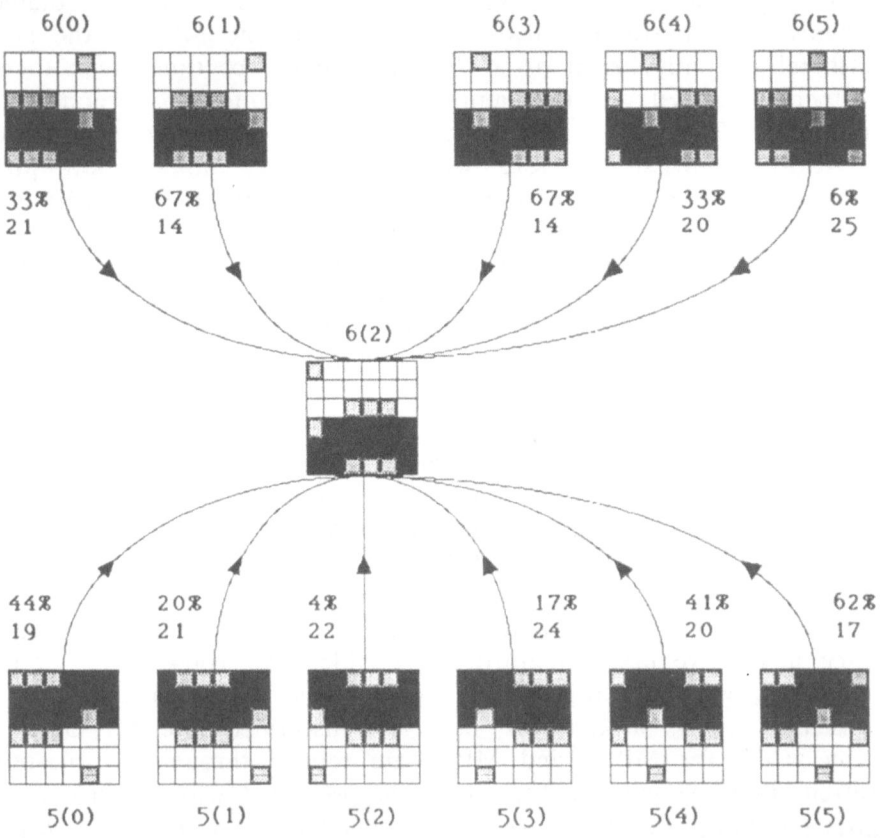

Figure 10. Example [7] of both rotation and time-reversal recognition in Monte Carlo calculations (see Fig. 9) after learning MP 6(2) . MPs 5 are related to MPs 6 by T.

302

We suggest that these "built-in" recognitions of MPs related by symmetries constitute the basic operations in the cortex for higher brain function.

Analytic Versus Creative Modes of Monte Carlo Evolutions

We [1] proposed a *neural* definition or description of analytic and creative modes. There are two types of Monte Carlo evolutions, Fig. 11, which we will identify as the *analytic mode* and the *creative mode*. The analytic mode corresponds to a sequential evolution of the MPs, see Fig. 11A, and has been used in our studies of memory recall and pattern recognition. With only somewhat different parameters than used for studying memory and pattern recognition, much more flowing and intriguing patterns emerged, see Fig. 11B, which we identify as the creative mode.

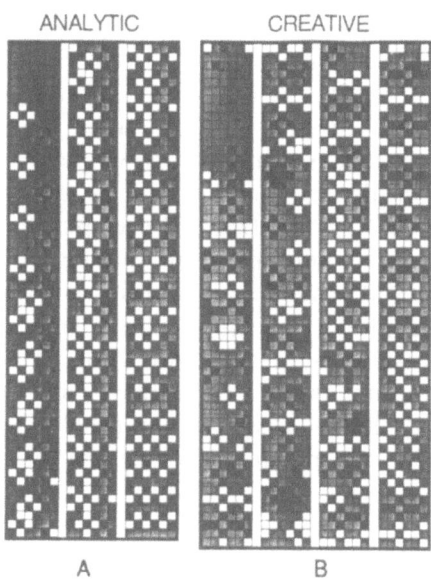

A   B

Figure 11. Probabilistic evolutions from the trion model of the firing activity of a columnar network of six trions giving examples of (A) the "analytic" mode and (B) the "creative" mode. Each square in a given row represents the firing levels of a trion at that time step. White, gray and black represent the firing levels above average, average and below average. The time steps (rows) are consecutively ordered, beginning at the top of the column at the left and continuing from the bottom of that column to the top of the next column, etc. (A). There are seven different inherent firing patterns which cycle two or more times. This is an example of a sequential or analytic evolution of the inherent firing patterns and has high symmetry in the connections. (B). This is an example of a creative evolution, in that specific inherent firing patterns appear and reappear in forms related by symmetry operations. This evolution has connections similar to that of (A) but with somewhat "broken" symmetry and a synaptic fluctuation parameter which is near the 1st "transition temperature" in (5), and presumably modifiable by the release of specific neuromodulators.

The inherent spatial-temporal firing patterns in the cortex are related by specific *symmetries*. For example, the "diamond" pattern starting at time step 8 in Fig. 11A can be seen in several spatially rotated forms in Fig. 11B. We suggest that computation by symmetry operations among the inherent patterns is a key property of higher brain function.

Consider now the analytic and creative modes in our Monte Carlo evolutions as in Fig. 11. In Fig. 11A, there is high symmetry in the connections and the B value is far from the phase transition B(1) in Eq. (5) leading to a Monte Carlo in which the MPs evolve in some natural sequence in what we define as the analytic mood. There will be a number of such sequences for each MP with different probabilities. Further, there are specific relationships (or symmetries) among these MPs which we are now studying. In Fig. 11B, there is broken symmetry in the connections and a fluctuation B value near the transition B(1) leading to a Monte Carlo in which MPs from a related set appear and reappear in what we define as the creative mode. It is as if the MPs evolve in related but novel ways in the creative mode and in related but more prescribed ways in the analytic mode. Thus we suggest that these columnar trion network descriptions are viable for the neural bases of analytic and creative thought.

We conclude this discussion with the idea that when the fluctuation parameter B is near a phase transition value in Eq. (5), the Monte Carlo evolutions are in the creative mode as compared to the analytic mode when the B value is far from these phase transitions. Certain neuromodulators [1-2,41] perhaps can induce this change in B and thus cause this important change in columnar network behavior back and forth from the analytic to the creative mode. Both modes are crucial to and complementary in a discussion of higher brain function in that it is usually necessary to analytically examine a creative solution for validity of the details. Behavioral tests of these ideas have been suggested [1-2] using music as a causal enhancement of spatial-temporal reasoning.

Trion Music

We proposed that the trion model patterns represent a candidate for the common internal language of the brain, that is, how one part of the brain communicates with the other parts. We then mapped [2,24] the Monte Carlo evolutions of the model, or translated this internal language of the brain, into music. *The results were striking when the Monte Carlo evolutions of the type in Fig. 11B of the creative mode were mapped onto pitches and instruments to produce music*

There are many, many possible choices of mappings! We presented in [24] three examples to illustrate the mapping procedures. As we demonstrated, they generate good approximations to different recognizable (western and eastern) human styles of music. (A cassette tape is available upon request.) It should be emphasized that pitches and instruments were the only two elements which were varied from one time step to the next (all other parameters, the length of the discrete time step and the volume are constant throughout each piece of music presented).

We then made the following generalizations [1-2]: Music composition, performance and listening all involve the evolution of this inherent repertoire of spatial-temporal patterns (MPs in the trion model). It is the creative genius of the composer such as Mozart or the brilliance of the conductor such as Szell which produces a magnificent performance that involves higher brain function at its highest level. *We suggested what is involved here is the perfect use of these inherent spatial-temporal patterns common to our species. As listeners, we need only appreciate the result of having these inherent patterns excited in our brains.* We can all appreciate music even though very few of us can compose it. As we have seen [24] different mappings of the same Monte Carlo gave different recognizable human styles of music (Western and Eastern).

The assumption that this inherent repertoire of patterns is essentially present at birth is perhaps a necessary condition to understand the appreciation of music (and particular pieces of music) by infants [14]. In addition, there are inherent structures in the brain which are devoted to music just as there is such a structure for language. Further, this structure is accessible for use almost from birth without substantial learning. In this sense, we might consider music as a sort of "pre-language."

Again the feature that millions of people over centuries will be captivated by certain composers and specific pieces of music speaks to the common universality of the repertoires of inherent spatial-temporal patterns.

These discussions gave the insight for us [1-2] to relate the neuronal processes involved in music and abstract spatial-temporal reasoning, and thus to propose that music can enhance spatial-temporal reasoning.

MUSIC ENHANCES SPATIAL-TEMPORAL REASONING

As discussed previously, Leng and I had proposed [1-2] that music training at an early age could enhance spatial-temporal reasoning in children. In particular, the neuronal processes of pattern development lasting some tens of seconds would employ in sequential manner the built-in basic ability of the cortex to recognize MPs related by symmetries. Music clearly involves this pattern development concept as does spatial-temporal reasoning Although cognitive abilities such as music and spatial-temporal reasoning crucially depend on specific, localized regions of the cortex, all higher cognitive abilities draw upon a wide range of cortical areas [27]. Musical abilities are evident in infants [30-31]. Music, then may serve as a "pre-language" [1] (with centers [32] distinct from language centers in the cortex), available at an early age, which can access inherent cortical spatial-temporal firing patterns and enhance the cortex's ability to accomplish pattern development.

In addition to this potential of long-term enhancement of spatial-temporal reasoning in young children through music training, Rauscher and I thought that even listening to certain types of music might prime the neuronal pathways for pattern development, specifically, spatial-temporal reasoning, leading to the "Mozart effect" short-term enhancement experiments [8-9].

Mozart Effect

We performed a experiment to determine if short-term causal enhancements of spatial-temporal reasoning could be invoked by merely listening to music [7].   Thirty-six UCI undergraduates listened to (the first) 10 minutes of Mozart's Sonata for Two Pianos, K. 448, and scored 8 to 9 points higher on the spatial IQ subtest of the Stanford-Binet Intelligence Scale than after they listened to taped relaxation instructions or silence. *This facilitation lasted only ten to fifteen minutes.*

In a follow-up study [8], seventy-nine UCI students participated for five consecutive days. We issued all students 16 Paper Folding and Cutting (PF&C) items on the first day of the experiment (Fig. 12), and then divided them into three groups with equivalent abilities (giving them16 different PF&C items on each following day).   Of the three sub-tests in the spatial reasoning portion of Stanford-Binet's Intelligence Scale, we chose the PF&C task because it best fit our concept of spatial-temporal pattern development [1].   On subsequent days, we showed that listening to the same Mozart sonata K.448 enhanced performance on the PF&C tasks over listening to i) silence, ii) a minimalist work by Philip Glass, iii) an audio-taped story; and iv) a dance (trance) piece.   Also, we showed that listening to the Mozart sonata did not enhance short-term memory.

These experiments established the causal enhancement of spatial-temporal reasoning by listening to specific music.

Our proposed mechanisms for this short-term enhancement of spatial-temporal reasoning by listening to music include the following: i) Listening to music helps "organize" the cortical firing patterns so that they do not wash out for other pattern development functions, in particular, the right hemisphere processes of spatial-temporal task performance. ii) Music acts as an "exercise" for exciting and priming the common repertoire and sequential flow of the cortical firing patterns responsible for higher brain functions. (We note that the potential mechanism ii) would imply that care in experimental design must be taken by researchers who want to extend our findings since the task can prime itself.)   iii) The cortical symmetry operations among the inherent patterns, are enhanced and facilitated by music.

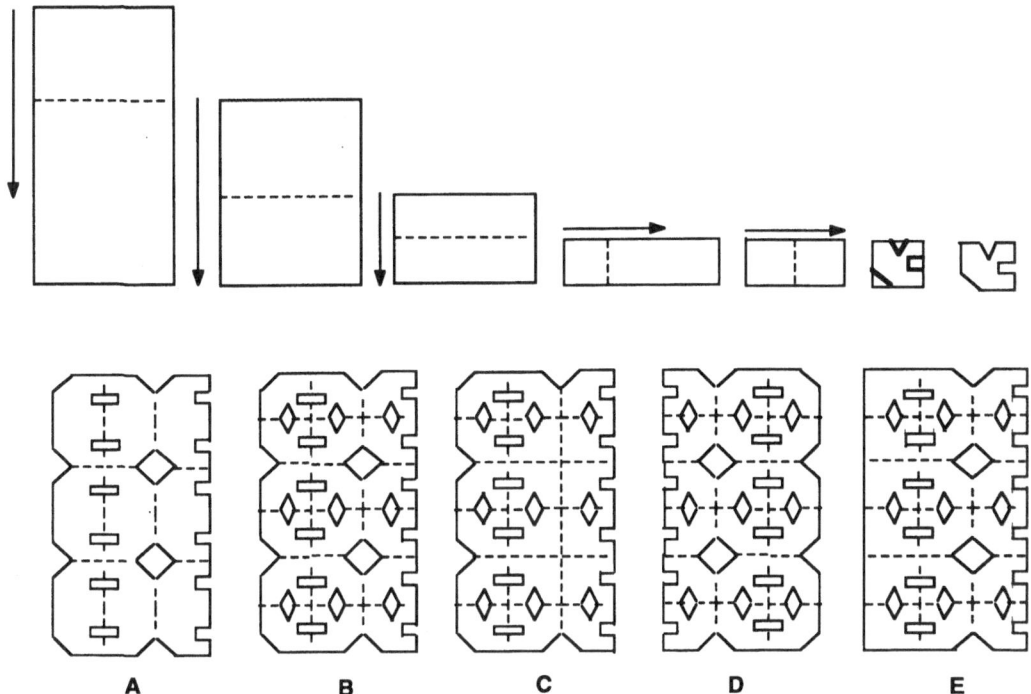

Figure 12. Example of the PF&C task used in our behavioral experiment. It depicts a picture of a paper before it was folded and cut (top left figure). The dotted lines and straight arrows represent the location and direction of folds. The solid lines represent cuts. Subjects were to choose which of the five choices below show how the paper would look unfolded.

We chose Mozart since he was composing at the age of four. Thus we expect that Mozart was exploiting the inherent repertoire of spatial-temporal firing patterns in the cortex. While one might explore the many possibilities with a large number of styles and composers, it would be perhaps more interesting to investigate the underlying neurophysiological bases using behavioral studies in conjunction with EEG studies. We have begun such a collaborative study at the University of Vienna Neurophysiological Laboratory, and the first results look very promising [42].

Music Training Enhances Reasoning in Preschool Children

Predictions from our structured neuronal model of the brain [1-2] led us to test the hypothesis that music training for young children enhances spatial-temporal reasoning. A substantial number of preschool children participated in the just completed study [10]: The Keyboard group of children were given private keyboard music lessons for 6 months, and there were three control groups of children. Four standard age-calibrated spatial reasoning tests were given at the beginning and at the end of the study; 1 test assessed spatial-temporal reasoning and 3 tests assessed spatial-recognition reasoning. A *highly significant improvement of large magnitude* was found for the Keyboard group in the spatial-temporal reasoning test (Object Assembly in which the child arranges pieces of a puzzle to create, e.g., a familiar animal by forming a mental image of the animal and rotating pieces to match the mental image; performance is helped by putting the pieces together in particular orders, defining the spatial-temporal nature of the task). No significant improvement was found on tests of spatial-recognition reasoning (such as matching, classifing and recognizing similarities among objects). The control groups did not improve significantly on any of the tests. We suggest that spatial-temporal reasoning is crucial for such adult endeavors as higher mathematics, engineering and chess. This experiment was not able to determine the temporal duration of the enhancement of spatial-temporal reasoning, but it was at least one day. *Thus there are enormous educational implications of these results.*

Discussion

Predictions [1,2] from our structured neural model [3-7] of the cortex led us to the hypothesis that music could causally enhance spatial-temporal reasoning. We have shown [8,9] that college students scored significantly higher on a spatial-temporal reasoning task after listening to the Mozart Sonata K.448, as compared to silence and to other listening conditions. This enhancement lasted roughly 10 minutes. Thus although this "Mozart effect" established the causal effect and has major scientific consequences, it does not have educational implications. We have shown [10] that preschool children who received private keyboard lessons for 6 months improved dramatically on a spatial-temporal reasoning task while appropriate control groups did not improve significantly. This enhancement lasts at least one day which is long enough to have major educational implications. Since this Volume is principally concerned with high temperature superconductivity, let me make the analogy with the reality that although the goal is to find a room temperature HTC, the present achievement of liquid nitrogen temperatures has enormous practical consequences. The Mozart effect establishes the causal short-term enhancement of spatial-temporal reasoning; the preschool music training study establishes an enhancement of at least one day (an increase by more than a factor of 100) making this of immediate educational use; clearly the goal would be to achieve a permanent enhancement in spatial-temporal reasoning.

Also it is unlikely that we have found the only type of enhancement, i.e., analytic spatial-temporal reasoning. Using the arguments above concerning analytic versus creative trion model evolutions in Fig. 11, we have proposed [9] how one might look for the enhancement of creative reasoning in chess tasks.

Finally, following our proposal that structured pattern development of neuronal firing in the human brain underlies abilities to perform higher brain functions such as mathematics and music, we inquire which are the key performance skills in reasoning and creativity and how can they be enhanced. Perhaps the standard tests of IQ do not adequately characterize the basic skills necessary to become a research mathematician, or even to properly understand a scientific curriculum [21]. For example, it has been extensively documented that *many* students have difficulty with the concepts of ratio and proportion: proportional reasoning [43]. Further, no key insight has been developed to teach proportional reasoning. Our brain theory model and our behavioral results on music training enhancing spatial-temporal reasoning in pre-school children have led us to design a program to exploit the general ability of children to search for patterns and relationships in space and time in order to overcome this difficulty in teaching proportional reasoning. We are just starting to implement this program at the pre-school level.

Acknowledgments

This paper, as noted in the references cited, represents the work of a large number of colleagues and students over many years. However, I want to specifically acknowledge the absolutely vital role of two researchers on the main theme of this paper: the influence of music on how we think and reason: Xiao Leng on the cortical theory and Fran Rauscher on the behavioral experiments.

I also would like to single out and thank several colleagues and friends who have encouraged me over the past years to pursue the *seemingly* soft science of music as a window into higher brain function: Bill Little, Frieda Belinfante, Leslie Brothers, Karl Bruhn, George Patera, and Fred Reines.

This research was supported by grants from the National Association of Music Merchants, the Ralph and Leona Gerard Foundation, The Seaver Institute, the Orange County Philharmonic Society, the National Academy of Recording Arts and Sciences, Walter Cruttenden and Associates, and the National Piano Foundation.

REFERENCES

1. X. Leng, and G.L. Shaw, Toward a neural theory of higher brain function using music as a window, *Concepts Neurosci.* 2:229 (1991).
2. X. Leng, Investigation of higher brain functions in music composition using models of the cortex based on physical system analogies. Ph.D. Thesis, University of California, Irvine (1990)
3. G.L. Shaw, D.J. Silverman, and J.C. Pearson, Model of cortical organization embodying a basis for a theory of information processing and memory recall, *Proc. Natl. Acad. Science, USA,* 82: 2364(1985).
4. D.J. Silverman, G.L. Shaw, and J.C. Pearson, Associative recall properties of the trion model of cortical organization, *Biol. Cybern.* 53:259(1986).
5. J.V. McGrann, G.L. Shaw, D.J. Silverman, D.J., and J.C. Pearson, Higher temperature phases of a structured model of cortical organization, *Physical Review* A43:5678 (1991).

6. K.V. Shenoy, J. Kaufman, J.V. McGrann and Shaw, G.L., Learning by selection in the trion model of cortical organization *Cerebral Cortex* 3:239 (1993).
7. J.V. McGrann, G.L. Shaw, K.V. Shenoy, X. Leng, and R.B. Mathews, Computation by symmetry operations in a structured model of the brain, *Physical Review* E49:5830 (1994).
8. F.H. Rauscher, G.L. Shaw, and K.N. Ky, Music and spatial task performance, *Nature*, 365:611 (1993).
9. F.H. Rauscher, G.L. Shaw, and K.N. Ky, Listening to Mozart enhances spatial-temporal reasoning:towards a neurophysiological basis, *Neuroscience Letters* 185:44 (1995).
10. F.H. Rauscher, G.L. Shaw, L.J. Levine, E.L. Wright, W.R. Dennis and R.L. Newcomb, Evidence that music training yields long-term enhancement of preschool children's reasoning abilities, submitted for publication.
11. J.J. Hopfield, Neural networks and physical systems with emergent collective computational behavior, *Proc. Natl. Acad. Science, USA,* 79: 2554(1982).
12. W.A. Little, Existence of persistent states in the brain, *Math. Biosci.* 19:101 (1974).
13. B. Katz, *The Release of Neural Transmitter Substances.*, Thomas, Springfield (1969).
14. G.L. Shaw and R. Vasudevan, Persistent states of neural networks and the random nature of synaptic transmission, *Math. Biosci.* 21:207 (1974).
15. W.A. Little and G.L. Shaw, A statistical theory of short and long-term memory *Behav. Biol.* 14:115(1975).
16. W.A. Little and G.L. Shaw, Analytic study of the storage capacity of a neural network, *Math. Biosc.* 39:281 (1978).
17. J.V. McGrann, Further theoretical investigations of the trion model of cortical organization. Ph.D. Thesis, University of California, Irvine (1992).
18. G.J. Allman, *Greek Geometry from Thales to Euclid*" Arno, New York1976).
19. M. Hassler, N. Birbaumer and A. Feil, Musical talent and visual-spatial abilities: a longitudinal study, *Psychology of Music* 13:99(1985).
20. L.D. Cranberg and M.L. Albert, The chess mind, In *The exceptional brain* , L.K. Obler and D. Fein, eds., Guilford, New York (1988).
21. W.S. Boettcher, S.S. Hahn, and G.L. Shaw, Mathematics and music: a search for insight into higher brain function, *Leonardo Music J.* 4:53 (1994).
22. V.B. Mountcastle, An organizing principle for cerebral function: the unit module and the distributed system. In *The Mindful Brain* , G. M. Edelman and V.B. Mountcastle, eds., MIT, Cambridge (1978).
23. D.O. Hebb, *Organization of Behavior* , Wiley, New York, (1949).
24. X. Leng, G.L. Shaw, and E.L. Wright, Coding of musical structure and the trion model of cortex, *Music Perception* 8:49 (1990).
25. L. Brothers, G.L. Shaw, and E.L. Wright, Durations of extended mental rehearsals are remarkably reproducible in higher level human performance, *Neurological Research* 15:413(1993).
26. W.G. Chase and H.A. Simon, The mind's eye in chess., In *Visual information processing* , W. G. Chase (Ed.), Academic Press, New York (1973).
27. H. Petsche, P. Richter, A. von Stein, S. Etlinger and O. Filz, EEG coherence and musical thinking, *Music Perception* 11:117 (1993).
28. K. Wynn, Addition and subtraction by human infants, *Nature* 358:749(1992).
29. E.S. Spelke, Principles of object perception, *Cognitive Science* 14: 29 (1990).
30. S.A. Trehub, Infants' perception of music patterns, *Perception and Psychophysics* 41:635 (1987).
31. C.L. Krumhansl and P.W. Jusczyk, Infants' perception of phrase structure in music, *Psychological Science* 1:70(1990).
32. I. Peretz, R. Kolinsky, M. Tramo, R. Labrecque, C. Hublet, G. Demeurisse, S. Belleville, Functional dissociations following bilateral lesions of auditory cortex, *Brain* 117:1283 (1994).
33. P.S. Goldman-Rakic, Introduction:the frontal lobes: uncharted provinces of the brain, *Trends Neurosci.* 7:425 (1984).
34. T. Bonhoeffer. and A. Grinvald, Iso-orientation domains in cat visual cortex arranged in pinwheel-like patterns, *Nature* 353:429 (1991).

35. M.E. Fisher and W. Selke, Infinitely many commensurate phases in a simple Ising model, *Phys. Rev Lett.* 44:1502 (1980).
36. G.L. Shaw and K.J. Roney, Analytic solution of a neural network theory based on an Ising spin system analogy *Phys. Lett.* 74A:146 (1979).
37. S.H. Shanbhag, Robotic motion and the trion model of the cortex, unpublished report (1991)
38. G.M. Edelman, *Neural Darwinism: The Theory of Neuronal Group Selection,* Basic, New York (1987).
39. D.E. Rumelhart, J.E. McClelland and PDP Research Group, eds., *Parallel Distributed Processing,* MIT, Cambridge (1986).
40. R.N. Shepard and J. Metzler, Mental rotation of three-dimensional objects, *Science* 171:701 (1971).
41. I. Mintz and H. Korn, Serontonergic facilitation of quantal release at central inhibitory synapses, *J. of neuroscience* 11:3359 (1991).
42. J. Sarthein, A. von Stein, H. Petsche, G.L. Shaw and F.R. Rauscher, EEG investigation of the positive effect of music on spatial-temporal reasoning, manuscript in preparation.
43. K.R. Karplus, S. Pulos and E.K. Stage, Early adolescent's proportional reasoning on 'rate' problems, *Educational Studies in Math* 14:219 (1983).

CHALLENGES IN TEACHING AND LEARNING INTRODUCTORY PHYSICS

Roger A. Freedman

Department of Physics and College of Creative Studies
University of California, Santa Barbara
Santa Barbara CA 93106–9530

INTRODUCTION

In a volume intended to celebrate Bill Little's many contributions to the research frontiers of fundamental and applied physics, an article about teaching the most elementary aspects of physics may seem somewhat out of place. But a substantial part of Bill's well–deserved reputation is based on his excellence as a teacher, especially for the introductory courses for engineers and biological science majors. This article is offered in homage to his tradition of excellence in education.

During his distinguished career, Bill has served as a mentor to a great number of graduate teaching assistants. In large part, I owe what success I have had as a physics educator to what I learned from Bill during my own teaching assistant days at Stanford. My goal in this article is to attempt, however feebly, to carry on his tradition.

In particular, I will discuss some aspects of the introductory physics course that will be of interest (and perhaps surprising) to new college physics teachers and new teaching assistants. These include recent research into student misconceptions and student difficulties with formal mathematical reasoning, as well as recent innovations that hold promise for improving both student understanding of physics concepts and student proficiency at solving physics problems. I hope that this article will also be of use to the "old hands" who, like me, discover something new about teaching every time they go into a classroom or talk with a student.

The challenges involved in teaching an introductory course in physics are legion. My goal in this brief article is not to be comprehensive — merely provocative! References 1 and 2 include a variety of other suggestions for physics faculty and teaching assistants.

WHY INTRODUCTORY PHYSICS EDUCATION IS IMPORTANT

The relative importance of teaching in the physics enterprise has increased dramatically in recent years. As federal funding for research decreases, the golden era in which newly–minted physics Ph. D.s were guaranteed at least a postdoctoral research position is becoming an ever more distant memory. A larger fraction of the available academic positions emphasize teaching, especially at four–year and two–year colleges.

From High Temperature Superconductivity to Microminiature Refrigeration
Edited by Cabrera *et al.*, Plenum Press, New York, 1996

Even at research universities, teaching is now playing a larger role in promotions and tenure decisions. In this brave new world, a physics graduate student who aspires to an academic career dare not neglect the teaching side of her or his graduate training.

While teaching physics at all postsecondary levels — beginning undergraduate, advanced undergraduate, and graduate — is important, the greatest importance attaches to the introductory courses taken by students in their first two years of college. These courses are the ones in which the budding physics major gets his or her first taste of the subject and decides whether or not to pursue the bachelor's degree in physics. The basic understanding achieved in these courses is the foundation for all subsequent study in physics.

The real importance of the introductory courses, however, lies in those students who are *not* physics majors. Indeed, the vast majority of students in introductory courses are likely to be engineers (in a calculus–based course), premedical students (in an algebra–based course), or humanities majors (in a "conceptual physics" course). These students constitute the educated electorate of the future, and their introductory physics courses are the only chance that we physicists have to plead our case with them.

The dominant public perception of physics is that it is tedious, abstract, and fundamentally irrelevant; the challenge in an introductory course is to convince the audience that physics is rewarding, fun, useful, and most of all a *worthwhile* endeavor. If we fail in this, and the public perception of physics does not change, there is little chance that future physics research will be funded at anything more than a token level. In this sense, introductory physics teaching is the foundation not only of a physics education, but of the physics enterprise as a whole. We neglect the teaching of these courses at our own grave peril.

WHY JOHNNY CAN'T ANALYZE CIRCUITS

College students begin most of their courses in a state of nearly perfect *tabula rasa*. Before they take their first course in world history, in economics, or in psychology, they know little or nothing about those subjects. The instructor can then help the students to implant fresh knowledge upon the palimpsests of their minds.

The situation in an introductory physics course is quite different. Although they would be shocked to hear you say it, students arrive in their first physics course with a set of physical theories that they have tested and refined over years of repeated experimentation. How can this be? The reason is that students have spent some eighteen years exploring mechanical phenomena by walking, running, throwing baseballs, catching footballs, and riding in accelerating vehicles. They have also some more limited experience with electrical phenomena, garnered from using electric circuits in the home, and with the behavior of light, lenses, and mirrors. Based on their observations, students have pieced together a set of "common sense" ideas about how the physical universe works.

Unfortunately, research carried out by physicists has shown these "common sense" ideas are in the main *incompatible* with correct physics. Worse still, these erroneous ideas are robust and difficult to dislodge from students' minds, in large measure because these ideas are not addressed by conventional physics instruction.

As an example, Fig. 1 illustrates some representative "common sense" ideas about electric circuits. In part (a) of the figure, a battery is connected to two identical light bulbs A and B in series. In part (b), the battery is connected to a single bulb C which is identical to bulbs A and B. McDermott and Shaffer[3] asked students in introductory physics courses to compare the brightnesses of bulbs A and B in circuit (a) and to compare these with the brightness of bulb C in circuit (b).

The results of this investigation were incredibly disappointing. The correct answer, that bulbs A and B in circuit (a) are equally bright and that bulb C in circuit (b) is brighter

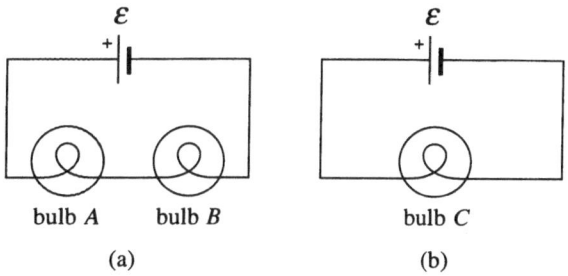

Figure 1. (a) A battery connected to two identical light bulbs A and B in series. (b) The battery is now connected to a single bulb C which is identical to bulbs A and B. When asked to compare the brightnesses of the bulbs in these circuits, only an embarrassingly small number of students gave the correct answer even after instruction in circuit theory.

still, was given by only about 10% of the students in algebra–based courses and by only about 15% of the students in calculus–based courses.

The most remarkable result of McDermott's and Shaffer's study is that the types of student errors made on this question are *unrelated* to, and *unaffected* by, conventional instruction. One common student error is the belief that in circuit (a), bulb A will be brighter than bulb B because bulb A "'uses up' the current first." Another common error is that the brightness of each bulb will be the same in either circuit because the battery provides a constant *current* in all cases. Neither of these incorrect ideas are learned from an introductory course, but neither are they *discredited* in a standard introductory course. Indeed, McDermott and Shaffer found that student performance on this question was nearly independent of whether the question was posed before or after instruction on electric circuits. Similarly disquieting results have been found regarding "common sense" ideas in mechanics[4, 5, 6] and in optics.[7]

Investigations of this sort show that it is not enough to merely teach students the *right* way to think about physics. Rather, the challenges to the instructor are to identify possible student misconceptions, to confront these misconceptions head–on, and to help students to *unlearn* these misconceptions at the same time that they are learning correct physics. Failure to do this will invariably leave students with their erroneous "common sense" ideas intact.

In order to rise to these challenges, an essential tool is an introductory physics textbook that addresses "common sense" ideas explicitly. Sadly, most contemporary textbooks are severely deficient in this respect. But some very recent textbooks make extensive use of research into student misconceptions,[8, 9] and these should be given consideration by instructors who are serious about helping students overcome their "common sense" ideas about physics.

"I UNDERSTAND THE CONCEPTS, I JUST CAN'T DO THE PROBLEMS"

Every physics instructor has heard this complaint from students at one time or another. All too often, however, what the student really means is the converse:

"I can do (some of) the problems, I just don't understand the concepts."

Students can usually handle problems that are akin to the worked examples in their textbook, especially if there are "special equations" that they can use. Problems that require using fundamental concepts, along the lines of how we might expect a physicist to think, are another matter altogether.

The proof of this statement is the difference between student performance on "standard" physics problems that require computation and calculation and their performance on purely conceptual, qualitative problems. As an example, McDermott and Shaffer[3] found that even students who performed well on standard numerical problems in circuit analysis, and even students with near–perfect scores on such problems, performed poorly on the conceptual question depicted in Fig. 1.

Part of the difficulty that students have with conceptual questions stems from the kind of problems that students are most often assigned. Instructors commonly assign homework and exam problems that involve computation or calculation, in the belief that these are "real" physics problems. A corollary to this belief is the assumption that a student's ability to successfully solve such problems is evidence of complete understanding. Alas, research shows that such is not the case. One example is an investigation of student understanding of the Newtonian concept of force carried out by Hestenes, Wells, and Swackhamer.[6] By comparing student performance on a set of conceptual questions posed both before and after a first course on mechanics, they found that conventional instruction (including the assignment of conventional homework problems) produces only marginal gains in conceptual understanding.

If we truly want students to learn about the *ideas* of physics, we must require them to *use* these ideas in their homework and then hold them accountable for these ideas in examinations. Most introductory textbooks include a wealth of conceptual questions, and questions of this sort these should be assigned regularly. My own students regularly comment that they find conceptual questions to be much more difficult than the "ordinary" problems; such comments convince me that conceptual questions are very useful tools for teaching and learning physics.

WHAT DOES THAT EQUATION MEAN?

A related issue is the question of how students deal with *formal,* mathematical expressions of physical concepts. Two examples are Newton's second law and the work–energy theorem:

$$\Sigma \vec{F} = m\vec{a} , \tag{1}$$

$$W_{net} = \Delta K = \frac{1}{2} mv_f^2 - \frac{1}{2} mv_i^2 . \tag{2}$$

It is very common for students to interpret Eq. (1) to mean that the product of a body's mass and its acceleration *is itself a force*. In other words, they fail to realize that a mathematical equality between two quantities does not imply that the two quantities are conceptually distinct. As a result, they do not appreciate that acceleration is the *consequence* of the presence of a net force. Thus students frequently make reference to such chimeras as "the force due to acceleration" or "the force due to momentum."

A similar confusion arises concerning the work–energy theorem, Eq. (2). When students are asked to explain what kinetic energy *means*, the most common response is that it is "one–half the mass times the speed squared." By fixating on the *mathematical* definition, they fail to grasp the essence of the work–energy theorem: that the kinetic energy of a particle is equal to the total work that was done to accelerate it from rest to its present speed, and equal to the total work that the particle can do in the process of being brought to rest.

This tendency to focus on mathematical definitions rather than physical meanings was shown convincingly by Lawson and McDermott.[10] They presented students with a simple question concerning the work–energy theorem. As depicted in Fig. 2, an object of mass m

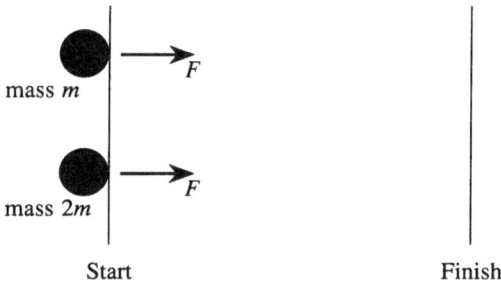

Figure 2. A top view of a "race" between two objects on a frictionless horizontal surface. The two objects are of different mass but are subjected to the same net force. When asked which object crosses the finish line with greater kinetic energy, only a few students were able to give the correct answer.

and another object of mass $2m$ are initially at rest on a frictionless horizontal surface. The same constant force of magnitude F is then applied to each object. The question to be answered is "Which object crosses the finish line with greater kinetic energy?"

Using the work–energy theorem, and keeping in mind the *physical* meaning of kinetic energy, it can easily be seen that each object has the same kinetic energy upon reaching the finish line. Yet in interviews with 28 students taken from two classes at the University of Washington, an honors section of calculus–based physics and a regular section of algebra–based physics, Lawson and McDermott found that only a few honors students were able to supply the correct answer and the correct reasoning without coaching. While most of the remaining honors students were able to eventually achieve success with guidance from the interviewer, almost none of the students from the algebra–based course were able to do so. No less disappointing results were obtained with a written version of the question presented to a regular section of calculus–based physics. I have had similar experiences with my own students: Their performance on conventional homework–type problems shows that they can *compute* quantities such as work and kinetic energy, but their performance on conceptual questions shows that they have much more difficulty *explaining* or *interpreting* their results.

This example shows again that emphasis on numerical problem–solving can obscure major conceptual deficiencies in students. It underscores the importance of requiring students to apply the fundamental concepts of physics in a variety of different situations, as well as requiring them to explain the logic that they use in solving physics problems of all kinds.

RETHINKING THE LECTURE AND DISCUSSION SECTION

I have described research showing that conventional physics instruction tends to be ineffective in helping students to develop a real understanding of physics. How, then, should the nature of physics instruction be changed? A number of different approaches have been suggested and explored; I will summarize the approaches that I believe to be the most promising.

The Misuses of the Lecture

The lecture is one of the most ancient of teaching methods. In the teaching of physics, it is typically used to demonstrate physical phenomena, to present derivations; and to show examples of how to solve problems. The first of these uses of the lecture is an important one, and is often neglected by instructors who feel compelled to "cover more material" or who regard the demonstrations as a distraction. My own experience is that good lecture demonstrations are absolutely indispensable as tools for helping students to relate physical concepts to the real world. Good lecture demonstrations also have the strength of being memorable. I have had students come to me a decade after taking one of my classes and tell me how they still remember a certain demonstration and the physics that they learned from it. The title "Lecture Demonstrator" is still in use at certain British universities to denote a science lecturer; the title alone speaks volumes about the importance of lecture demonstrations.

By contrast, the use of lecture time to present derivations is typically ineffective. (I have yet to have a former student tell me how vividly they remember my derivation of the thin–lens formula.) A derivation presented on the blackboard is less useful to the student than the same derivation presented in the textbook, where it can be traced through repeatedly at the student's leisure. My suspicion is that instructors tend to present derivations in lecture because they doubt that their students read the book. While this is indeed a valid concern, it would seem that using the lecture to reiterate the contents of the book is ultimately counterproductive; it merely helps to *ensure* that the students won't read the book.

Far and away, however, the least effective use of lecture time is for presenting the solutions to physics problems. The essential difficulty here is that physics problem–solving is a *skill* that has to be learned by repeated practice. In learning a skill, it can be useful to first watch an expert exercise that skill, but that is by no means the most important part of the learning process. If it were, the millions who watch professional sports would themselves naturally develop into top–notch players; avid movie–goers would inexorably turn into accomplished actors (who really want to direct); and the poor souls who watch televised court proceedings would slowly but surely mutate into highly paid defense attorneys. Of course, none of these evolutions really take place. In the same way, students who watch their instructor (an expert problem–solver) work out a solution on the board may be impressed by the instructor's prowess, but they will augment their own problem–solving skills only marginally. The disappointing problem–solving performance of students who have had such conventional instruction, referred to earlier, is testimony to this.

A Lecture Model with "Active Learning"

Numerous instructors, myself included, have found that lectures become more useful when students are forced to become *active participants* in the lecture.[11] In my own classes, I speak briefly about each new topic (proceeding under the assumption that students have read the required material from the textbook *before* class), and do a lecture demonstration or two as appropriate. I then give the students an exercise to work out. They then spend several minutes working out this exercise, which is chosen to be specific to the topic at hand: it may involve tasks such as drawing free–body diagrams, writing down (but not necessarily solving) the key equations for a group of related but distinct situations, or making graphs of different types of motion. While this is going on, I roam around the classroom inspecting the students' work. I then instruct the students to confer with their neighbor to compare their responses and to resolve any discrepancies. Remarkably, this works very well even in a large lecture hall; the sound level from the discussions among 300 students can be quite impressive! Finally, I discuss with the students the correct way to tackle the exercise, being careful to point out common errors to the students. I typically do two or three sequences of instructor description — student work — instructor discussion during a typical lecture.

This technique has several merits. First, the students have something constructive to do during the lecture; it is a sure–fire cure for the torpor that grips students midway through a conventional lecture. Second, students are forced to discuss physics with their peers and to defend their ideas. Third, students get immediate feedback as to whether or not they understand a concept that has been presented in class, and any points of confusion can be corrected at an early stage in the students' apprehension of the concept. Last, but by no means least, the instructor can learn a great deal about her or his students' understanding of the material. This last point was brought home to me vividly during a lecture when I asked students to draw the free–body diagram for a car rounding a banked curve; many of the diagrams I saw while walking around the lecture hall included a number of creative and wholly imaginary forces that I had never dreamed existed!

When conducting the lecture in this way, it is best if the students have a printed sheet with the exercise on it. These can be time–consuming to develop and to prepare in printed form, however. I have relied heavily on Alan Van Heuvelen's *ALPS Kit* (an acronym for *Active Learning Problem Sheets*), which is a workbook containing several hundred exercises and activities which are expressly designed for student use during lecture.[12] Students purchase this inexpensive workbook at the campus bookstore, and are required to bring it with them to lecture; happily, I find that almost all of them do so religiously.

Some will no doubt complain that this technique of "active learning" forces the lecturer to cover less material. It is indeed true that the lecturer *talks about* less material with this approach; the challenge to the lecturer is to choose between the material that is worthy of discussion during the lecture and the easier material that the students can learn adequately on their own from the textbook. Thus this technique does not require that any material be deleted from the course syllabus.

Employing "active learning" in the lecture keeps students engaged in the lecture. More importantly, it yields substantially better student performance on exams than does conventional instruction.[11, 12]

Discussion Sections, Teaching Problem–Solving, and "Cooperative Learning"

Most introductory physics courses have both a lecture component and a discussion section (or "recitation section") component. The discussion section, typically led by a teaching assistant (TA), is intended principally to be a forum in which students gain insight into problem–solving technique by observing the discussion leader, by practicing solving problems, and by discussions with other students.

Unfortunately, physics discussion sections very often fail to live up to this intent. Too many students come to discussion sections with the intent that they will get their weekly homework "done for them" by the TA. As a result, despite the earnest efforts of hard–working and talented TAs, it is difficult to cajole students into actually doing problem–solving work in a discussion section. Furthermore, it is next to impossible to initiate and sustain any real student discussion within the unstructured format of a typical discussion section. The upshot is that few students are able to move beyond the "formulaic" approach to problem–solving, which consists of hunting through the textbook for a likely–looking equation or set of equations into which they can plug the values stated in the problem. This sad state of affairs is especially frustrating for TAs who, after working diligently with a group of students for an entire term, must grade those students' disappointing work on exams.

A very promising effort to rectify these shortcomings of the discussion section has been described by Heller, Keith, Anderson, and Hollabaugh.[13, 14] They reorganized the discussion sections in two rather different physics courses, the first quarter of a large algebra–based introductory course at the University of Minnesota and a sophomore modern physics course with a dozen students at Normandale Community College. In both courses, students were taught a general problem–solving strategy based on the methods used by expert problem–solvers, and were required to write up their problem solutions in a way that explicitly reflects the use of that strategy. (In addition to shaping the students' approach to problem–solving, this technique helps to clarify for the grader what conceptual ideas the students are using.) To discourage "formulaic" problem–solving, students were assigned so–called "context–rich" problems. Such problems do not always explicitly identify the unknown variable, may include extraneous information, and may require reasonable assumptions (e. g., the acceleration is constant) or estimation (e. g., the mass of a typical cat). In other words, they are less like standard textbook problems and more like the problems encountered by real scientists and engineers. The following is an example of such a "context–rich" problem, taken from Ref. 14:

While visiting a friend in San Francisco, you decide to drive around the city. You turn a corner and find yourself going up a steep hill. Suddenly a small boy runs out on the street chasing a ball. You slam on the brakes and skid to a stop, leaving a skid mark 50 ft long on the street. The boy calmly walks away, but a policeman watching from the sidewalk comes over and gives you a ticket for speeding. You are still shaking from the experience when he points out that the speed limit on this street is 25 mph.

After you recover your wits, you examine the situation more closely. You determine that the street makes an angle of 20° with the horizontal and that the coefficient of static friction between your tires and the street is 0.80. You also find that the coefficient of kinetic friction between your tires and the street is 0.60. Your car's information book tells you that the mass of your car is 1570 kg. You weigh 130 lb, and a witness tells you that the boy had a weight of about 60 lbs and took 3.0 s to cross the 15–ft wide street. Will you fight the ticket in court?

Such "context–rich" problems are of the kind that we would like our students to be able to solve, but which are usually thought too difficult and challenging for students to solve on their own. Remarkably, students in both of the test groups described in Refs. 13 and 14 were able to solve such problems when each discussion section was organized into *cooperative groups* of three students. The students in each group were required to work together to produce a group solution to the assigned problem, using the problem–solving strategy that they had been taught. All students in the group received the same grade for their group assignment. The students in each group were assigned the roles of Manager (who keeps the group on task and manages the sequence of steps), Skeptic (who helps the group to avoid overly quick agreement and asks questions like "Are there other possibilities?"), and Checker/Recorder (who checks for consensus among the group and who writes up and hands in the group solution). These roles were rotated among the students each week. The use of such definite roles, and the challenging nature of the assigned "context–rich" problems, kept the students from simply working independently.

To reinforce the use of the problem–solving strategy and of the skills used in the cooperative groups, each course exam included a "context–rich" problem that had to be solved by the students in their cooperative group during the discussion section. (More conventional individual exams were given during lecture.)

Heller et al. found that over two quarters of using these methods, the problem–solving technique of students of all ability levels improved.[13] It may not be surprising that this proved to be the case for students in the lowest third and middle third of the class. The structured problem–solving strategy and the requirement to discuss ideas with other students seems well–suited to helping students whose understanding of problem–solving was initially only fair or poor. What is remarkable is that participation in cooperative groups also helped the *best* students in the class to improve their problem–solving skills, and that these students improved at about the same rate as the students in the lowest and middle thirds. For example, the percentage of students in the lowest third of the class whose individual solutions followed a logical mathematical progression improved from

20% to 50% over two quarters; this percentage for students in the upper third improved from 60% to 90%. Furthermore, this improvement of all students was found in both group problem–solving and individual problem–solving.

The use of cooperative groups and "context–rich" problems can have a very beneficial effect on student problem–solving skills. We have just begun to implement these innovations in the introductory calculus–based physics course at UC Santa Barbara, and the preliminary results look encouraging. This method is not a panacea, however; Heller at al. found that their innovations did not have much beneficial effect on students' understanding of the conceptual aspects of physics.[13] This suggests that these aspects are best addressed in the lecture using the "active learning" technique described previously.

CONCLUSION

Teaching and learning introductory physics are both challenging tasks. While traditional methods have led to frequently disappointing results, I have tried to show that there *is* hope. As instructors, we should heed the lessons about our students' thought processes learned from research into "common sense" ideas about physics and into students' difficulty with formal mathematical reasoning. We must see to it that students truly learn how to use the *concepts* of physics, in order that they may learn how to think like a scientist or engineer. And we should be willing to consider new forms and new approaches for the time–honored lecture and discussion section. Whatever gains we can make in improving student understanding and appreciation of physics cannot help but improve the public perception of physics as a useful, interesting, and above all *comprehensible* human activity.

ACKNOWLEDGMENTS

I thank my many teachers at Stanford for helping to cultivate my interest in the field of physics education. In particular, I thank Alan Schwettman, Dirk Walecka, and especially Bill Little for his service as a role model to generations of Stanford graduate student TAs.

REFERENCES

1. A. B. Arons, *A Guide to Introductory Physics Teaching,* Wiley, New York (1990).
2. M. J. Bozack, Tips for TA's: The role of the physics teaching assistant, *Phys. Teach.* **21,** 21 (1983).
3. L. C. McDermott and P. S. Shaffer, Research as a guide for curriculum development: an example from introductory electricity, Part I: Investigation of student understanding, *Am. J. Phys.* **60,** 994 (1992).
4. I. Halloun and D. Hestenes, The initial knowledge state of college physics students, *Am. J. Phys.* **53,** 1043 (1985).
5. I. Halloun and D. Hestenes, Common–sense concepts about motion, *Am. J. Phys.* **53,** 1056 (1985).
6. D. Hestenes, M. Wells, and G. Swackhamer, Force concept inventory, *Phys. Teach.* **30,** 141 (1992).
7. F. M. Goldberg and L. C. McDermott, An investigation of student understanding of the real image formed by a converging lens or concave mirror, *Am. J. Phys.* **55,** 503 (1987).
8. L. C. McDermott, *Physics by Inquiry,* Wiley, New York (1996).
9. H. D. Young and R. A. Freedman, *University Physics,* 9th edition, Addison–Wesley, Reading, MA (1996).

10. R. A. Lawson and L. C. McDermott, Student understanding of the work–energy and impulse–momentum theorems, *Am. J. Phys.* **55,** 811 (1987).
11. A. Van Heuvelen, Learning to think like a physicist: A review of research–based instructional strategies, *Am. J. Phys.* **59,** 891 (1991).
12. A. Van Heuvelen, Overview, Case Study Physics, *Am. J. Phys.* **59,** 898 (1991). The ALPS Kit is available from Professor Alan Van Heuvelen, Department of Physics, Ohio State University, Columbus OH 43210–1106.
13. P. Heller, R. Keith, and S. Anderson, Teaching problem solving through cooperative grouping, Part 1: Group versus individual problem solving, *Am. J. Phys.* **60,** 627 (1992).
14. P. Heller and M. Hollabaugh, Teaching problem solving through cooperative grouping, Part 2: Designing problems and structuring groups, *Am. J. Phys.* **60,** 637 (1992).

GENES OF WILLIAM LITTLE

Excerpts from an after-dinner talk on September 30, 1995, by
W.E. Meyerhof

... I want to begin with the origin of Bill's genes. I won't start with the Big Bang, nor the apes, but jump right to 1833 when one of Bill's four great grandfathers, Thomas Bowler, left England. He sailed to South Africa to work at the Royal Observatory in Capetown. Bowler was an astronomer, but he was so contentious, litigious, and fiercely temperamental that he had to leave his prior job in England. He did not last too long in his position at Capetown either, and soon concentrated on his artistic, rather than scientific, talents. He became a painter. Some of his many paintings still hang in museums in London and Capetown. He had a daughter whom grandfather Little had previously met and fetched to marry, when he was on his way to India.

Grandfather Little sailed to India from England via South Africa. Obviously, there is a travel gene in the Little family, which has settled in Bill , too. Grandfather Little studied law in Calcutta and became a judge. He had six children, the youngest of which was Bill's father. Like all the Little ancestors, Grandfather Little would have lived to a ripe old age, but he was poisoned by the son of a man he had condemned to death. Thereafter, grandmother Little decided to return to her father Thomas Bowler in South Africa, together with her six children. That was around 1880, when Bill's father was still a baby.

Father Little grew up in South Africa and got involved in the Boer War. Bill's father fought on the British side, of course. Following the war, like his father, he studied law and became a magistrate. He was sent to the Dutch part of South Africa, but as a Britisher was not too popular there. After convicting a Dutch farmer of some misdeed, he was transferred to a lesser position.

Father Little also inherited the travel gene. Several times, he sailed to Europe and got caught up in the First World War also. After that went to Scotland to marry a girl he had met. When he got there, he found her sister more to his liking and married her instead. She was a teacher, and she was over twenty years younger than her husband.

Bill was born in 1930. Looking at his forebears, we see that the genes available to him were, besides the travel gene and the artistic gene, an independent-spirit gene from his father's side and an incentive towards learning from his mother's side.

As a nine-year old, Bill started to hang out around a radio repair store. To get rid of him, the owner gave him a few parts with which Bill started to build a radio. He managed to get signals from the London BBC and from the San Francisco Treasure Island station. With the travel gene stirring in his mind, he was fired up by adventures in the wider world. On a more practical side, he learned to repair radios and make some much needed pocket money.

Bill's education followed the British system. At the end of high school, he completed the School Certificate with such high marks that the city of East London provided him with a scholarship: all of $300 per year. Throughout his university studies Bill survived thanks to scholarships and loans.

For undergraduate studies, Bill went to Rhodes university, not far from home, and then worked for a master's degree. He did an experimental thesis on the lifetimes of organic fluorescent materials. He invented a phaseshift method of measuring nanosecond lifetimes, which is still being used in commercial devices today. The work was so good that his professor told him he could get a Ph.D. degree from Rhodes University, if he would only add a bit of work on the theory of the effect.

Meanwhile, though, Bill had applied for fellowships to study in England. He was a finalist for a Rhodes fellowship at Oxford, but did not get it because of some political shenanigans in the Selection Committee. I am afraid, this soured him on Oxford forever. Luckily, he was awarded a fellowship from the Royal Dutch-Shell Oil Companies. Bill's professor at Rhodes suggested Bill should go to Glasgow University, where the professor knew the nuclear physicist Dee.

When Bill arrived in Glasgow, Dee wanted him to build Geiger counters for nuclear experiments. But that was not to Bill's liking and the independent-spirit gene started to stir. Bill retired to the library for six months to figure out what he *really* wanted to do in the two years he had at Glasgow. In his reading, he came across the Overhauser Effect. Overhauser had shown that one could polarize atomic nuclei indirectly by polarizing the atomic electrons in a magnetic field. He had used liquid samples. Bill thought it would be fun to polarize atomic nuclei. Because Dee would not support this work, he had to scrounge around for apparatus. But, he soon mamanged to get the experiment going and found that one could polarize atomic nuclei in solid materials, also. This was quite an advance, because in experiments with polarized nuclei, solids are much easier to handle than liquids.

I think, around this time Bill's love affair with electrons began, but before I get into that I want to mention a couple of other love affairs. It is part of British higher education to encourage sports and to encourage service in the Reserve Officers Training Corps. Bill excelled in both. Still in South Africa, he became a high-jump champion: at age 23, he was the third best high jumper in the entire British empire. Also, he enlisted in the aviation wing of the ROTC and already around age 20 obtained a commercial pilots license. He loved flying and at first continued doing so in Glasgow.

Two years after coming to Glasgow, Bill married, but flying then was out. During those two years, Bill worked toward a Ph.D. degree at Glasgow. Bill obtained a fellowship which allowed him to work for two years at any Canadian university. He chose the University of British Columbia at Vancouver. The low-temperature physicist Jim Daniels was there, and Bill wanted to try out another method of polarizing atomic nuclei, this time using low-temperature techniques. The work led to the question how heat is transferred at low temperatures and resulted in Bill's most frequently cited research paper, on the Kapitza heat resistance.

Meanwhile Bill's fame spread, even to Stanford University. Felix Bloch read about Bill's work, and together with Leonard Schiff proposed to the Department to offer Bill a position as Assistant Professor. This was early in 1957. But, with the independent-spirit gene lurking in the back, Bill decided he first wanted to complete the two-year fellowship at Vancouver. He stayed at Stanford for nearly forty years.

Bill came to Stanford with the idea of producing low temperatures, in particular to continue studies of heat transfer and superconductivity. Shortly after Bill came to Stanford, Bill Fairbank arrived and started talking to him about his most admired physicist, Fritz London. London was a low-temperature physicist with wide interests which included biology and the origin of life. All this fascinated Bill, and with a couple of students he started a reading seminar in biophysics. The inpendent-spirit gene soon turned Bill's thought processes toward neural networks and the basic structure of the brain. He developed a model of the brain which is closely related to a model of superconductivity he had previously invented. Both are based on strong interactions between pairs of atomic electrons. Graduate students were always involved in these developments, as well as postgraduate researchers. Among the latter, I mention in particular Hanoch Gutfreund and Gordon Shaw. In all, Bill has supervised nearly forty graduate students.

At the undergraduate level, Bill was a stimulating teacher, able to communicate his enthusiasm for physics to his audience. He received two teaching awards, which is quite exceptional: one from the Dean of Humanities and Sciences, the other from the University. Bill noticed also that before the teaching laboratory periods began, students would linger around in the corridor in front of the labs. He had the brilliant idea that one should set up small demonstration experiments in the corridor, with which the students could play and at the same time learn something about physics. He applied for funding, and now, twenty years later, these demonstration experiments are still giving pleasure and stimulation to students waiting for their lab. periods to begin.

To stimulate the undergraduate students in other ways, Bill invented a course called "Modern Physics Through Science Fiction". This course was so popular and successful that Bill published an article about it, together with Roger Freedman who was the Head Teaching Assistant in the course. As a result, similar courses are now being taught at other universities.

Turning once more to Bill's scientific interests, I mentioned earlier his love affair with electrons. Bill started to study their interference properties which are very marked at low temperatures. Later, he used interference devices, known as SQUIDS. These are physically very small devices. This set Bill thinking how one could cool small devices to low temperatures without disturbing their electrical properties by noise, such as occurs in ordinary refrigerating machinery. He thought of using the cooling effect of a gas when it is forced through a small opening, known as the Joule-Thomson effect.

By the end of the 1970's, Bill, together with students, had developed a working miniature refrigerator in which a gas is forced through tiny channels etched into a microscope slide. After a talk about this refrigerator at a physics conference, he was surrounded by a hundred people who all wanted to know more about this device. This gave Bill the idea to start a company where the refrigerators could be produced commercially. The company, called Micro-Miniature Refrigerators, or MMR in short, has now been going successfully for over 15 years. Bill is rightfully proud that a few years ago, MMR received a coveted industrial research award known as IR-100 for its ingenious products.

Bill's mind has been so fertile that is not possible for me to tell you about all the developments he has stimulated, but I give you one more example. In the 1960's he proposed a model of superconductivity, which indicated that in principle this effect could be produced even at room temperature. This was at a time when superconductivity was observed at best at a temperature of -260°C , so the superconducting establishment soundly pooh-poohed Bill's model. But, the independent-spirit gene has sustained Bill over all these years, and the properties of the recently discovered high-temperature superconductors indicate that he may yet be correct in some of the basic ideas of his model.

In any case, in the 1960s, Marvin Chodorow in Physics, Carl Djerassi in Chemistry, and Ed Ginzton at Varian got interested in the possibility of producing room temperature superconducting materials. The company SYNVAR was founded for this purpose. Although in the end, a room temperature superconductor was not developed, the research led to the invention of the most sensitive test for drugs. This is now a 400-million dollar a year business and Bill wished he had a small part of it...

INDEX